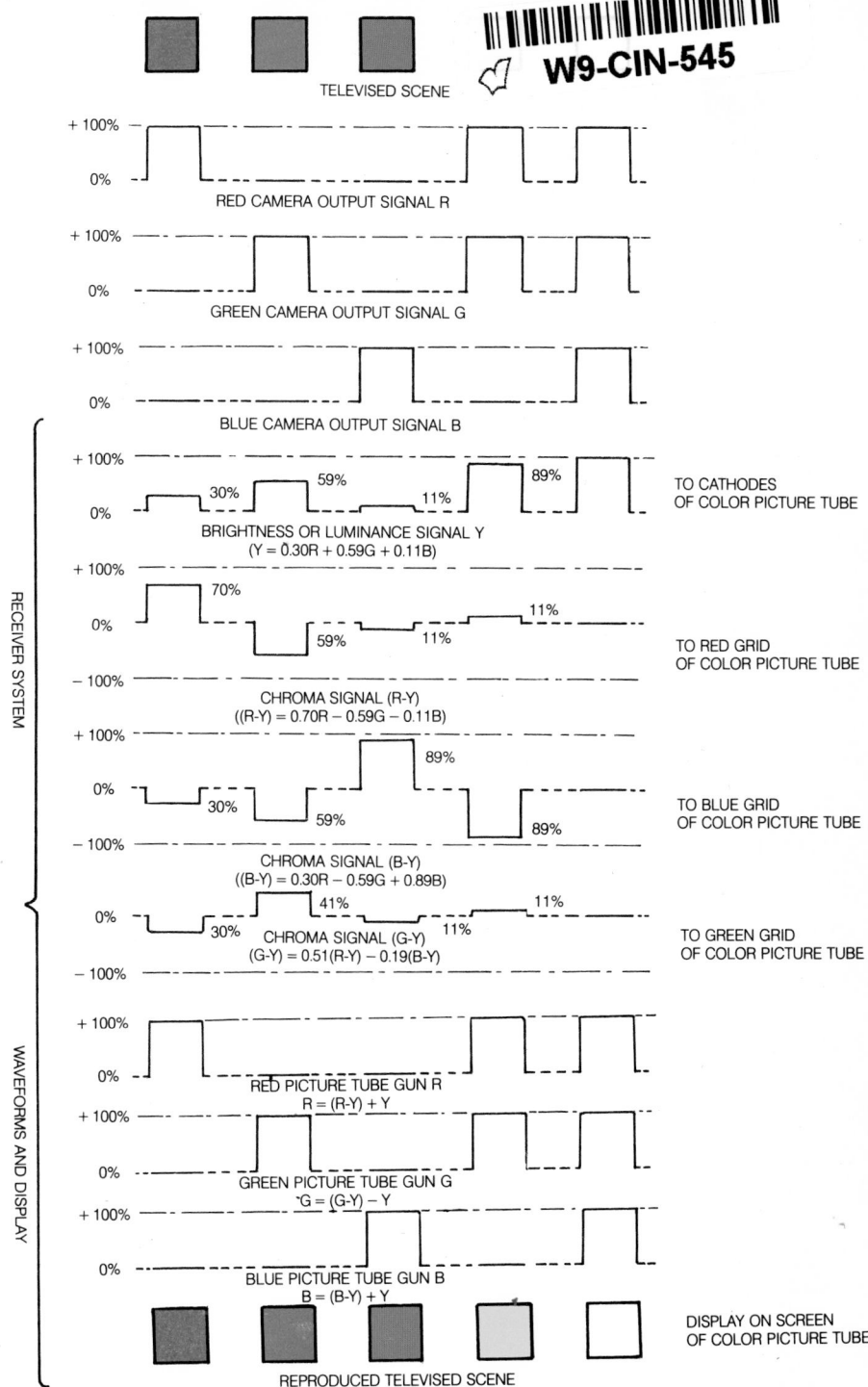

FIGURE 3-12

System operation with a color-bar signal

Color Television:
Theory
and
Servicing

Second Edition

Color Television: Theory and Servicing

Clyde N. Herrick

RESTON PUBLISHING COMPANY, INC., Reston, Virginia
A Prentice-Hall Company

Library of Congress Cataloging in Publication Data

Herrick, Clyde N
    Color television, theory and servicing.

    Includes index.
    1.   Color television.   I.   Title.
TK6670.H47   1976   621.388'04   75-34282
ISBN 0-87909-123-1

©1977 by
Reston Publishing Company, Inc.
A Prentice-Hall Company
Reston, Virginia 22090

10   9   8   7   6   5   4   3   2   1

Printed in the United States of America

Contents

Preface

Color-television technology is now dominated by solid-state devices such as bipolar and field-effect transistors, silicon controlled rectifiers, and integrated circuits. The trend has been to all solid-state receivers, although thousands of hybrid receivers (using both transistors and electron tubes) are still in use. In turn, the need for a state-of-the-art book covering the theory and troubleshooting of solid-state, color television receivers is evident. This coverage necessitates application data for color television test equipment to the extent required in setup and basic servicing procedures.

It is assumed that the reader has either completed courses in electricity, electronics, radio communication, and black-and-white television or has attained a practical background in these areas. Relevancy is stressed throughout the profusely illustrated text. Mathematical treatment has been minimized, and equations are employed only in topics that have a basic quantitative context. A prerequisite background in arithmetic, algebra, geometry, and trigonometry is also necessary. It is essential that the reader have an elementary understanding of analytic geometry, to the extent that he can work with curves and graphs.

In this new edition, discussions have been added concerning the slotted-mask color picture tube, all-electronic tuner, troubleshooting SCR horizontal-sweep circuits, troubleshooting horizontal-section modules, troubleshooting vertical-sweep modules, and digital display of the prevailing channel number with the time. Thus, the student is provided with a relevant state-of-the-art text that meets the needs of both college-prep and vocational curricula.

Acknowledgement is made to those who have preceded the author by their development of other books on color television, and to the faculty of San Jose City College, who have made many helpful suggestions and criticisms. This book can be properly described as a team effort, although the individual members would choose to minimize the measure of their own contributions. It is appropriate that this book be dedicated as a teaching tool to instructors as well as students involved in television technology.

Clyde N. Herrick

Chapter 1

Color Fundamentals And Colorimetry

1.1 LIGHT AND COLOR VISION

Visible light consists of a small interval in the spectrum of electromagnetic radiation, as shown in Fig. 1-1. All visible colors are contained in this spectrum. The wavelengths of visible electromagnetic radiation extend from 400 to 700 millimicrons, or from 16 to 28 millionths of an inch. Although we name the colors of the rainbow red, orange, yellow, green, blue, indigo, and violet, these are merely the most obvious hues. Hundreds of variations in colors can be distinguished, even by a comparatively untrained observer. Physics teaches us that a beam of white light passed through a prism, as depicted in Fig. 1-2, splits up into a color spectrum. Conversely, if a color spectrum is passed through a prism, the reverse process takes place—spectral colors are recombined to form a beam of white light.

A beam of white light can also be formed by combining a minimum of three colored lights, which are called *additive primaries*. The most useful additive primaries are red, green, and blue; these are the primary hues that are utilized in color television. Additive primaries must not be confused with the *subtractive primaries* used in printing processes and in paintings. Subtractive primaries are viewed by reflected light, and the most useful subtractive primaries are red, yellow, and blue. To anticipate subsequent discussion, a color

1

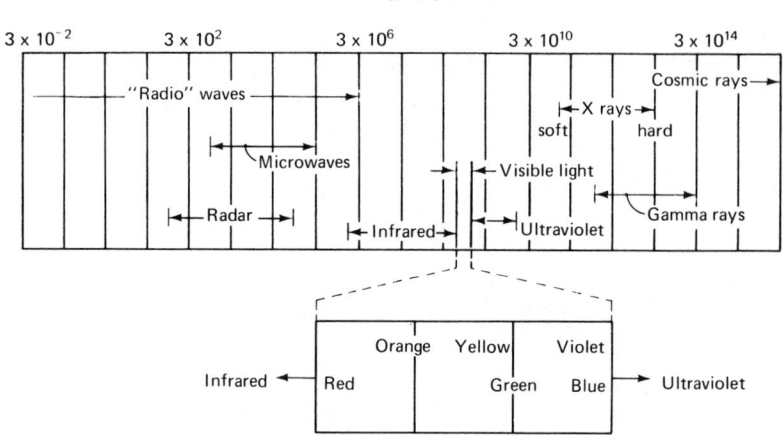

FIGURE 1-1   Spectrum of electromagnetic radiation.

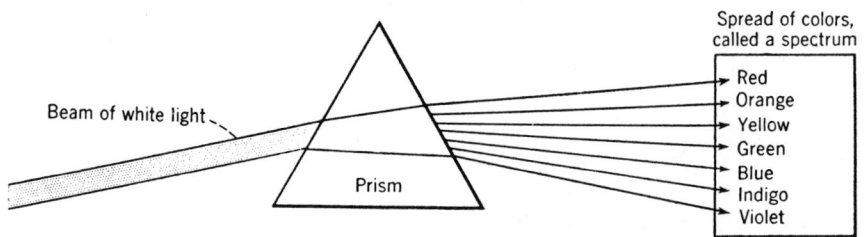

FIGURE 1-2   A prism refracts electromagnetic waves. White light is split up into a color spectrum (a color spectrum can be recombined into white light).

picture tube is a source of colored lights that are viewed directly. A color picture tube employs red, green, and blue phosphors.

Red, green, and blue are the most useful additive primaries because they allow the formation of the maximum range or gamut of hues when blended in various proportions. Figure 1-3 shows how the primary colors blend in pairs to form the complementary colors and how all three primary colors blend to form white. Thus, red and blue blend to form their complementary color magenta; blue and green blend to form their complementary color, cyan; and red and green blend to form their complementary color, yellow. Technically, white is not a color, but rather is light that is free of color. Black is defined as the absence of light. Note that the subtractive complementary colors are cyan, magenta, and orange.

FIGURE 1-3   The additive primary and complementary colors. (See inside front cover of book for artwork.)

1.2   PRINCIPLES OF COLORIMETRY

Colorimetry is the science of specifying colors. Any color utilized in a technical process such as color television can be produced by some suitable blending of the three primary colors. This fact is shown to good advantage by means of a *chromaticity diagram*. Figure 1-4 exemplifies a useful form of chromaticity diagram with wavelengths along the *boundary* (or range) noted in millimicrons. (A micron is equal to one-millionth of a meter.) Any color on the chromaticity diagram can be obtained by suitable blending of red, green, and blue primary hues. Note that white occurs at a point on the diagram that corresponds to equal proportions of red, green, and blue. Note also that colors in the whitish area are very pale or *unsaturated*, whereas colors near the boundary of the diagram are intense or highly *saturated*.

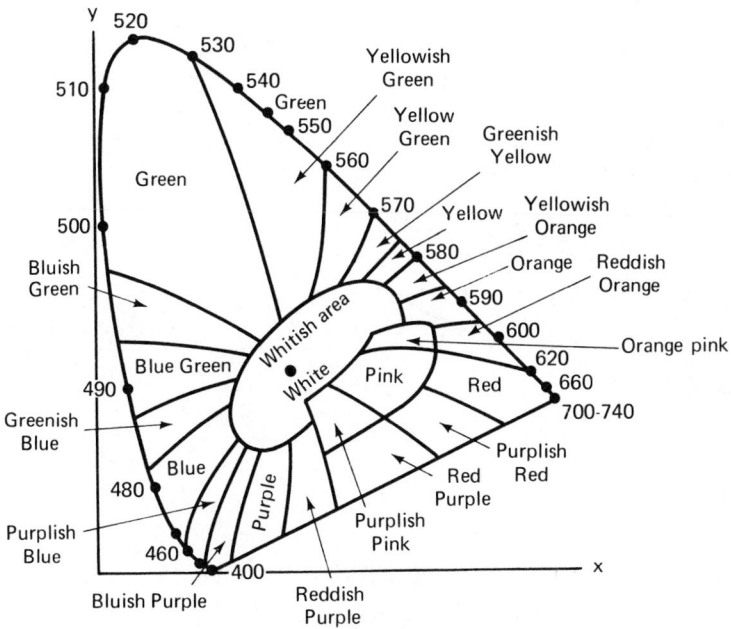

FIGURE 1-4   A standard chromaticity diagram.

Colors utilized in technical processes have a smaller gamut than those enclosed by the chromaticity diagram. For example, a color picture tube uses the primaries denoted by R_1, B_1, and G_1 in Fig. 1-5. Thus, the range of a color-TV image is enclosed by the associated triangle. Note that this color-TV range is somewhat greater than the range utilized in color printing processes. In summary, a color-TV system can reproduce any color that occurs in nature, although certain hues such as green and purple cannot be reproduced at maximum saturation. However, this is a minor limitation because it is very seldom that any hue occurs in nature in maximum saturation.

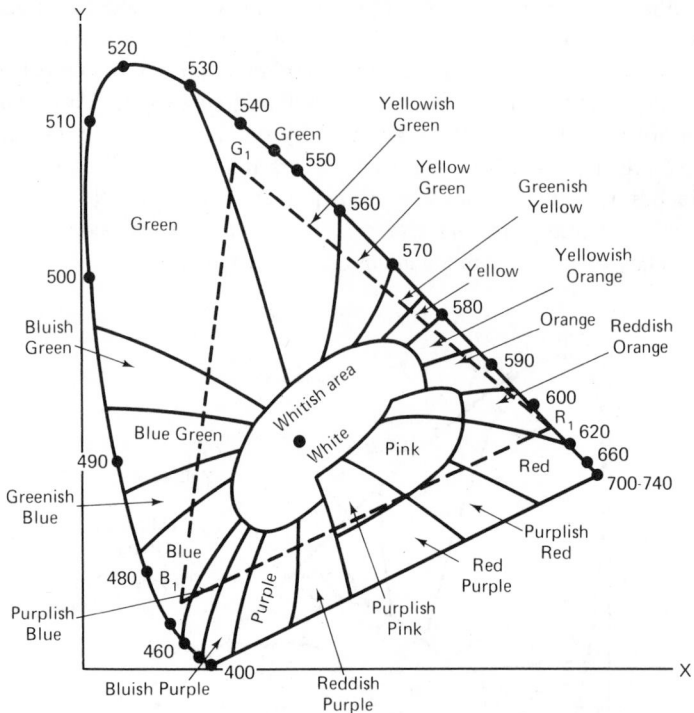

FIGURE 1-5   Chromaticity diagram showing range employed in color television.

1.3   COLOR VISION VERSUS BLACK-AND-WHITE VISION

Some of the fundamental characteristics of the color-TV system are based on principles of color vision. For example, black-and-white vision is more acute than color vision. An important aspect of this distinction is the inability of the eye to perceive very small colored areas in a scene. Thus, although the eye has three-primary vision for large colored areas, it has only two-primary vision for small areas. Finally, the eye has only black-and-white vision for even smaller areas. Purple and green-yellow hues become indistinguishable from gray in small areas of an image, although red and cyan colors remain visible. As a colored area is further reduced in size, the red and cyan hues also become indistinguishable from gray. The technical importance of these facts is that a highly detailed reproduction of black, white, and gray areas is required for TV, whereas colored areas do not need to be as sharply defined. Also, less definition is required for the blue hues than for the orange hues.

The eye is also more critical of some off-hue situations than of others. For example, a color-TV viewer tends to be very critical of variation in flesh tones, whereas variation in brown and green tones is likely to be accepted. In other words, the orange tints need to be reproduced with the minimum departure from the original scene, whereas green tints can vary appreciably from the original without recognition of this variance by the viewer. This requirement places stringent demands on the color-TV system, as is evident from inspection of Fig. 1-5. That is, as we proceed around the triangle $R_1 G_1 B_1$, the orange hues occupy a comparatively small interval. On the other hand, the green hues occupy an extended interval. Technically, this means that orange hues (and flesh tones in particular) must be reproduced with minimum error. Special circuits are often included in color receivers to contend with this problem.

Colors are technically described in terms of brightness, hue, and saturation. *Brightness* denotes relative light levels; for example, a sunlit scene is brighter at noonday than at twilight. *Hue* denotes the basic distinction between different kinds of colors; thus, red is a hue different from green. *Saturation* denotes the extent to which a hue is diluted by white light; for example, a red traffic light is highly saturated, and is often described as a *vivid* red light. On the other hand, a pastel pink color is considerably less saturated, and is said to be a weak shade of red. Note that a vivid shade of red and a weak shade of red have the same hue. They may also have the same brightness. In summary, brightness, hue, and saturation are independent variables in the science of colorimetry.

1.4   THE TRANSMISSION PRIMARIES

Although only three primary colors are required to reproduce any desired hue, these three primaries are not the only colors usable. Four primaries may be employed if we choose to do so. In fact, four transmission primaries are used in operation of a color-TV transmitter. This might seem to be an unnecessary complication, but it is mandatory because of the technical requirements that are placed on the color-TV system. These requirements will be subsequently explained in greater detail. At this time, merely note the four basic transmission primaries depicted in Fig. 1-6. These primaries are named +(R−Y), −(R−Y), +(B−Y), and −(B−Y). Observe that +(R−Y) corresponds to a purplish-red hue, −(R−Y) corresponds to a bluish-green hue, −(B−Y) corresponds to a greenish-yellow hue and +(B−Y) corresponds to a purplish-blue hue.

Note that green, for example, is transmitted by a combination of −(R−Y) and −(B−Y) signals. Cyan is transmitted by a combination of +(B−Y) and −(R−Y) signals. Similarly, any one of the three primaries or

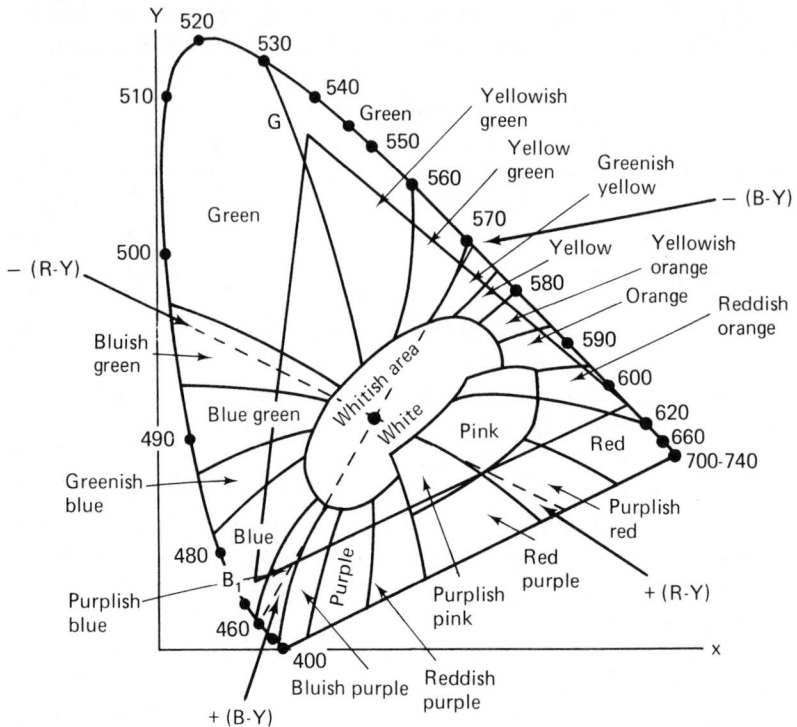

FIGURE 1-6   The four primaries utilized in color-television transmission.

complementaries is transmitted by a combination of two of the transmission primaries. It is instructive to consider the technical properties of the four transmission primaries. In the first place, the transmission primaries are *color-difference signals*. This term seems puzzling until we consider what this means in respect to brightness, hue, and saturation. Any color can be specified in terms of its brightness, hue and saturation and each of these three properties of color can be treated separately.

Since a black-and-white picture is described entirely in terms of its brightness from one area to another and from one point to another, it follows that a black-and-white picture has no hue and no saturation. Therefore, if a black-and-white picture is to be changed to a color picture, hue and saturation information must be added to the brightness information. This is basically what is accomplished by the color-TV system—an ordinary black-and-white television image is changed into a color image by addition of hue and saturation information. To summarize briefly, a black-and-white image is produced by adding color-difference signals to the brightness signal. As its name indicates, a color-difference signal provides the distinction between a black-and-white image and a color image.

R–Y and B–Y are examples of color-difference signals. Consider how a color-difference signal is formed. White is formed from a suitable blend of red, green, and blue primary hues. Technically, a *white signal* (an ordinary television signal) is called a *Y signal*. Now, to form an R–Y signal, a process in which the Y signal is subtracted from a red signal is employed. In turn, an *R minus Y signal*, or R–Y, signal is obtained. Since a Y signal is the sum of red, green, and blue signals, proceed as follows:

$$Y = R + G + B \qquad (1.1)$$

Equation (1.1) is not entirely complete because the red primary is 30% as bright as reference (standard) white, green is 59% as bright as reference white, and blue is 11% as bright as reference white. In turn, to make the equation "balance" and mathematically correct, these brightness factors must be taken into account to make it as follows:

$$1.00Y = 0.30R + 0.59G + 0.11B \qquad (1.2)$$

Note that the Y signal, which has been formed from red, green, and blue signals, is positive. This is the familiar black-and-white television signal, often called a *video signal*. To form a –Y signal, the Y signal is passed through a phase inverter. Thus, the Y signal is "turned upside down," or its polarity is reversed. To obtain an R–Y signal, subtract the Y signal from a red signal. This is the same as adding a –Y signal to a red signal. Thus, proceed as follows:

$$R - Y = 1.00R - 0.30R - 0.59G - 0.11B \qquad (1.3)$$

$$\text{or } R - Y = 0.70R - 0.59G - 0.11B \qquad (1.4)$$

Equation (1.4) states that to obtain an R–Y signal, add red, green, blue signals together in certain proportions and polarities. This forms the color-difference or transmission primary signals from red, green, and blue primary signals. Just how this is done by equipment at the color-TV transmitter will subsequently be explained in greater detail. At this point, the important fact to be noted is that when a Y signal is added to an R–Y signal, an R signal will be obtained. This is done at the color-TV receiver so that a red primary signal may be obtained from the red color-difference signal. This method of processing the primary color signals has been standardized by the National Television Systems Committee, a group whose name is generally abbreviated to NTSC. To anticipate subsequent discussion, the NTSC system enables a black-and-white TV receiver to reproduce a color signal as a normal black-and-white image. Conversely, a color-TV receiver reproduces a black-and-white signal as a normal black-and-white image, or, if tuned to a color signal, the receiver reproduces a color image. This aspect of the NTSC system is called *compatibility*.

Although a color-difference signal has no brightness component, it corresponds to a certain hue, as noted previously. Thus, a +(R–Y) signal may be described in general terms as corresponding to a "red" hue. However, this is not an entirely accurate description, as shown in Fig. 1-6. A +(R−Y) signal corresponds to a red hue that is tinged with purple. This fact is apparent from Fig. 1-7, which illustrates a true red hue beside a +(R−Y) hue. Similarly, a +(B−Y) signal may be described in general terms as corresponding to a "blue" hue. However, to be strictly accurate, a +(B−Y) signal corresponds to a blue hue that is tinged with purple, as seen in Fig. 1-6. Again, this fact is apparent in Fig. 1-7, which illustrates a true blue hue beside a +(B−Y) hue.

> **FIGURE 1-7**   Comparison of red and blue hues with +(R−Y) and +(B−Y) hues. (See inside front cover of book for artwork.)

As we proceed in our study of color television, it will be helpful to correlate hues with signal phases. This representation may take more than one form, and one useful form consists of a "color wheel," an example of which is illustrated in Fig. 1-8. This is fundamentally another method of presenting the color spectrum shown in Fig. 1-2 and has the advantage of depicting progressive blends of the primary colors. For example, as green light is blended with red light, a yellow-red hue results. A larger proportion of green yields the complementary color yellow. A still larger proportion of green results in a green-yellow hue, which is the basic −(B−Y) transmission primary. Finally, a pure green light is the primary green hue. There are other transmission primaries besides those indicated in Fig. 1-8, and these will be subsequently explaned in greater detail.

> **FIGURE 1-8**   A color wheel depicting the primary, complementary, and basic transmission primary hues. (See inside front cover of book for artwork.)

1.5   HISTORICAL NOTES

Although color television is popularly considered to be a comparatively recent development, it actually has a long history and is almost as old as black-and-white television. Experimental color-television systems were in use before the introduction of the cathode-ray tube. These early systems were very crude, but they demonstrated the fact that transmission of moving pictures in color by means of electromagnetic waves was physically possible. Historically, the first color-TV system that was practical for commercial use became available soon after the establishment of black-and-white television broadcasting following World War II. Television receivers had already been perfected almost to their present form, and cathode-ray tubes had been developed into large-screen picture tubes.

To obtain color image transmission and reception, a *color scanning disk* was employed, as is shown in Fig. 1-9. This is a circular disk comprising three transparent color filters arranged in sequential segments. These were primary-color transmission filters. Thus, if a scene were viewed through the red segment, only the red rays would be passed. If the scene were viewed through the green segment, only the green rays would be passed. Therefore, when the color scanning disk was rotated in front of an ordinary TV camera, as depicted in Fig. 1-10, red, green, and blue images were sequentially focussed by the camera on the screen of the camera tube. Thus, the scene was broken down into its three primary colors, and each color image was changed into a corresponding video signal. These three video signals were then transmitted sequentially.

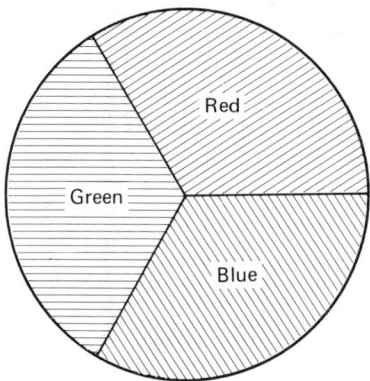

FIGURE 1-9   A color scanning disk.

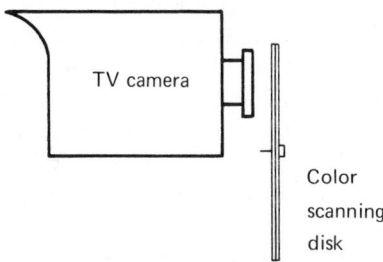

FIGURE 1-10   Arrangement at color-television transmitter.

At the receiver of this early color-TV system, a similar color scanning disk was rotated in front of the picture tube, as depicted in Fig. 1-11. It is evident that when the rotation of the receiving disk was synchronized with that of the transmitting disk, the primary color images were displayed in correct sequence. Since the color images followed one another at a rate that provided persistence of vision, a full-color picture was observed by the viewer. This historical color-TV system was entirely practical, although it had certain disadvantages. First, it was not compatible with existing black-and-white receivers. That is, a conventional black-and-white receiver required circuit modifications before it could produce a normal image when tuned to a color-TV station. Second, the color scanning disk was large, clumsy, and subject to mechanical wear, vibration, and noisy operation. Third, rapid movements in the scene being scanned resulted in blurred or jerky images, which compared unfavorably with the quality of conventional black-and-white reception.

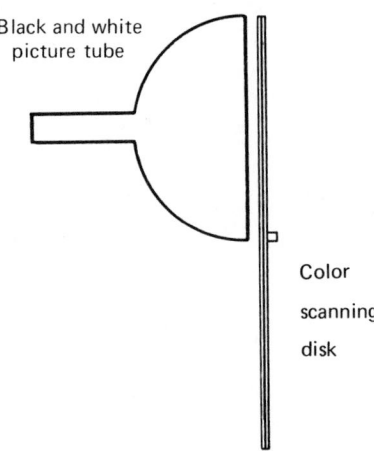

FIGURE 1-11   Arrangement of color scanning disk receiving system.

Because of the inherent disadvantages of the color scanning disk system of transmission and reception, it was soon displaced by the NTSC (National Television Systems Committee) system. Although the NTSC system is comparatively elaborate and complex, it provides the technical means whereby compatibility is achieved in an all-electronic system that provides high-quality color reproduction without perceptible jerkiness or blurring of rapidly-moving action scenes. How these objectives were realized is explained in Chapters 2 through 15.

It should be noted that the color scanning disk system is still in use for certain specialized applications. For example, this system was employed in Apollo flights to the moon. Some commercial applications also utilize the color scanning disk system.

EXERCISES

Questions

1. What is the ratio of the visible electromagnetic spectrum to the total electromagnetic spectrum?

2. State the wavelength limits of the visible electromagnetic spectrum.

3. Name the additive primary colors; the additive complementary colors.

4. Describe the physical characteristics of black and white.

5. Briefly discuss the meaning of the term *saturation*.

6. Do orange hues require more or less definition than blue hues in the reproduction of a color image?

7. State the three basic characteristics of a color.

8. Can four primary colors be employed in reproduction of a color image?

9. Name the basic transmission primaries.

10. How is the term *color-difference signal* defined?

11. State the comparative brightnesses of the red, green, and blue hues.

12. What is the brightness of white? of black?

13. Write the equation for $R-Y$ in terms of R, G and B.

14. Define the term *compatibility*.

15. To what general hue does an $R-Y$ signal correspond; a $-(R-Y)$ signal?

16. To what general hue does a $B-Y$ signal correspond; a $-(B-Y)$ signal?

17. Briefly describe a *color scanning disk*.

18. What are the disadvantages of a color scanning disk system?

19. State the objectives that were realized by the NTSC system.

20. Note a present-day application for the color scanning disk system.

True-False

1. Visible light has a wavelength near the low-frequency end of the electromagnetic spectrum.

2. X-rays have a higher frequency than visible light waves.

3. A micron is equal to one-millionth of a meter.

4. All wavelengths of light are equally diffracted by a glass prism.

5. The additive primaries are red, yellow, and blue.

6. Subtractive complementary colors are the same as additive primary colors.

7. A color picture tube employs red, green, and blue phosphors.

8. No primary hue, such as green, can be produced by combining the other two primary hues, such as red and blue.

9. Black is the complement of white.

10. Any color that occurs in nature can be reproduced by the color-TV system.

11. Highly saturated hues exist that cannot be reproduced by the color-TV system at full saturation.

12. Many colors that occur in nature have maximum saturation.

13. Color vision is more acute than black-and-white vision.

14. Very small areas in a color image are perceived as grays by the human eye.

15. Although the eye is very critical of green tints, it is quite tolerant of off-hue orange tints.

16. Brightness, hue and saturation are independent variables.

17. The transmission primaries are $\pm(R-Y)$ and $\pm(B-Y)$.

18. A color-difference signal has zero brightness.

19. White light is formed by blending suitable proportions of red, green and blue hues.

20. A fully saturated red hue is 30% as bright as white; a fully saturated green hue is 59% as bright as white; and a fully saturated blue hue is 11% as bright as white.

21. If a Y signal is reversed in polarity, a $-Y$ signal is obtained.

22. When a red signal voltage is added to a $-Y$ signal voltage, an $R-Y$ signal voltage is obtained.

23. A $-Y$ signal is also called a color-difference signal.

24. The term *NTSC* denotes the National Television Systems Committee.

25. *Compatibility* denotes that a color receiver reproduces a color broadcast in a color, a black-and-white broadcast in black-and-white, and that a black-and-white receiver reproduces either a black-and-white broadcast or a color broadcast in black-and-white.

Multiple Choice

1. Television waves have a/an _____ frequency, compared to visible light.
   (a) higher
   (b) lower
   (c) identical
   (d) imaginary

2. The visible wavelengths of electromagnetic radiation are in the vicinity of _____ .
   (a) 550 millimicrons
   (b) 150 millimicrons
   (c) 950 millimicrons
   (d) 360 meters

3. A ray of red light is diffracted _____ than a ray of blue light by a glass prism.
   (a) more
   (b) less
   (c) more at low temperatures, less at high temperatures
   (d) less at low temperatures, more at high temperatures

4. Additive primary colors are _____ .
   (a) brighter than subtractive primary colors
   (b) highly saturated, compared to the subtractive primary colors
   (c) highly saturated, compared to the additive complementary colors
   (d) red, green, and blue

5. Yellow, an additive complementary color, is formed by blending _____ .
   (a) red and green
   (b) red, white, and blue
   (c) red and blue
   (d) red, green, and blue

6. A chromaticity diagram can be characterized as a _____ .
   (a) schematic diagram
   (b) block diagram
   (c) color map
   (d) flow chart

7. As a color area is progressively reduced in size, purple and green-yellow hues appear as grays before _____ hues appear as grays.

(a) black and white
(b) red and cyan
(c) yellow and magenta
(d) purple and blue-green

8. More definition is required for _____ hues than for blue hues.
   (a) gray
   (b) colorless
   (c) orange
   (d) invisible

9. Brightness, hue and saturation are _____ variables.
   (a) dependent
   (b) independent
   (c) fixed
   (d) unpredictable

10. The transmission primaries are _____ .
    (a) red, green, and blue
    (b) red, yellow, and blue
    (c) red, white, and blue
    (d) $\pm(R-Y)$ and $\pm(B-Y)$

11. $R-Y$ corresponds to a _____ hue.
    (a) purplish-red
    (b) reddish-purple
    (c) bluish-green
    (d) greenish-blue

12. $B-Y$ corresponds to a _____ hue.
    (a) purplish-blue
    (b) reddish-purple
    (c) bluish-green
    (d) greenish-blue

13. $-(R-Y)$ corresponds to a _____ hue.
    (a) bluish-green
    (b) greenish-blue
    (c) reddish-purple
    (d) purplish-red

14. $-(B-Y)$ corresponds to a _____ hue.
    (a) bluish-purple
    (b) purplish-blue
    (c) greenish-yellow
    (d) bluish-green

15. A green hue is transmitted by a combination of _____ signals.
    (a) (R–Y) and (B–Y)
    (b) –(R–Y) and –(B–Y)
    (c) (R–Y) and –(B–Y)
    (d) –(R–Y) and (B–Y)

16. A cyan hue is transmitted by a combination of _____ signals.
    (a) (R–Y) and (B–Y)
    (b) –(R–Y) and –(B–Y)
    (c) (R–Y) and –(B–Y)
    (d) –(R–Y) and (B–Y)

17. Color-difference signal voltages are produced by subtracting the _____ signal voltage from a hue signal voltage.
    (a) brightness
    (b) saturation
    (c) opposite
    (d) same

18. Black is a/an _____ .
    (a) blend of –R, –G and –B signal voltages
    (b) blend of R, G and B signal voltages
    (c) –Y signal voltage
    (d) absence of light

19. White is a blend of _____ .
    (a) –R, –G and –B signal voltages
    (b) R, G and B signal voltages
    (c) R–Y and B–Y signal voltages
    (d) –(R–Y) and –(B–Y) signal voltages

20. An R–Y signal voltage can be formed from a combination of _____ signal voltages.
    (a) R, –G and –B
    (b) –R, G and B
    (c) R, G and –B
    (d) R, –G and B

Problems

1. If a red hue has a wavelength of 700 millimicrons, what is its wavelength expressed in millionths of an inch?

2. What is the ratio of the wavelengths for long radio waves to gamma rays?

3. State the frequency of a red hue that has a wavelength of 700 millimicrons.

4. What is the ratio of the frequencies for long radio waves to gamma rays?

5. A B—Y signal has an optical wavelength of 460 millimicrons; to what frequency does this wavelength correspond?

6. State the optical wavelength of a —(B—Y) signal.

7. If a 0.30R signal voltage is added to a 0.59G signal voltage and to an 0.11B signal voltage, what is the brightness value of the combination?

8. Calculate the brightness value of the resultant combination if a 0.70R signal voltage is added to a —0.59G signal voltage and to a —0.11B signal voltage.

9. If an R—Y signal voltage is added to another 0.5(R—Y) signal voltage, what is the brightness value of the combination?

10. Calculate the brightness value of a combination of an R—Y signal voltage with a Y signal voltage.

Chapter 2

Survey Of Color-Television Technology

2.1 BASIC COLOR-TELEVISION SYSTEM

Although the technical details of a television system are comparatively elaborate, its basic plan is simple (Fig. 2-1). A black-and-white TV system can be arranged by connecting the output from an ordinary TV camera to the electron gun in a picture tube. A closed-circuit TV system, for example, operates in this manner. A color-TV system can be arranged by connecting the outputs from three color-TV cameras to the three electron guns in a color picture tube. To anticipate subsequent discussion, a color picture tube produces a red image when its red gun is energized, a green image, when its green gun is energized, and a blue image, when its blue gun is energized. Note that insofar as black-and-white pictures are concerned, both of the arrangements depicted in Fig. 2-1 serve equally well. In other words since a suitable combination of red, green, and blue hues produces white or gray, the color-TV arrangement provides a black-and-white image that is indistinguishable from that provided by the black-and-white TV arrangement. This is the most fundamental aspect of *compatibility*.

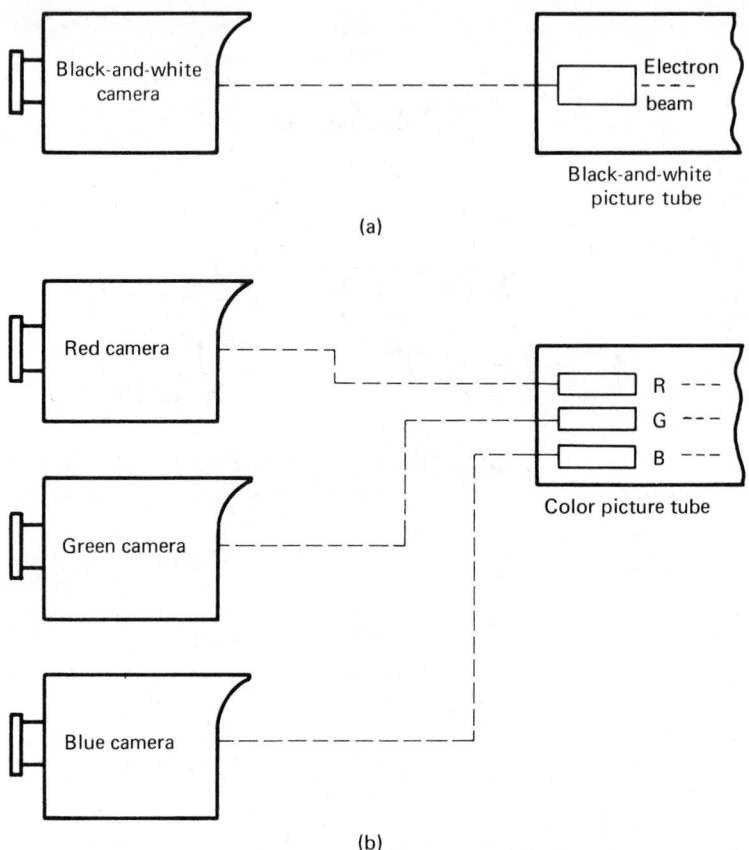

FIGURE 2-1 Fundamental television systems. (a) Basic black-and-white television system; (b) basic color television system.

In practice, the three color-TV cameras depicted in Fig. 2-1 are built into a unit that employs a single lens, as shown in Fig. 2-2. The image formed by the lens is split into three images by means of glass prisms. These prisms are designed as dichroic mirrors. A dichroic mirror passes one wavelength and rejects other wavelengths (colors) of light. Thus, red, green, and blue color images are formed. The rays from each of the light splitters also pass through color filters, called *trimming filters*. These filters provide highly precise primary color images, which are converted into video signals by image-orthicon or vidicon camera tubes. Thus, three primary color signals are generated. They are commonly called the R, G, and B signals.

To transmit these R, G, and B signals by means of electromagnetic waves, three television channels could be employed. This was the method

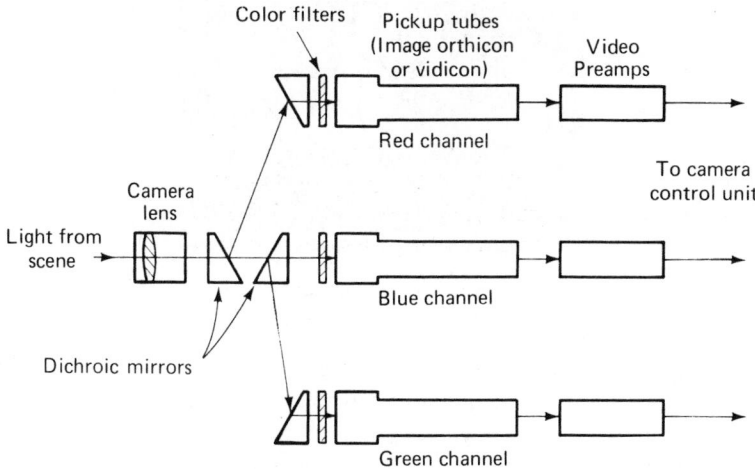

FIGURE 2-2   Plan of a color-television camera.

used by the pioneer workers. However, if three 6-MHz television channels were allocated to each color-TV transmitter, only one-third as many TV broadcast stations could have been accommodated in the VHF and UHF bands. Moreover, a three-channel transmission system would not have been compatible with existing black-and-white TV receivers. Therefore, a more sophisticated system was developed with the objective of providing complete compatibility between color and black-and-white receiver operation. As noted previously, the basic plan of the NTSC system is to add hue and saturation information to the brightness information contained in a standard black-and-white TV signal.

2.2   ANALYSIS OF THE BLACK-AND-WHITE TELEVISION SIGNAL

Figure 2-3 illustrates the waveform of a Y (black-and-white) signal. This video signal is said to occupy a bandwidth of 4 MHz. This is true as far as it goes. It can be shown, however, that the signal occupies its spectrum in clusters, with "empty" spaces between clusters, as depicted in Fig. 2-4(a). Therefore, hue and saturation information, or the color-difference signal, can be added to best advantage by its insertion in these "empty" spaces, as shown in Fig. 2-4(b). The color-difference signal is generally called a *chroma signal*. When the clusters of chroma signal energy and the clusters of Y signal energy are staggered in the video frequency spectrum, the signals are said to be *frequency interleaved*. This process of signal interleaving is also called *frequency interlacing.*

Frequency interleaving is important because of compatibility requirements. That is, when a black-and-white TV receiver is tuned to a color

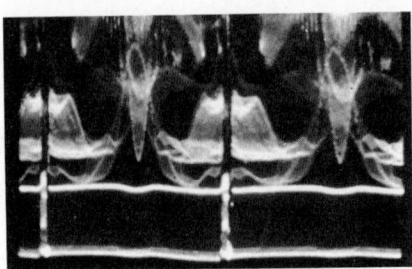

FIGURE 2-3   Waveform of a black-and-white video signal at vertical rate (dark background).

transmission, the chroma signal must be rejected insofar as possible. Figure 2-5 shows the signal relations that are involved in the frequency-interleaving process. Note that the Y signal repeats in phase on successive fields, as depicted in Fig. 2-5(a). The interleaved chroma signal repeats out of phase on successive fields, as depicted in Fig. 2-5(b). Therefore, the chroma signal effectively cancels itself out when processed by a black-and-white receiver. We will find that a color-TV receiver has additional circuits that separate the chroma signal from the Y signal, and reconstitute the R, G, and B signals before they are applied to the color picture tube.

To interleave the chroma signal with the Y signal, a *color subcarrier* must be employed. This color subcarrier will automatically interleave with the Y signal, provided that the subcarrier's frequency is an odd harmonic of one-half the horizontal scanning frequency. Thus, there is a very large number of possible frequencies for the color subcarrier. However, there are two technical reasons for assigning a high frequency to the color subcarrier. In the first place, the interleaving process is not 100% efficient, particularly in the darker portion of the image. Consequently, a low-visibility dot pattern may appear in the image when a black-and-white receiver is tuned to a color transmission. The visibility of this residual dot pattern can be minimized by making the dots very small. This is accomplished by assigning a high video frequency to the color subcarrier.

In the second place, the Y signal spectrum contains most of its energy at the low-frequency end, as shown in Fig. 2-6. Similarly, the chroma signal spectrum contains most of its energy in the vicinity of the color sub-carrier. Accordingly, residual beat interference (line crawl) between the chroma signal and the Y signal can be minimized by placing the chroma signal at the high-frequency end of the Y spectrum. This is accomplished by assigning a high video frequency to the color subcarrier. Note in passing that beat interference would not occur if the receiver (including the picture tube) were a perfectly linear system. However, practical receiver circuitry and picture tubes have residual amplitude nonlinearities. Therefore, suitable design

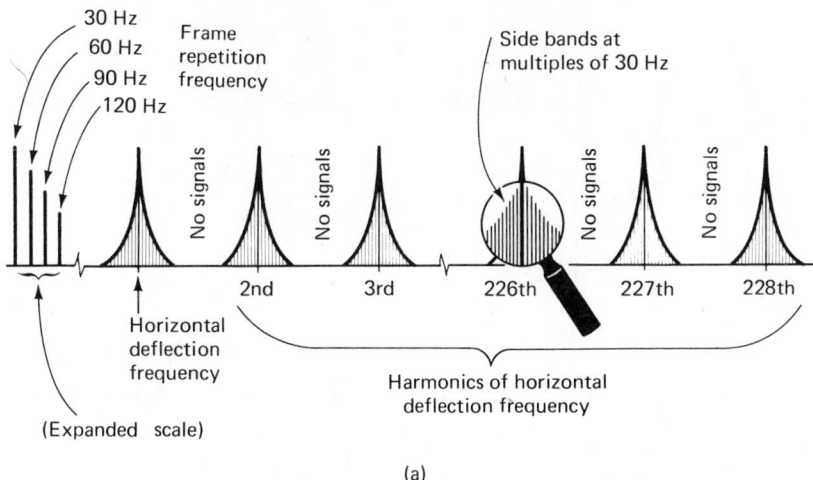

(a)

Light shaded areas indicate distribution of the chroma
energy for a subcarrier whose frequency is an odd multiple
of one-half the line frequency.

Dark shaded areas indicate distribution of brightness or Y energy
at whole multiples of line frequency.

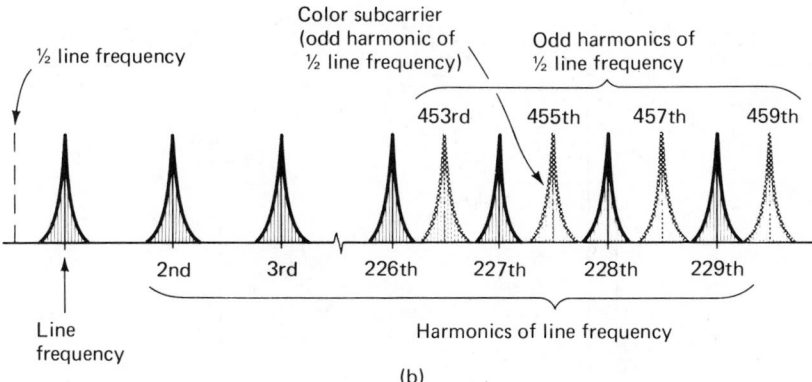

(b)

FIGURE 2-4   Distribution of signal energy in the video-frequency
spectrum. (a) Cluster distribution by a black-and-
white signal; (b) interleaving of clusters by color-
subcarrier and Y signals.

measures must be employed to minimize residual beat interference between
the Y and chroma signals.

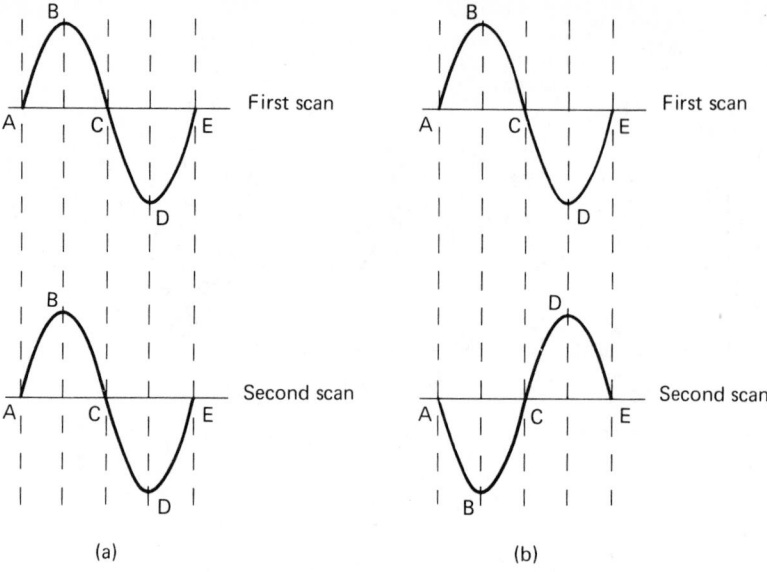

FIGURE 2-5   Signal relations in the frequency interleaving process. (a) Y signal repeats in phase on successive fields; (b) chroma signal repeats out of phase on successive fields.

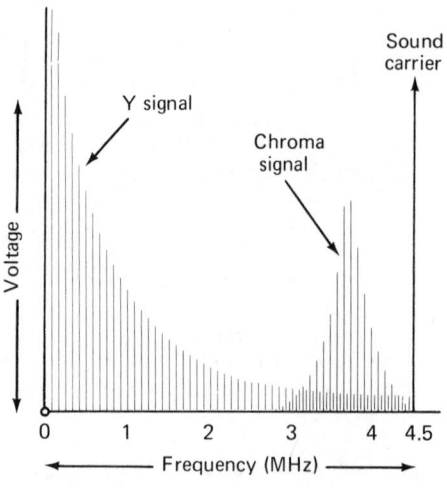

FIGURE 2-6   Y and chroma signal energy distributions.

2.3   COLOR-SUBCARRIER FREQUENCY AND RELATED OPERATING FREQUENCIES

Since the video-frequency channel has a bandwidth of approximately 4.2 MHz, the color-subcarrier frequency could theoretically have this value. During development of the NTSC system, color-subcarrier frequencies as high as 4.1 MHz were investigated; however, difficulties were encountered. First, as depicted in Fig. 2-6, the sound carrier has a frequency of 4.5 MHz. It was found impractical to design traps for separating the 4.1-MHz subcarrier from the 4.5-MHz sound carrier. Second, single-sideband transmission of the chroma signal is required with a 4.1-MHz color subcarrier, and this requirement involves increased receiver complexity with its attendant high production costs. Third, it is technically difficult to obtain a reasonably linear phase characteristic near the cutoff point of the video-frequency channel. Nonlinear phase response results in distortion of complex waveforms.

Accordingly, the NTSC specifications assign a frequency of 3.579545 MHz to the color subcarrier. As Fig. 2-6 shows, this choice provides double-sideband transmission of the basic chroma signal and is also so high in frequency that residual interference effects are unobjectionable. Since the subcarrier frequency is separated from the sound carrier frequency by approximately 0.9 MHz, efficient trapping can be provided by simple circuitry, and an acceptably linear phase characteristic can be realized through the chroma-signal region. Note that the 3.579545-MHz subcarrier frequency is interleaved not only with the Y signal, but also with the 4.5-MHz sound carrier. This interleaving provides minimum visibility of any 900-kHz dot pattern that might result from beating between the chroma and sound signals in circuits that are not entirely linear.

Since the color-subcarrier frequency must be an odd multiple of both the horizontal and vertical scanning frequencies, the NTSC standards stipulate a horizontal scanning frequency of 15.734264 kHz, and a vertical scanning frequency of 59.94 Hz. These scanning frequencies are very close to the 15.75-kHz and 60-Hz frequencies utilized in black-and-white transmissions. Therefore, a color receiver will automatically "lock in" when tuned to a black-and-white transmission, and a   black-and-white receiver will automatically "lock in" when tuned to a color transmission. Accordingly, compatibility is not sacrificed in providing for complete system frequency interleaving. Note in passing that the color-subcarrier frequency is usually given as 3.58 MHz in the interest of brevity.

2.4   ENCODING OF THE COLOR SIGNAL INTO THE Y SIGNAL

A *color encoder* (also termed a *color modulator* or a *Colorplexer*) is a subsystem that processes the outputs from three color cameras to form the NTSC color signal. In the first analysis, the encoder arrangement processes the outputs from the preamplifiers depicted in Fig. 2-2 to form the interleaved Y and chroma signals shown in Fig. 2-6. To *encode* means to translate—as to prepare a program for a computer. One basic function of a color encoder was noted previously, viz., to change the red, green, and blue primary signals into ±(R—Y) and ±(B—Y) transmission primary signals. Figure 2-7 shows the basic plan of a color encoder. We will now consider the function of each section in this simplified color encoder.

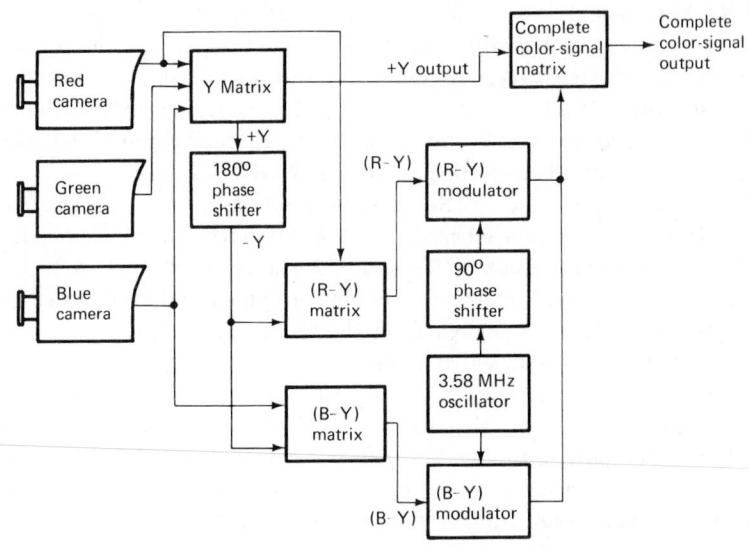

FIGURE 2-7   Basic plan of a color encoder.*

If a color card consisting of red, green, and blue strips is being scanned, as depicted in Fig. 2-8, each of the color cameras produces a square-wave output voltage. These square waves do not occur at the same time, but follow one another in accordance with the colors of the strips on the card. In scanning this arrangement we observe the following facts:

1. There is 100% output from the red camera, while the color cameras are scanning the red strip. Conversely, there is zero output from the red camera while scanning the green and blue strips.

2. There is 100% output from the green camera, while the color cameras are scanning the green strip. Conversely, there is zero output from the green camera while scanning the red and blue strips.

---

*As explained in greater detail subsequently, the 3.58-MHz subcarrier is suppressed in the complete color signal, in order to minimize chroma-B/W interference.

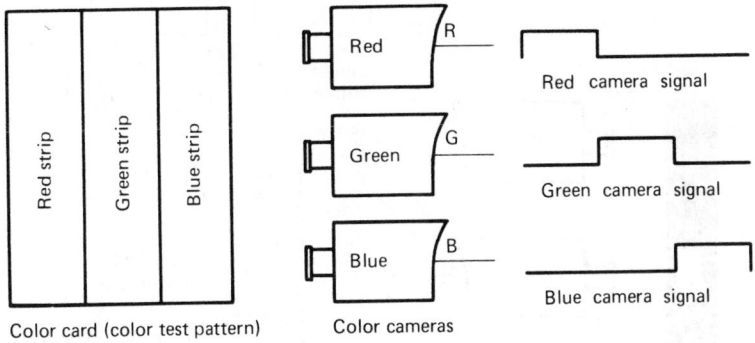

Color card (color test pattern)   Color cameras

FIGURE 2-8   Output voltages from the red, green, and blue color
cameras while scanning a color card.

3. There is 100% output from the blue camera, while the color cameras
are scanning the blue strip. Conversely, there is zero output from the
blue camera while scanning the red and green strips.

As shown in Fig. 2-7, the outputs from the three color cameras are
fed into the *Y matrix*. This Y matrix is a resistive network that combines the
outputs from the cameras in certain proportions. Figure 2-9 shows the basic
action of the Y matrix. In this example, a card that has black and white strips
is being scanned. We know that the red, green, and blue primary colors must
be blended in certain proportions to produce white. Similarly, the output
voltages from the red, green, and blue cameras must be mixed in certain proportions to form a Y signal. While scanning the white strip, the Y matrix combines 30% of the red-camera signal with 59% of the green-camera signal, and
11% of the blue-camera signal to form the Y signal. Thus, the Y matrix develops 100% output while scanning a white strip. Of course, there is zero output
from the Y matrix while scanning a black strip. Note that the output from
the Y matrix is the same as it would have been if a black-and-white TV camera were used.

Observe the outputs from the color cameras and from the Y matrix
while scanning a red strip, as depicted in Fig. 2-10. It is evident from Fig. 2-7
that there will be 30% output from the Y matrix. To form the R—Y signal,
the output from the Y matrix is passed through a unity-gain amplifier, thereby inverting the Y signal and forming a —Y signal. Then, the 100% R signal
and the 30% —Y signal are combined in the R—Y matrix. Accordingly, the
R—Y signal is formed and is fed to the R—Y modulator. Note in Fig. 2-11
that this R—Y signal has a relative level of 70%, inasmuch as it is the difference between the R and the —Y voltage levels. To summarize briefly, the
R—Y signal is simply a square-wave voltage with a certain level relative to the
red-camera and Y-matrix output voltages.

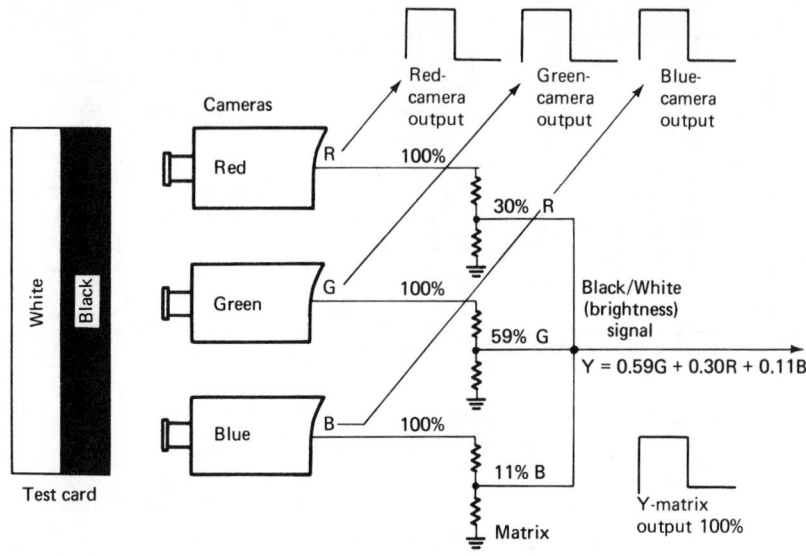

FIGURE 2-9 Basic operation of the Y matrix.

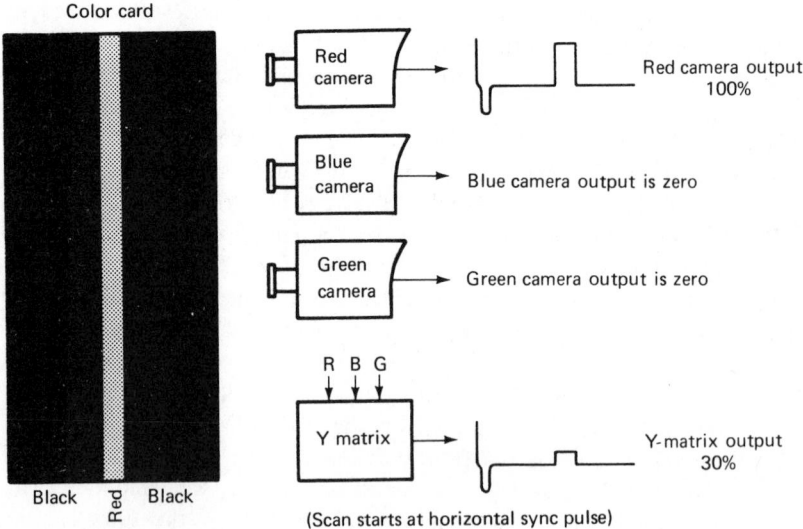

FIGURE 2-10 Outputs from the color cameras and the Y matrix while scanning a red strip.

Meanwhile, although there is no output from the blue camera in Fig. 2-10, there is a 30% −Y output from the Y matrix that is fed to the B−Y matrix, as seen in Fig. 2-7. Accordingly, there is a square-wave voltage with a

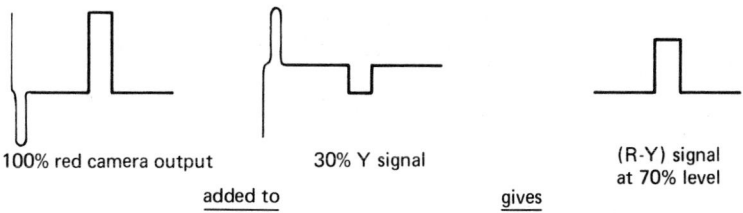

100% red camera output   30% Y signal   (R-Y) signal
                                                                at 70% level

          added to   gives

FIGURE 2-11   Formation of the R–Y signal while the color cam-
                   eras are scanning a red strip.

30% level fed from the B–Y matrix to the B–Y modulator. Observe that the
3.58-MHz subcarrier voltages in the R–Y and B–Y modulators are 90° out of
phase. Therefore, the outputs from the R–Y and B–Y modulators combine
to form a color-difference signal with a 76% amplitude, as shown in Fig. 2-12.
With reference to Fig. 2-13, the foregoing process is summarized. Observe also
in Fig. 2-7 that the complete color-signal matrix is energized by the 76%
color-difference or chroma signal, and by a 30% Y signal from the Y matrix.
Therefore, the output from the complete color-signal matrix is a 76% 3.58-
MHz sine wave centered on the 30% Y signal, as depicted in Fig. 2-13.

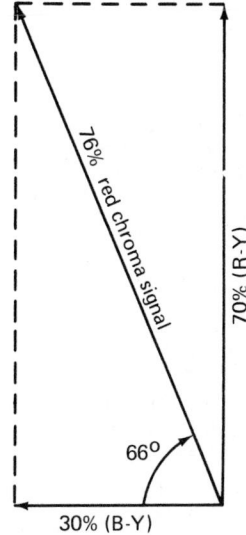

76% red chroma signal

70% (R-Y)

66°

30% (B-Y)

FIGURE 2-12   A 70 per cent R–Y voltage combines with a 30
                   per cent B–Y voltage to form a 76 per cent resul-
                   tant with a phase angle of 67°.

The encoding process for a red strip or bar signal is now complete,
and an NTSC color signal has been formed. This red chroma signal has a phase

1. Output voltage from the red camera is 100 per cent of maximum amplitude.

2. Output voltage from the green camera is zero.

3. Output voltage from the blue camera is zero.

4. Output from the Y matrix is +30 per cent of maximum amplitude.

5. Output from the 180° phase shifter is -30 per cent of maximum amplitude.

6. Output from the (R-Y) matrix is +70 per cent of maximum amplitude.

7. Output from the (B-Y) matrix is -30 per cent of maximum amplitude.

8. Combined outputs from the (R-Y) and (B-Y) modulators is a 3.58 MHz signal at 76 per cent peak value.

9. Output from the complete color-signal matrix is the 76 per cent chroma signal, centered on the +30 per cent Y signal.

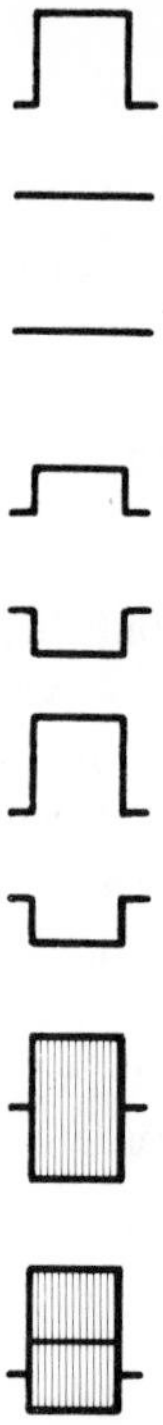

The 30 per cent Y level corresponds to the brightness of the red hue. The 76 per cent chroma level corresponds to a fully saturated red hue. The two modulators have a resultant output signal phase corresponding to the phase of a red hue.

FIGURE 2-13   Step-by-step encoding process for a red bar signal.

angle of 67°, as was shown in Fig. 2-12. In summary, the red chroma signal is simply a 3.58-MHz sine-wave voltage with a relative amplitude of 76%, and a phase angle of 67°. If the encoding process for green and blue bar signals is analyzed, the chroma-signal diagram shown in Fig. 2-14 will be arrived at. If the encoding process for yellow, magenta, and cyan bar signals are analyzed, these complementary colors will be found to correspond to equal, but oppositely phased voltages compared to the primary color signals, as shown in Fig. 2-15. The chroma-signal values indicated in Fig. 2-15 are called *unadjusted chroma values*, for reasons that will be considered in the next paragraph.

When the unadjusted chroma values noted in Fig. 2-15 are utilized to form a color-bar signal, the waveform shown in Fig. 2-16(a) is obtained. This is the basic NTSC color-bar waveform. Observe, however, that it is not prac-

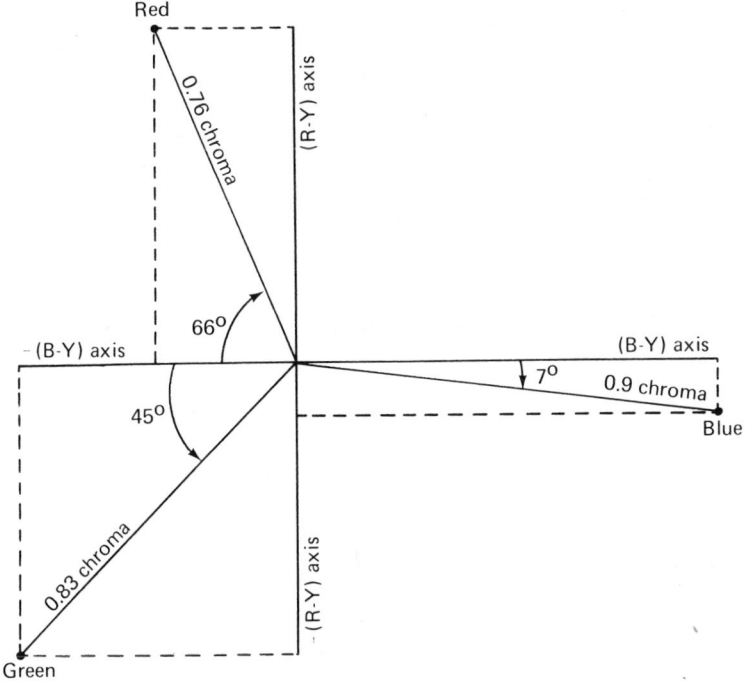

FIGURE 2-14   Chroma-signal diagram showing relative amplitudes
and phase angles for the primary colors.

tical to transmit this waveform. The chroma-signal peaks would exceed both the maximum sync-tip and white levels. Accordingly, overmodulation would be produced, and the reproduced colors would be objectionably distorted. To avoid overmodulation, it is necessary to reduce the chroma-signal values at the transmitter. The transmitter radiates readjusted chroma-signal values which must then be increased to the unadjusted chroma values at the color-TV receiver.

Figure 2-17 shows how readjusted chroma values are produced at the transmitter. Note that the R—Y signal is reduced to 0.877 of its original value, and the B—Y signal is reduced to 0.493 of its original value. No reduction is made in the amplitude of the Y signal. Thus the relative amplitude of red and cyan becomes 0.63, of green and magenta 0.59, and of blue and yellow 0.45. Observe also that readjustment of chroma values results in a change of the chroma phase angles. Thus, the readjusted phase angle for red and cyan is 76.5°, for green and magenta, 61.1°, and for blue and yellow, 12°. Observe in Fig. 2-18 that although these readjusted chroma values result in a transmitted color-bar pattern that still exceeds the white level, the chroma-signal is approximately within limits.

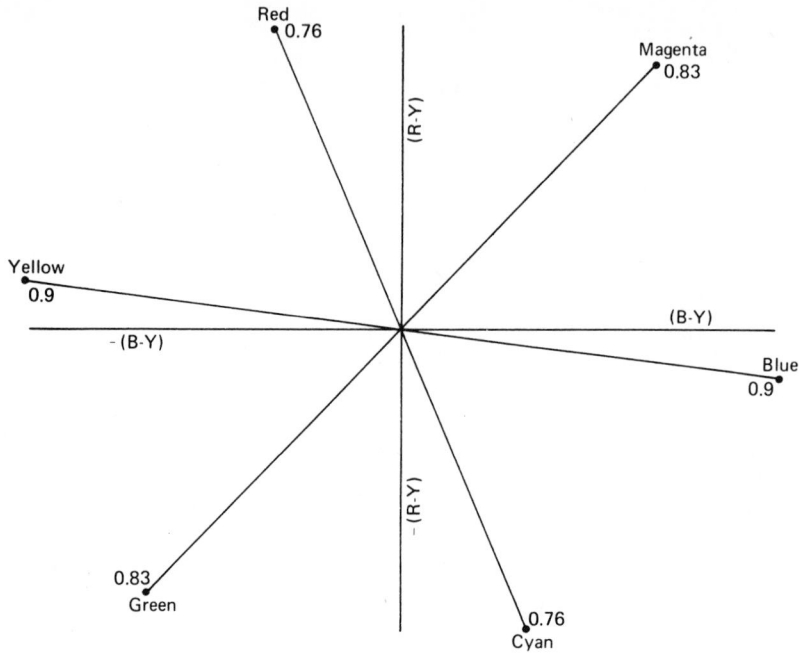

FIGURE 2-15   Chroma-signal diagram for the primary and complementary colors. (Unadjusted chroma values.)

It is instructive to note that the color-bar pattern illustrated in Fig. 2-19 corresponds to hues with full brightness and full saturation. In practice, the saturation of hues in natural and staged scenes seldom exceeds 75%. Since the amplitude of the chroma signal is proportional to the saturation of a hue, maximum chroma-signal amplitudes are seldom encountered in color-TV transmission. Therefore, employment of readjusted chroma values results in a complete color signal that will rarely, if ever, overmodulate the picture carrier at the color-TV transmitter. Finally, note that a color-burst signal appears on the back porch of the horizontal sync pulse in Fig. 2-19. The burst signal will be discussed in detail.

Figure 2-20 shows the saturation waveforms for two color-bar patterns and a summary of the color-B/W compatibility characteristics for the NTSC system is depicted in Fig. 2-21.

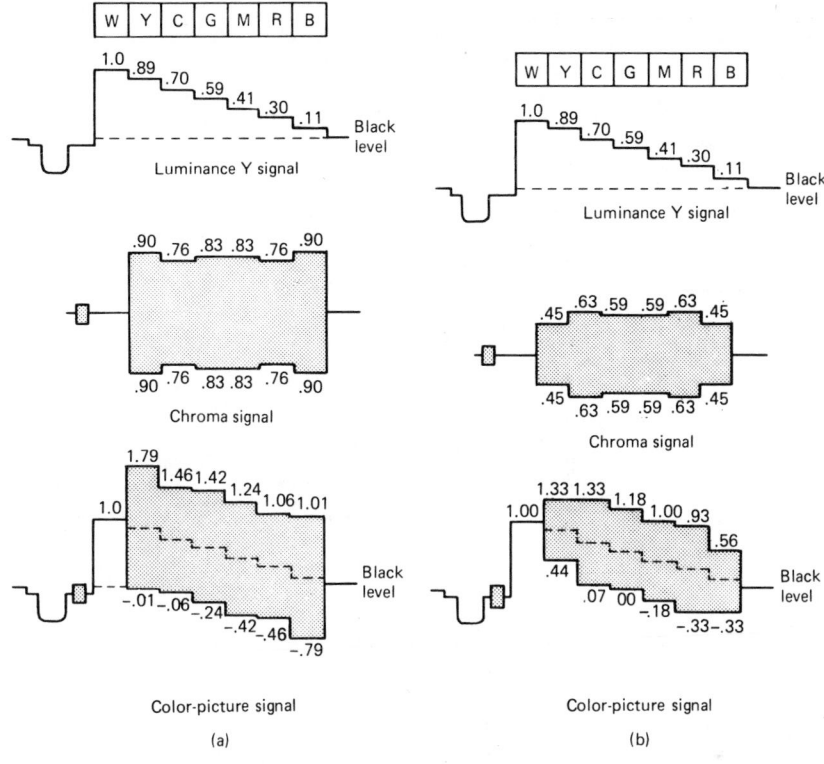

FIGURE 2-16 Color-bar waveforms with unadjusted and read-justed chroma values. (a) Color-bar waveform utilizing the unadjusted chroma-signal values shown in Fig. 2-15; (b) color-bar waveform utilizing the readjusted chroma-signal values shown in Fig. 2-18.

EXERCISES

Questions

1. How many color-TV camera tubes does a color-TV transmitter employ?

2. What is a *trimming filter*?

3. Briefly describe the basic plan of the NTSC system.

4. Describe the *clusters* that are present in a video signal.

5. Define the term *frequency interleaving*.

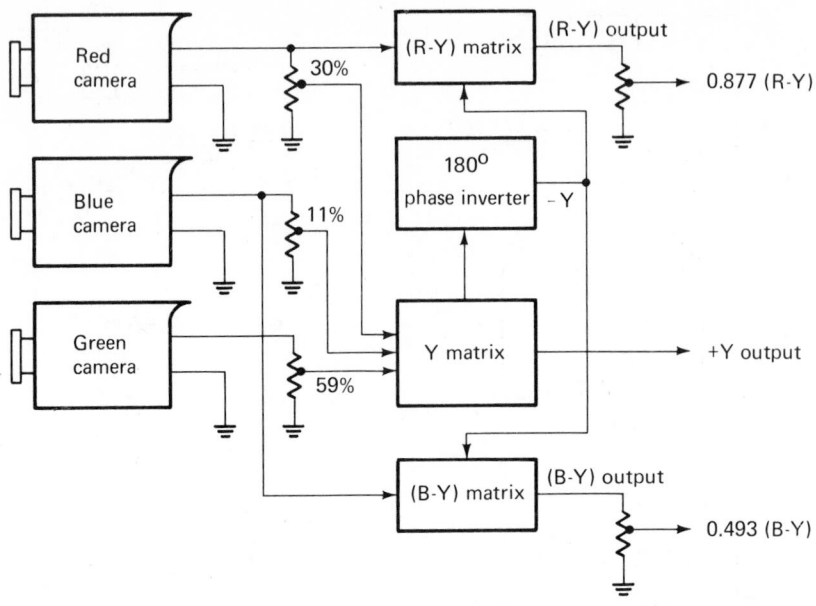

FIGURE 2-17   How readjusted chroma values are produced at the transmitter.

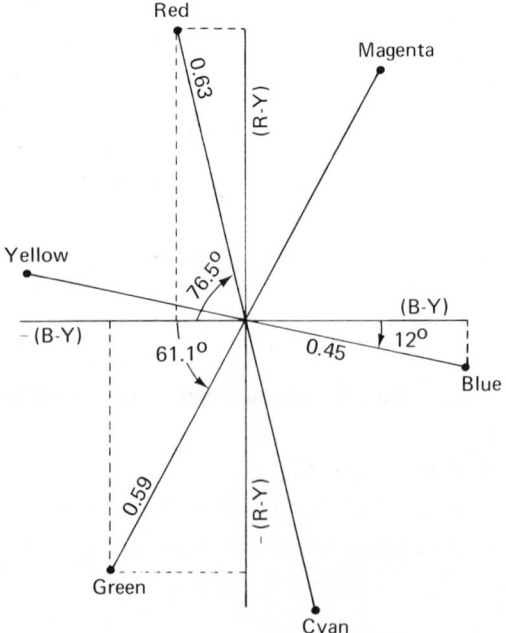

FIGURE 2-18   Chroma-signal diagram for the primary and complementary colors using readjusted chroma values.

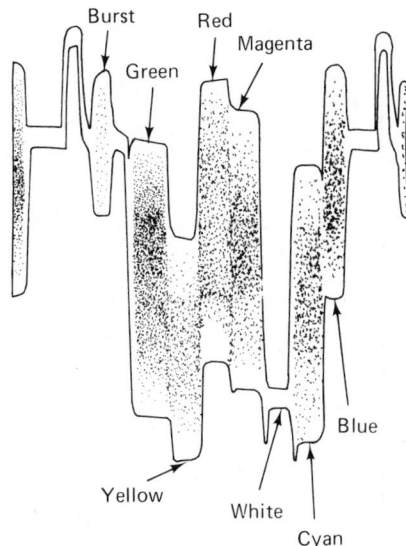

FIGURE 2-19   Standard color-bar signal with readjusted chroma values.

6. What is the color-subcarrier frequency?

7. Explain the relation of the color-subcarrier frequency to the horizontal scanning frequency.

8. Does the Y signal contain most of its energy at low frequencies or at high frequencies?

9. Does the chroma signal contain most of its energy in the vicinity of the color subcarrier, or at some other frequency or frequencies?

10. What is the approximate bandwidth of the video-frequency channel?

11. Is the basic chroma signal a double-sideband signal or a single-sideband signal?

12. State the difference between the color-subcarrier frequency and the intercarrier-sound frequency.

13. How does the vertical-scanning frequency in a color system differ from the vertical-scanning frequency in a black-and-white system?

14. How does the horizontal-scanning frequency in a color system differ from the horizontal-scanning frequency in a black-and-white system?

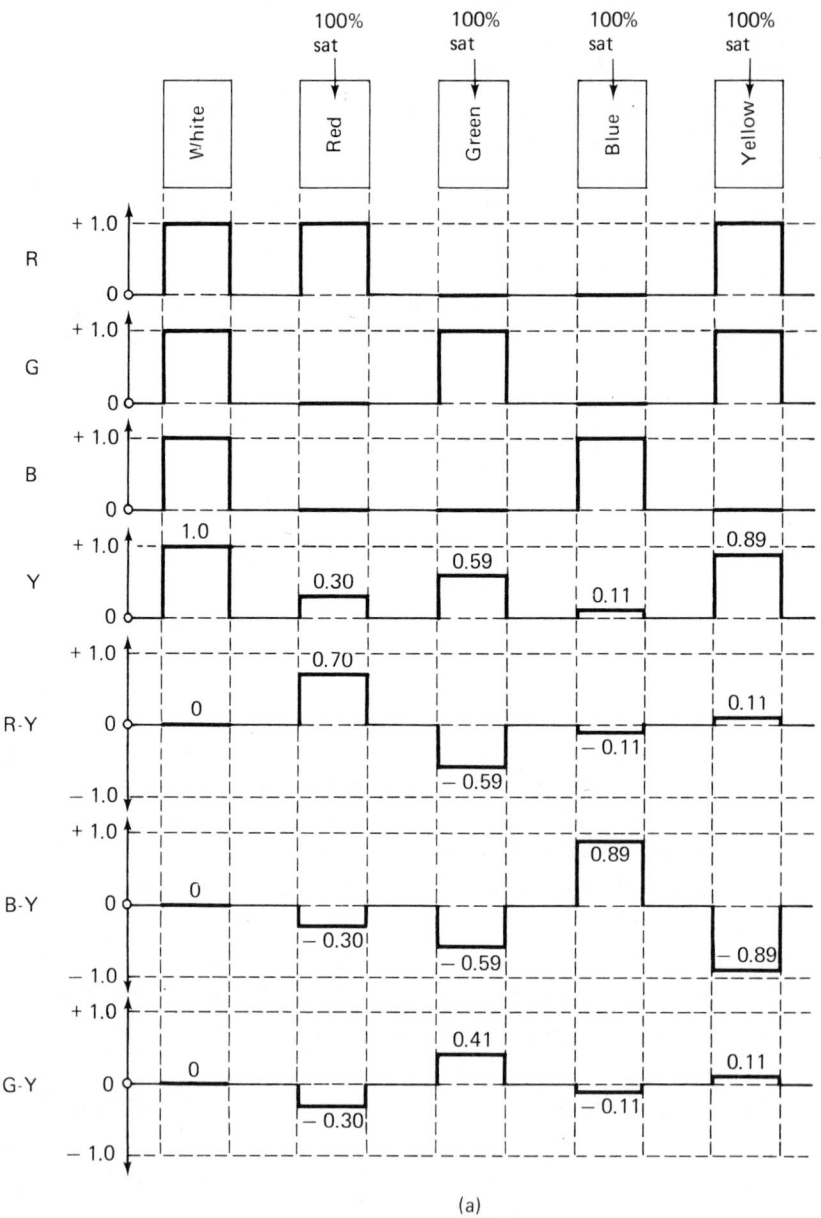

(a)

FIGURE 2-20 Waveforms for 100% saturated and 50% saturated color-bar patterns. (a) Waveforms for 100% saturated colors; (b) waveforms for 50% saturated red, blue, and green colors.

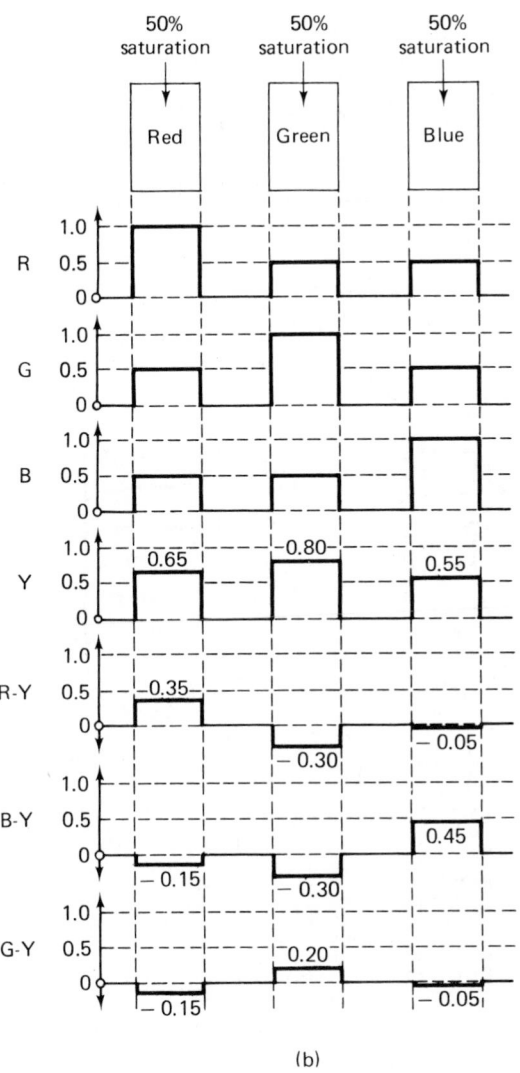

(b)

FIGURE 2-20 (continued)

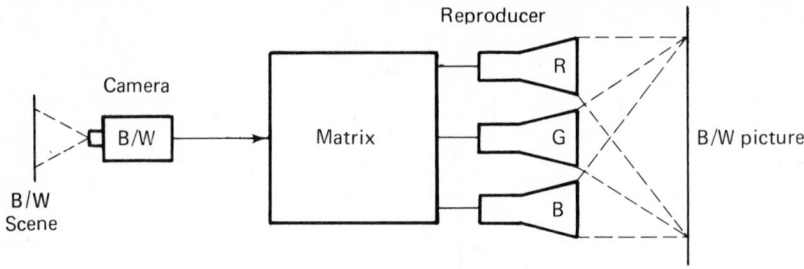

B/W to color

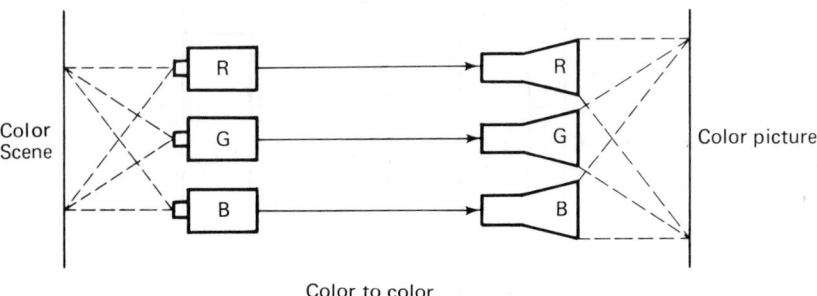

Color to color

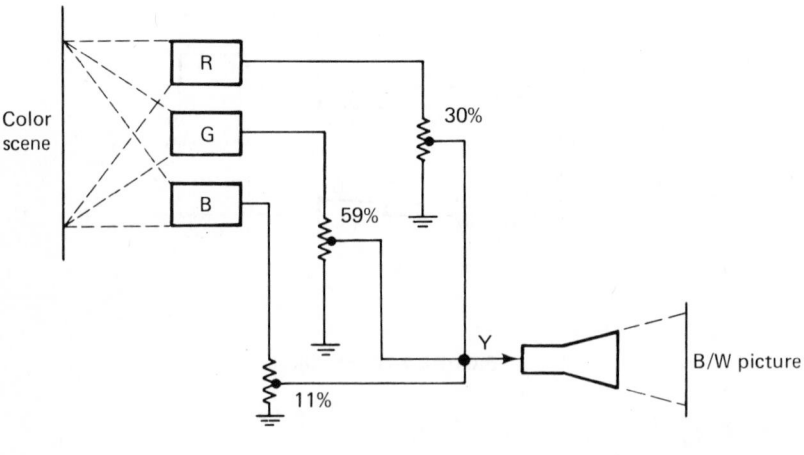

Color to B/W

FIGURE 2-21   Summary of color-B/W compatibility characteristics for the NTSC system.

15. Define the term *encode.*

16. If red, green, and blue color cameras are scanning a pattern comprising red, green, and blue strips at full saturation, what is the value of the red-camera output while scanning the red strip? The green strip? The blue strip?

17. Briefly describe the operation of a Y matrix.

18. What is the phase angle (with respect to burst) of the red chroma signal?

19. Explain the meaning of *unadjusted chroma values.*

20. Why is it impractical to transmit unadjusted chroma values?

21. How are readjusted chroma values produced at the transmitter?

22. Does readjustment of chroma values result in change of chroma phase angles?

23. When desaturated colors are being transmitted, do the chroma amplitudes change? The Y amplitudes?

24. Are readjusted chroma values converted to unadjusted chroma values at the color receiver?

25. Briefly describe the *color-burst signal.*

True-False

1. White, gray, and black shades consist of various proportions of red, green and blue hues.

2. A dichroic mirror passes one wavelength of light and rejects other wavelengths.

3. Each primary color requires 6 MHz of channel space.

4. Each complementary color requires 1 MHz of channel space.

5. A color-TV transmission requires only one-third as much bandwidth in the UHF spectrum as in the VHF spectrum.

6. Chroma signals and color-difference signals are identical.

7. Frequency interlacing is the same process as frequency interleaving.

8. The color subcarrier must have a frequency of 3.579545 MHz in order to transmit a color signal.

9. If the sound signal beats with the chroma signal, a 920-kHz dot pattern will be displayed on the picture-tube screen.

10. A reasonably linear phase characteristic must be maintained through the chroma-signal region of the picture channel.

11. Interleaving is provided only with respect to the horizontal-scanning frequency.

12. A black-and-white receiver will not automatically "lock in" when tuned to a color-TV transmission.

13. Color-TV transmission is sequential for the Y signal, but is simultaneous for the chroma signal.

14. A Y matrix forms white and gray shades from R, G and B camera outputs.

15. A gray shade is the same as a white level with reduced amplitude.

16. Black has zero brightness.

17. If a Y signal is passed through a CE amplifier stage, the output will be a $-Y$ signal.

18. The chroma signal contributes no energy to the complete color signal.

19. Readjustment of chroma values is effected by reducing Y to 0.877 of its original value, and reducing $-Y$ to 0.493 of its original value.

20. Overmodulation of the color-TV transmitter is impossible when readjusted chroma values are employed.

Multiple Choice

1. A closed-circuit color-TV system is _____ a color-TV broadcast system.
   (a) the same as
   (b) simpler than
   (c) more complex than
   (d) completely different from

2. Color-TV cameras utilize _____ tubes.
   (a) image-orthicon or vidicon
   (b) mercury-vapor or neon
   (c) exhaust or intake
   (d) X-ray or violet-ray

3. A color-TV camera comprises _____ cameras.
   (a) red, yellow, and blue
   (b) red, white, and blue
   (c) color and black-and-white
   (d) red, green, and blue

4. *Frequency interleaving* denotes _____.
   (a) staggering the clusters of Y and chroma signals
   (b) merging the clusters of Y and chroma signals
   (c) readjustment of chroma values
   (d) desaturation of high video frequencies

5. The advantage of frequency interleaving is _____.
   (a) avoidance of overmodulation
   (b) reduction of interference between the chroma and Y signals
   (c) rejection of adjacent-channel interference
   (d) trapping of intercarrier-sound interference

6. An odd multiple of half the horizontal-scanning frequency is employed for the _____ frequency.
   (a) vertical-scanning
   (b) intercarrier-sound
   (c) local-oscillator
   (d) color-subcarrier

7. A separation of approximately _____ exists between the color-subcarrier frequency and the intercarrier-sound frequency.
   (a) 4.5 MHz
   (b) 3.58 MHz
   (c) 15,750 Hz
   (d) 920 kHz

8. Color-TV transmissions employ a horizontal-scanning frequency of _____.
   (a) 15,750 Hz
   (b) 60 Hz
   (c) 15.734264 kHz
   (d) 59.94 Hz

9. Color-TV transmissions employ a vertical-scanning frequency of _____ Hz.
   (a) 15,750
   (b) 60
   (c) 15.734264
   (d) 59.94

10. Frequency interleaving occurs between the color subcarrier and the _____ signal.
    (a) intercarrier-sound
    (b) audio
    (c) local-oscillator
    (d) AFC error

11. A *color encoder* processes the outputs from _____.
    (a) three microphones
    (b) three color cameras
    (c) three microphones and three cameras
    (d) the FM sound section

12. An encoder changes _____ signals into _____ signals.

(a) RGB; ±(R−Y) and ±(B−Y)
(b) Y; RGB
(c) RGB; Y
(d) Y; −Y

13. If the red camera is scanning a fully-saturated red strip, there is
_____ output from the B−Y matrix.
    (a) 100%
    (b) 30%
    (c) zero
    (d) infinite

14. While a color-TV camera is scanning fully-saturated red, green, and
blue strips, the output from the Y matrix is _____.
    (a) 100% at all times
    (b) zero at all times
    (c) 30% at all times
    (d) changing with each change in color

15. A Y signal voltage consists of _____ signal voltages.
    (a) 0.59R, 0.30G, 0.11B
    (b) 0.59G, 0.30R, 0.11B
    (c) 0.59B, 0.30G, 0.11R
    (d) 0.59R, 0.30B, 0.11G

16. R−Y and B−Y voltages are _____ .
    (a) in phase
    (b) 180° out of phase
    (c) 90° out of phase
    (d) changing in phase

17. Unadjusted and readjusted blue signal voltages have a ratio of
_____.
    (a) 0.9 to 0.45
    (b) 1 to 1
    (c) 0.877 to 0.493
    (d) 15,750 to 60

18. Unadjusted and readjusted red signal phases have a ratio of _____.
    (a) 1 to 1
    (b) 67° to 76.5°
    (c) 0° to 90°
    (d) 3.58 to 1

19. A color burst is located _____.
    (a) on top of the horizontal sync pulse
    (b) on the front porch of the horizontal sync pulse
    (c) on the back porch of the horizontal sync pulse
    (d) at the end of the vertical sync pulse

20. The color burst has a frequency of _____ .
    (a) 15,750 Hz
    (b) 4.5 MHz
    (c) 920 kHz
    (d) 3.579545 MHz

Problems

1. Calculate the frequencies of the 227th and 228th harmonics of 15,734.264 Hz.

2. Calculate the frequency of the 455th harmonic of 7,867.132 Hz.

3. What is the difference between the frequency of the 455th harmonic of 7,867.132 Hz and the midfrequency between the 227th and 228th harmonics of 15,734.264 Hz?

4. If a 70% R–Y signal is added to a 30% B–Y signal, what is the amplitude of the resultant? What is the phase angle of the resultant with respect to burst?

5. What is the ratio of the unadjusted green-signal amplitude to the readjusted green-signal amplitude, for a fully-saturated hue?

6. What is the ratio of the unadjusted green-signal phase angle to the readjusted green-signal phase angle?

7. Power is proportional to voltage squared. Calculate the power ratio of the unadjusted blue-signal voltage to the readjusted blue-signal voltage, for a fully saturated hue.

8. If two hues are blended, the brightness of the blend is equal to the sum of the brightnesses of the hues. Calculate the brightness of a fully saturated yellow hue.

9. A color blend is comprised of one full part of red and one-half part of green. What is the brightness of the blend, assuming that the red and green hues are fully saturated?

10. A color blend consists of one full part of green and one-half part of red. What is the brightness of the blend, assuming that the red and green hues are fully saturated?

Chapter 3

Color Signal Characteristics And System Operation

3.1   THE COMPLETE COLOR SIGNAL

A complete color signal consists of a Y component and a chroma component as shown in Fig. 3-1. Note that the color burst is a 3.58-MHz signal, with the $-(B-Y)$ phase. This burst signal is employed for color synchronization in the color-TV receiver.

Observe that the complete color signal depicted in Fig. 3-1(c) can be separated into its Y and chroma components as shown in Fig. 3-1(a) and (b). This separation is accomplished in a color-TV receiver as indicated in Fig. 3-2, the color receiver system consisting of two basic signal sections. The black-and-white section passes the Y signal to the color picture tube, but does not pass the chroma signal. However, the chroma section picks up the chroma signal that is rejected by the black-and-white section. In turn, the chroma section decodes the chroma signal and feeds this color information to the color picture tube.

The horizontal sync pulse has a repetition rate of 15.75 kHz, whereas the color burst has a frequency of 3.58 MHz. Therefore, these two signal components are processed near the low-frequency and the high-frequency

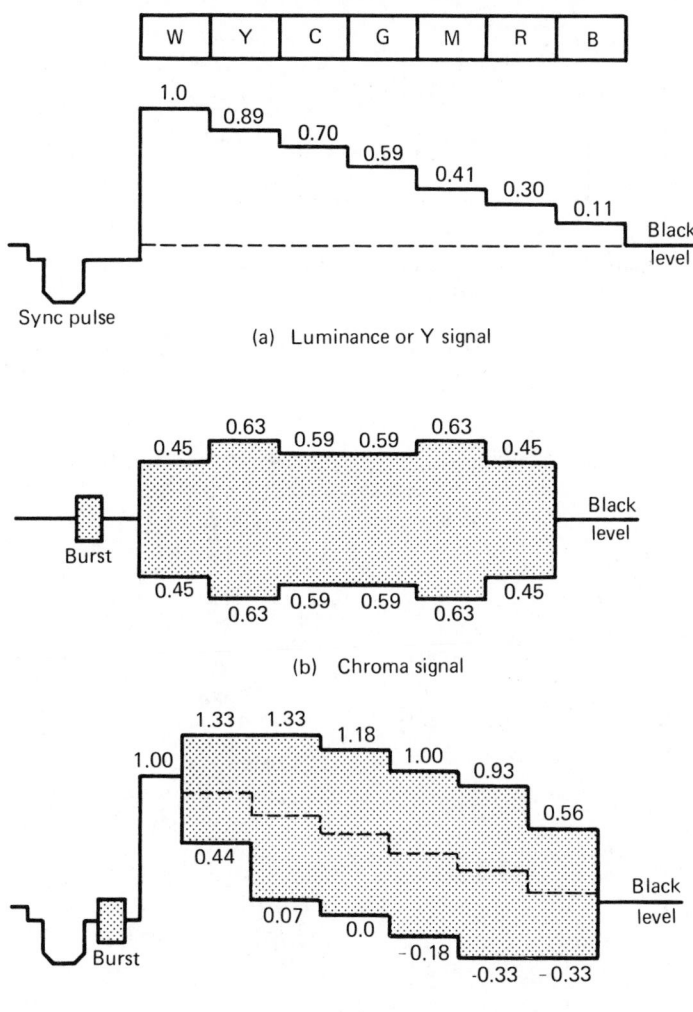

(a) Luminance or Y signal

(b) Chroma signal

(c) Complete color-bar signal

FIGURE 3-1   Y and chroma voltage amplitudes in the complete color signal.

ends of the receiver signal channel, as Fig. 3-3 shows. Video-amplifier response curves generally have a greater bandwidth in a color receiver than in a black-and-white receiver. For example, a black-and-white receiver may have reduced high-frequency response, as indicated by the dotted line in Fig. 3-3. When a black-and-white receiver is tuned to a color signal, the color burst (and chroma signal) is substantially attenuated in most cases. Thus, the 3.58-MHz signal component is largely rejected before it is applied to the picture

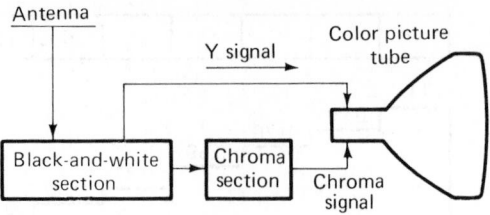

FIGURE 3-2 The two basic signal sections of a color-television receiver.

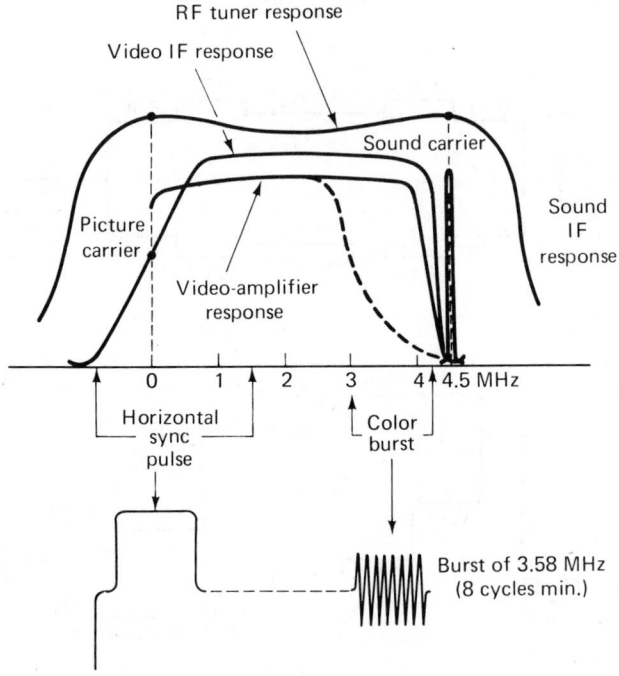

FIGURE 3-3 Relation of the color burst and the sync pulse to the frequency-response curves in a receiver.

tube. The residual 3.58-MHz signal is then canceled out as a result of the frequency-interleaving process.

It is instructive to observe the frequency band that is occupied by the chroma signal in a typical color-TV receiver. Figure 3-4 shows the chroma-section passband compared with the IF passband for a typical color-TV receiver. The chroma-signal section generally has a frequency response of from 3.1 to 4.1 MHz. That means that the IF channel is four times as wide as the chroma channel. This bandwidth limitation is practical for chroma-signal processing because color information does not have to be reproduced in fine detail, as has been previously explained. The chief advantage of bandwidth limitation is in the reduction of residual beat interference between the chroma and Y signals. Note also that since the chroma signal has a high center frequency (3.58 MHz), any dot patterns that it might produce will be small and, accordingly, of low visibility.

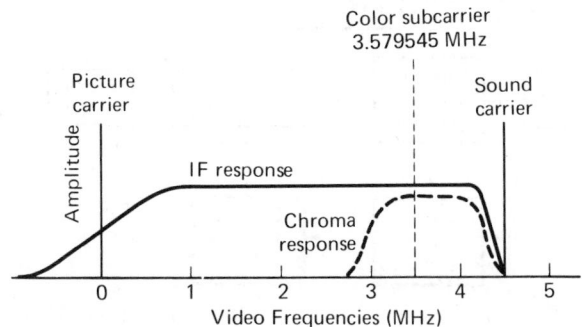

FIGURE 3-4   Chroma-section passband compared with IF section passband.

Possible residual beat interference between the Y and chroma signals is further minimized by suppressing the 3.58-MHz color subcarrier at the transmitter, only the chroma sidebands being transmitted. Therefore, the color subcarrier must be reinserted at the receiver. (Details are reserved for subsequent discussion.) Suppression of the color subcarrier reduces the energy in the chroma signal so that any beat patterns that might be produced will have minimum visibility. Since all of the color information is contained in the chroma signal, the chroma signal will disappear whenever a black-and-white scene is being scanned by the color cameras. For example, if a black-and-white chart, drawing, scoreboard, or piece of printed material is being scanned, a Y signal is transmitted without any chroma signal. In practice, this feature of the NTSC system further reduces the possibility of visible residual beat interference.

3.2   BASIC COLOR-TELEVISION RECEIVER ARRANGEMENT

Consider the block diagram for a color-TV receiver shown in Fig. 3-5. Some of the sections are the same as those utilized in black-and-white receivers. However, the shaded sections are employed only in color receivers. Observe that the black-and-white signal channel is the same as in a black-and-white receiver, except for the inclusion of a delay line in the former. This delay line slows up the Y signal by almost 1 microsecond. Although the delay line has no visible effect on the black-and-white image, the line serves to provide correct registration of the color image with the black-and-white image, as depicted in Fig. 3-6. If the color signal and Y signal are incorrectly timed, the color image will be displayed on the picture-tube screen. This defect in system operation is called "poor color fit."

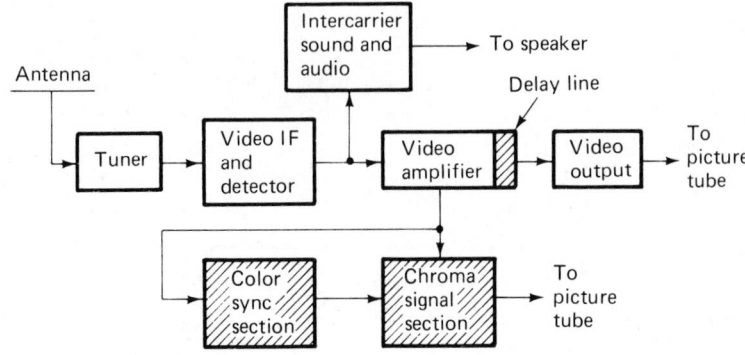

Note: Shaded blocks indicate stages used only in color receivers.

FIGURE 3-5   Basic block diagram for a color-television receiver.

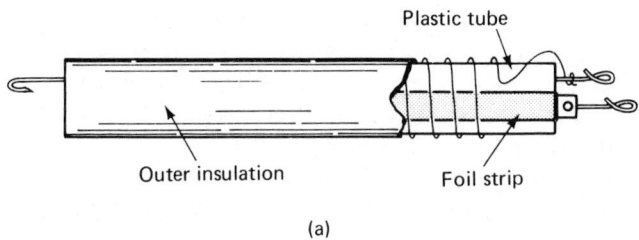

(a)

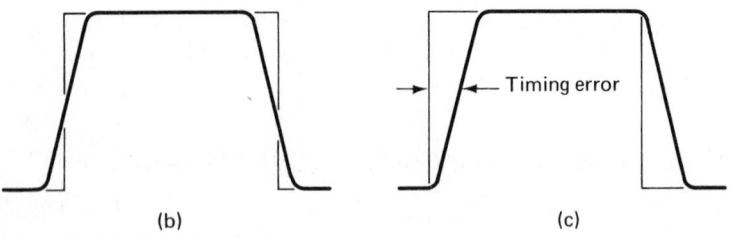

(b)   (c)

FIGURE 3-6   Circuit action of a Y-signal delay line. (a) Construction of a delay line; (b) Y and chroma signals correctly timed; (c) timing error in absence of a delay line.

Figure 3-4 shows why a Y-signal delay line is required: the chroma-signal section has only one-quarter of the bandwidth utilized by the Y-signal section. When a waveform passes through a narrow-band channel, the leading edge rises more slowly than when the waveform passes through a wide-band channel. In effect, the signal in the narrow-band channel is delayed in comparison to the signal in the wide-band channel. Therefore, a compensating delay must be provided in the wide-band channel. The delay-line construction depicted in Fig. 3-6 is employed to obtain the required delay with negligible distortion of the Y signal. Although the Y signal is delayed, its rise time is practically unimpaired.

Next, observe that the chroma-signal section in Fig. 3-5 operates to amplify the chroma signal and decode the signal into its R–Y and B–Y components, as shown in Fig. 3-7. These R–Y and B–Y chroma signals combine with the Y signal at the picture tube to form the red and blue primary hues. In addition, a G–Y chroma signal is required to combine with the Y signal at the picture tube to form the green primary hue. A chroma section is therefore included, which processes the R–Y and B–Y signals to produce the G–Y signal. In the example of the chroma signal of Fig. 3-7, a G–Y matrix is utilized. Some color receivers employ a G–Y demodulator instead of a matrix. In any case, the end result is the same, and the G–Y signal is decoded from the complete chroma signal. Note that the G–Y signal is not separately generated at the transmitter, because a G–Y signal can be formed by mixing –(R–Y) and –(B–Y) signals in suitable proportions.

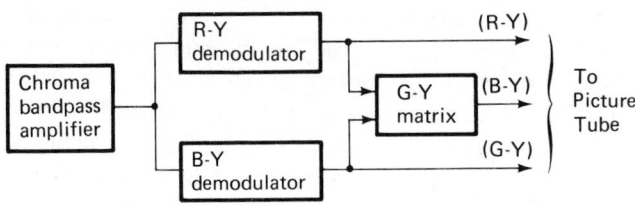

FIGURE 3-7   Plan of the chroma-signal section.

The relation of the G–Y signal to the R–Y and B–Y signals is shown in Fig. 3-8. Since the G–Y signal has components on the –(R–Y) and –(B–Y) axes, it is necessary only to mix the –(R–Y) and –(B–Y) signals in suitable proportions to decode the G–Y signal. This is the basic action of the G–Y matrix in Fig. 3-7. Also observe in Fig. 3-8 that the (G–Y)$\angle 90°$ chroma phase is at right angles, or in *quadrature*, to the G–Y chroma phase. Thus, these are quadrature signals, just as R–Y and B–Y are quadrature signals. We will return to consideration of quadrature signal relations when we progress to color-generator and oscilloscope tests of chroma circuitry.

With reference to Fig. 3-5, the basic function of the color-sync

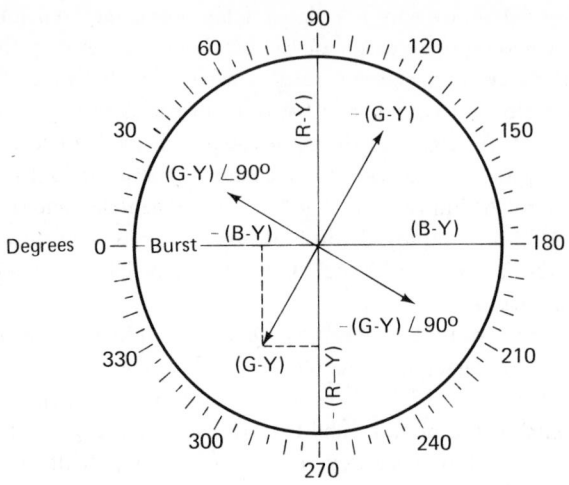

FIGURE 3-8   Relation of the G—Y signal to the R—Y and B—Y signals.

section is to keep the local color-subcarrier oscillator in step with the color-burst phase, as depicted in Fig. 3-9. First, the burst amplifier, which is tuned to 3.58 MHz, picks out the color burst from the complete color signal. Next, this burst signal is fed to the phase-comparator and AFC section, where a discriminator arrangement is energized by both the burst signal and a feedback loop from the subcarrier oscillator. In case the oscillator phase tends to drift away from the burst phase, the discriminator develops a corresponding control-voltage output. This control voltage operates a varactor diode (or some equivalent arrangement), which pulls the subcarrier oscillator back into correct phase. Finally, the reconstituted color-subcarrier voltage is fed to the chroma demodulators, where it is inserted with the chroma sidebands.

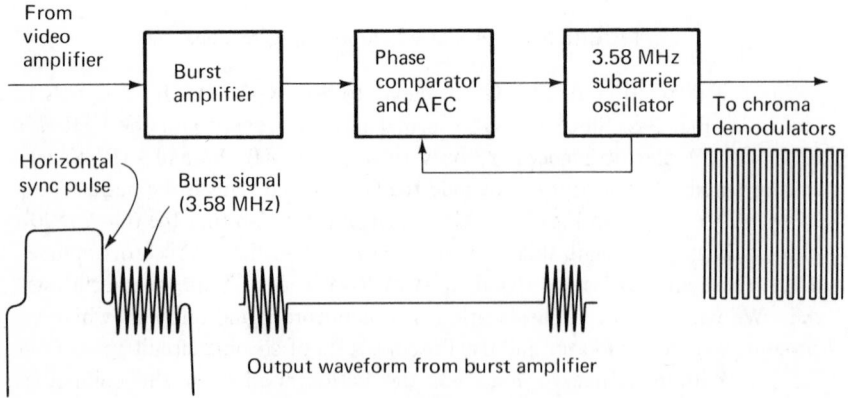

FIGURE 3-9   Plan of the color-sync section.

Now consider the block diagram of Fig. 3-10. Observe that a *color-killer* arrangement is included in the chroma section. This is an electronic switch automatically operated by the color burst. The color killer switches the chroma bandpass amplifier into operation when the receiver is tuned to a color-TV signal and cuts the bandpass amplifier off when the receiver is tuned to a black-and-white TV signal. It is desirable to disable the chroma channel during black-and-white reception because random noise pulses might otherwise produce spurious outputs from the R−Y, B−Y, and G−Y stages. Spurious chroma outputs result in color flashes and streaks on the screen of the picture tube. This type of interference is called *confetti*. If the color killer is operating normally, confetti is automatically eliminated.

As noted previously, the Y signal proceeds through the video amplifier to the color picture tube, and the chroma signal proceeds through the bandpass amplifier as depicted in Fig. 3-11. After the chroma signal has been reconstituted by insertion of the subcarrier, and decoded in the chroma-demodulation section, it is applied to the grids of the color picture tube. Since the Y signal is applied to the cathode of the picture tube, the Y and chroma signals are combined in the three electron beams. This constitutes a matrixing action in which the picture tube serves as the final decoding device—in addition to its operating as a display device. For example, if a color-bar pattern is being displayed, the grid and cathode waveforms appear as in Fig. 3-12. These waveforms can be observed with an oscilloscope.

At this point, it is helpful to compare the waveforms for a 100% saturated color-bar pattern with the waveforms for a 50% saturated color-bar pattern. These waveforms are shown in Fig. 3-13. It is apparent that the amplitude of the chroma signals depends upon the saturation of the color that is being scanned. In other words, a 100% saturated red color has an R−Y amplitude of 0.70, whereas a 50% saturated red color has an R−Y amplitude of 0.35. Again, a 100% saturated green color has an R−Y amplitude of −0.59, whereas a 50% saturated green color has an R−Y amplitude of −0.30. Similarly, a 100% saturated blue color has an R−Y amplitude of −0.11, whereas a 50% saturated blue color has an R−Y amplitude of −0.05. Proceeding to B−Y and G−Y amplitudes, it will be noted that these amplitudes are also proportional to the saturation of the color that is being scanned.

Figure 3-13 shows how the color picture tube operates as a Y-and-chroma matrix for a red bar signal. A negative Y signal is applied to the three cathodes, and the signal produces a 30% output from each of the three electron guns. A positive R−Y signal is applied to the red grid; the signal adds a 70% output from the red gun. Accordingly, there is 100% output from the red gun. At the same time, a negative B−Y signal is applied to the blue grid; this signal cancels the 30% output from the blue cathode. Therefore, there is zero output from the blue gun. Similarly, a negative G−Y signal is applied to the green grid; this G−Y signal cancels the 30% output from the green cathode. Thus, there is zero output from the green gun. Since only the red gun is operative in this example, a red hue will be displayed on the screen of the color picture tube.

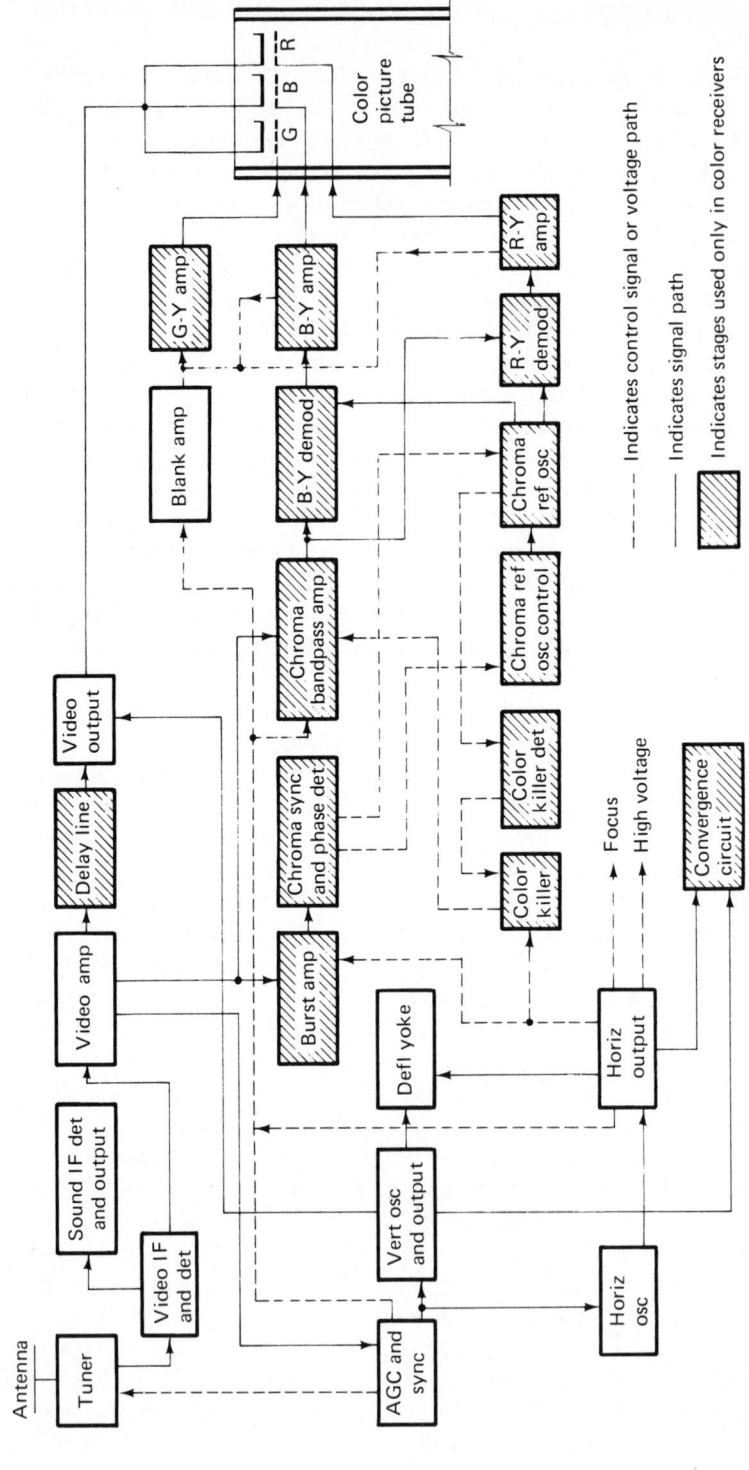

FIGURE 3-10   Block diagram of a color-television receiver.

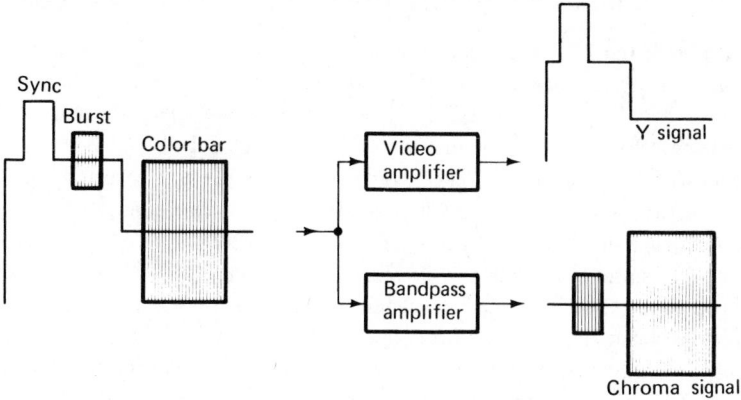

FIGURE 3-11   Separation of the Y and chroma signals from the complete color signal.

FIGURE 3-12   System operation with a color-bar signal (see inside front cover for artwork).

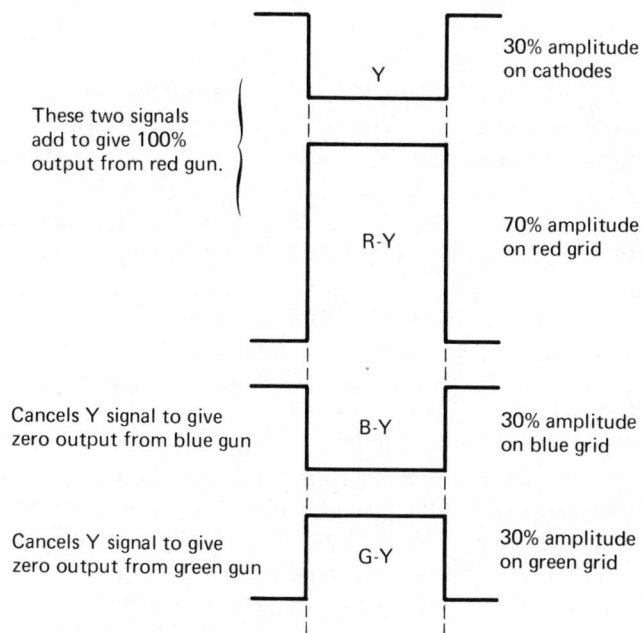

FIGURE 3-13   Waveform relations at color picture tube for display of a red bar.

3.3   CHROMA-DEMODULATOR AND MATRIX ARRANGEMENTS

Although the end result is always the same, various *chroma-demodulator* and *matrix arrangements* are utilized to obtain it. Six of the more basic arrangements are depicted in Fig. 3-14. One of the earliest methods, and one that is still used by a number of manufacturers, employs R–Y and B–Y demodulators with a G–Y matrix. This arrangement was previously noted. Demodulator actions are diagrammed in Fig. 3-15. This processing method is called *quadrature demodulation* or *synchronous demodulation*. A chroma demodulator is sometimes called a *product detector*. It operates in step with the injected 3.58-MHz subcarrier signal; that is, the R–Y demodulator device is normally cut off, but goes into conduction briefly on the positive peak of the injected subcarrier signal. Similarly, the B–Y demodulator device is normally cut off, but goes into conduction briefly on the positive peak of the injected subcarrier signal. Since there is a 90° phase difference between the subcarrier signals that are injected into the R–Y and B–Y demodulators, it is evident that they conduct alternately, as shown in Fig. 3-15.

A chroma demodulator is essentially a sampling device that produces an output in the form of pulses. However, since these pulses recur 3.58 million times per second, for all practical purposes the output waveform may be considered to be continuous and uninterrupted. The R–Y and B–Y signals are *multiplexed*; that is, they arrive together at the demodulators. However, because of the 90° phase difference between the R–Y and B–Y signals, the signals can be decoded and separated by quadrature demodulators. Observe in Fig. 3-15 that when the R–Y signal is going through its positive peak, the B–Y signal is going through zero. Therefore, the output from the R–Y demodulator contains samples of the R–Y signal only, and does not contain any B–Y signal. Similarly, the B–Y signal is going through its positive peak as the R–Y signal is passing through zero. Thus, the output from the B–Y demodulator contains samples of the B–Y signal only, and does not contain any R–Y signal.

A few receivers employ R–Y and G–Y demodulation with B–Y matrixing. This arrangement provides complete separation of the multiplexed signals, because although R–Y and G–Y are not in phase quadrature, they have R–Y and B–Y components, as Fig. 3-8 shows. An advantage to R–Y and G–Y demodulation is that the G–Y signal has a greater amplitude than the B–Y signal in most transmissions. G–Y demodulation provides a better signal-to-noise ratio than B–Y demodulation.

The twin-device demodulator and matrix arrangement is essentially an R–Y and B–Y demodulator arrangement with G–Y matrixing. However, the G–Y signal is typically matrixed in the common-emitter branch of a two-transistor arrangement that develops R–Y and B–Y outputs from the collector branches. The chief advantage of the twin-device arrangement is its economy in production, inasmuch as fewer components are required.

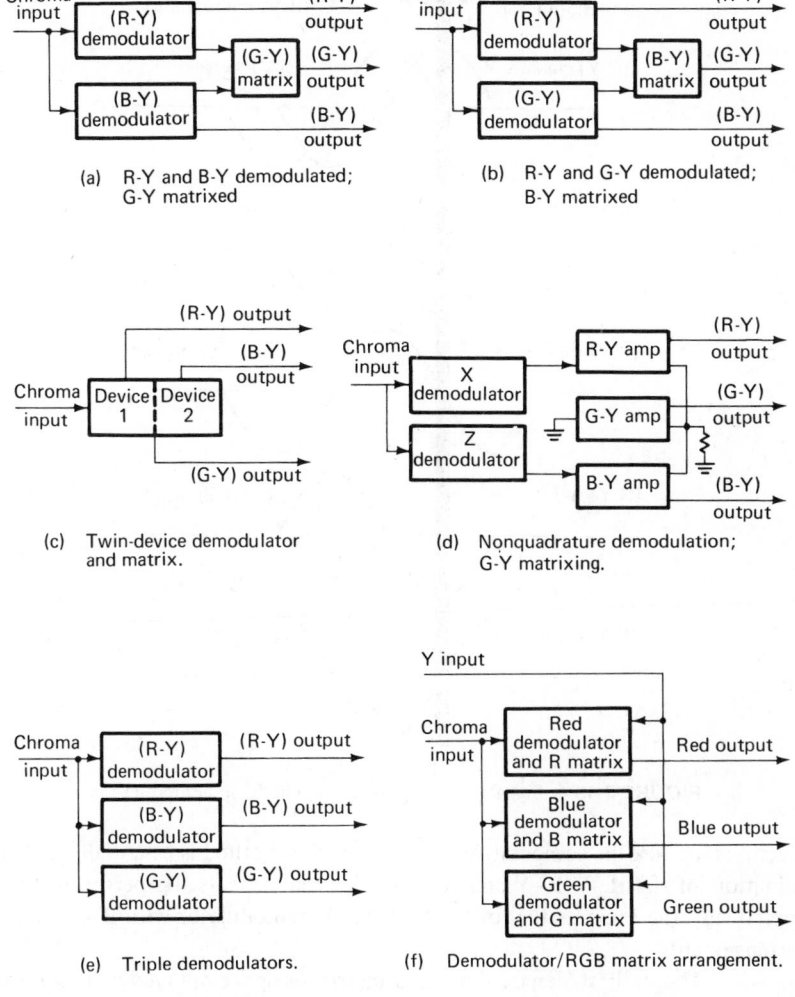

FIGURE 3-14   Six basic arrangements for chroma demodulation
and matrixing.

The X and Z demodulator arrangement with G—Y matrixing is
widely used. This is another example of non-quadrature demodulation, com-
parable to the R—Y and G—Y demodulation method, and having the same
general advantages and disadvantages as that method. Its production is also
economical. Note that the X and Z demodulators do not necessarily operate
along the R—Y and G—Y axes. Various manufacturers prefer demodulation
phase angles anywhere in the range of 50° to 90° (sometimes described as in
the range of 90° to 130°). Triple-demodulator arrangements without matrix

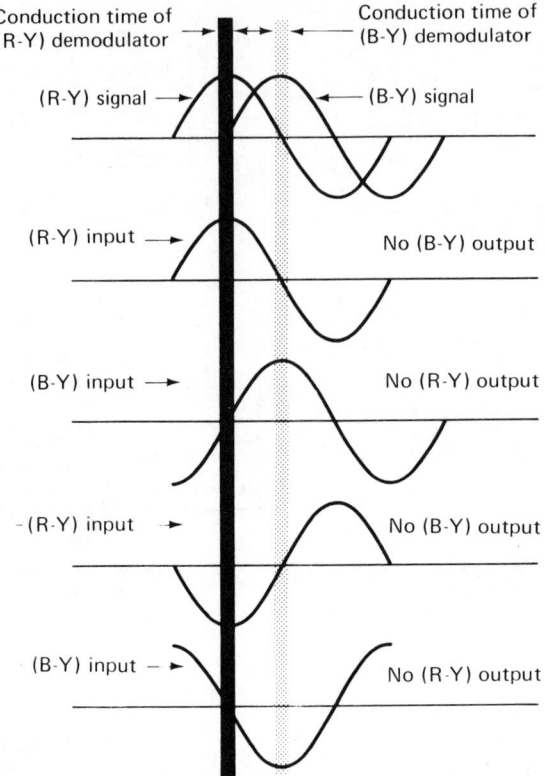

FIGURE 3-15   Signal sampling by R—Y and B—Y demodulators.

sections are also in comparatively wide use. This method is essentially a combination of the R—Y/B—Y and R—Y/G—Y arrangements. Its performance is practically the same as that of the R—Y/G—Y demodulator with B—Y matrix arrangement.

The earliest demodulator-and-matrix design employed two chroma demodulators, a chroma matrix, and an RGB matrix in which the Y signal was combined with the three chroma signals. In other words, the color picture tube was not employed as a final matrix. Although this arrangement was discarded for a number of years, it has been reintroduced in modified form, and is utilized in some of the most recent receiver designs. This demodulator/RGB matrix arrangement employs semiconductor devices fed by chroma and Y signal inputs. The chroma signals are demodulated in the R, G, and B phases, and simultaneously combined with the Y signal. Circuit actions will be subsequently explained in greater detail.

EXERCISES

Questions

1. What are the two chief components of a complete color signal?
2. How can the basic components of the complete color signal be separated?
3. Is the chroma channel in a color receiver narrower than the IF passband?
4. Why is the color subcarrier suppressed at the transmitter?
5. When a black-and-white scoreboard is being transmitted by a color-TV station, what is the value of the accompanying chroma signal?
6. Does the amplitude of the color burst change when the color-TV camera shifts from a color scene to a black-and-white scene?
7. Explain the function of a *delay line.*
8. State the approximate envelope (time) delay provided by a Y-amplifier delay line.
9. Define the meaning of *quadrature signals.*
10. Describe the output waveform from the burst amplifier.
11. How is a chroma signal *reconstituted* in a color receiver?
12. Why is the chroma channel automatically disabled during reception of a black-and-white TV broadcast?
13. Name the circuit sections and stages that are used only in color receivers.
14. Discuss the *decoding* of a chroma signal in a color receiver.
15. Explain how a color picture tube can operate as the final decoding device.
16. State the six classes of chroma-demodulator-and-matrix arrangements that are employed by designers of color receivers.
17. How are R–Y and B–Y signals separated by a pair of chroma demodulators?
18. What are the two methods used for development of the G–Y signal?
19. Explain how X-and-Z demodulation differs from (R–Y)-and-(B–Y) demodulation.
20. Briefly discuss the earliest demodulator-and-matrix arrangement.

True-False

1. Some of the most recent color-receiver designs employ a version of the earliest demodulator-and-matrix arrangement that was subsequently abandoned.

2. The repetition rate of the color burst is 15,734.264 Hz/sec.

3. Although the color burst has a frequency of 3.579545 MHz, this particular frequency is absent from the transmitted chroma signal.

4. A color burst has a greater bandwidth than a chroma signal.

5. Frequency interleaving (interlacing) cancels out all harmonics of the color-subcarrier frequency on the screen of a black-and-white picture tube.

6. Frequency interleaving (interlacing) cancels out all of the chroma-sideband frequencies on the screen of a black-and-white picture tube.

7. A chroma signal section in a color receiver has a typical frequency response from 3.1 to 4.1 MHz.

8. Narrow-band signal channels have the effect of delaying the envelope of a signal, compared with signal passage through a wide-band signal channel.

9. Waveforms processed through narrow-band signal channels have a slower rise time than waveforms processed through a wide-band signal channel.

10. A delay line is included in the chroma section to provide a delay of approximately 1 microsecond.

11. The G—Y signal component can be recovered by mixing suitable proportions of —(R—Y) and —(B—Y) signal voltages.

12. A third chroma demodulator may be employed to decode the G—Y signal from the chroma signal.

13. Correct frequency and phase of the subcarrier oscillator is maintained by a phase-comparator and AFC section.

14. An electronic switch, called the *color killer*, automatically disables the Y channel when a color receiver is tuned to a black-and-white broadcast.

15. *Confetti* is a term that denotes colored flashes and streaks on the screen of a color picture tube during reception of a black-and-white broadcast.

16. Unadjusted chroma values are applied to the grids in a color picture tube.

17. A color picture tube may be operated as a Y-and-chroma matrix.

18. A black-and-white picture tube operates as a Y-and-chroma matrix when the receiver is tuned to a color broadcast.

19. Positive grid signals and negative cathode signals both operate to increase the picture-tube beam current.

20. *Quadrature demodulator, synchronous demodulator* and *product detector* are equivalent terms.

21. Incoming R—Y and B—Y signals are said to be *multiplexed.*

22. Chroma demodulators operate by means of a peak-sampling process.

23. A G—Y demodulator operates by sampling suitable proportions of the R—Y and B—Y signals to form the G—Y signal voltage.

24. X and Z demodulators do not operate in phase quadrature.

25. R—Y, B—Y and G—Y signals are matrixed from X and Z signals following chroma demodulation.

Multiple Choice

1. A color burst has the _____ phase.
   (a) R—Y
   (b) B—Y
   (c) —(R—Y)
   (d) —(B—Y)

2. When fully saturated colors are being transmitted, red and cyan chroma signals have _____ amplitude than the other primary and complementary hues.
   (a) the same
   (b) less
   (d) greater
   (d) more variable

3. When fully saturated colors are being transmitted, yellow and blue chroma signals have _____ amplitude than the other primary and complementary colors.
   (a) the same
   (b) less
   (c) greater
   (d) more variable

4. In a fully-saturated color-bar signal at maximum brightness, the _____ signals have maximum downward modulation.
   (a) yellow and cyan

(b) green and magenta
(c) red and blue
(d) white and black

5. In a fully-saturated color-bar signal at maximum brightness, the
   _____ signals have maximum upward modulation.
   (a) yellow and cyan
   (b) green and magenta
   (c) red and blue
   (d) black and white

6. The chroma signal section in a color receiver generally has a bandwidth from _____ .
   (a) 60 Hz to 15,750 Hz
   (b) 41.25 MHz to 45.75 MHz
   (c) 88 to 108 MHz
   (d) 3.1 to 4.1 MHz

7. A delay line functions by _____ .
   (a) attenuating the Y signal
   (b) attenuating the chroma signal
   (c) attenuating the color burst
   (d) delaying the Y signal

8. Incorrect timing of the color signal with respect to the Y signal
   causes _____ .
   (a) confetti
   (b) distorted sound
   (c) poor color fit
   (d) poor linearity

9. A delay line has no practical effect upon the _____ of
   the Y signal.
   (a) rise time
   (b) envelope delay time
   (c) time of passage
   (d) input-output time relation

10. A $G-Y\underline{/90^\circ}$ signal is in phase quadrature to the _____ .
    (a) color burst
    (b) R—Y signal
    (c) B—Y signal
    (d) G—Y signal

11. The subcarrier-oscillator voltage is injected into the chroma signal
    _____ .
    (a) after demodulation

   (b) during demodulation
   (c) at the picture tube
   (d) at the burst amplifier

12. A chroma-bandpass amplifier is driven by _____ .
   (a) the burst amplifier
   (b) both chroma demodulators
   (c) R–Y and B–Y matrixes
   (d) the video amplifier

13. A color killer functions by _____ .
   (a) disabling the color-subcarrier oscillator
   (b) disabling the bandpass amplifier
   (c) disabling the burst amplifier
   (d) enabling the color picture tube

14. Chroma matrixes are located _____ .
   (a) before the bandpass amplifier
   (b) before the chroma demodulators
   (c) after the Y amplifier
   (d) after the chroma demodulators

15. The color burst has a repetition rate of _____ .
   (a) 3.58 MHz
   (b) 455 kHz
   (c) 15,734.264 Hz
   (d) 59.94 Hz

16. A video amplifier in a color receiver has _____ bandwidth than the video amplifier in a black-and-white receiver.
   (a) greater
   (b) less
   (c) more variable
   (d) less variable

17. IF amplifiers have approximately _____ times as much bandwidth as chroma channels in color receivers.
   (a) 0.877
   (b) 0.493
   (c) 3.1
   (d) 4

18. Beat interference between the color subcarrier and the intercarrier-IF signal has a frequency of _____ .
   (a) 3.579545 MHz
   (b) 455 kHz
   (c) 920 kHz
   (d) 4.5 MHz

19. Video amplifiers have approximately _____ times as much bandwidth as chroma channels in color receivers.
    (a) 0.877
    (b) 0.493
    (c) 3.1
    (d) 4

20. Y amplifiers have _____ bandwidth than video amplifiers.
    (a) more
    (b) less
    (c) more variable
    (d) less variable

Problems

1. Amplifier bandwidth is inversely proportional to the rise time of an output square wave. If a video amplifier with a bandwidth of 4.1 MHz has a rise time of 0.08 μs, calculate the rise time of an amplifier with a bandwidth of 2 MHz.

2. Calculate the rise time of a 1-MHz chroma amplifier.

3. Calculate the rise time of a 0.5-MHz chroma-demodulator channel.

4. An IF amplifier has a rise time of 0.1 μs. What is the bandwidth of the amplifier?

5. If there is a −25 volt signal on the cathodes of a color picture tube, what signal voltage will be required on the red grid to cancel the cathode-drive signal?

6. A circuit fault permits the color subcarrier to beat with the sound signal in the picture channel. Calculate the interference frequency that is displayed on the picture-tube screen.

7. If a color-subcarrier trap coil has an inductance of 20 μH, calculate the value of shunt capacitance required to resonate the trap at 3.58 MHz.

8. If a sound-trap coil has an inductance of 10 μH, calculate the value of shunt capacitance required to resonate the trap at 4.5 MHz.

9. A sound-takeoff coil has a center frequency of 4.5 MHz and a Q value of 80. Calculate the approximate bandwidth of the tuned circuit.

10. A subcarrier oscillator coil has a center frequency of 3.58 MHz and a bandwidth of 35.8 kHz. Calculate the approximate Q value of the tuned circuit.

Chapter 4

Color Picture Tube
Principles
And Operation

4.1 BASIC TYPES OF COLOR PICTURE TUBES

Although several types of color picture tubes have been developed, only one general design has been adopted commercially. This design is called the *shadow-mask arrangement*, and it is utilized in two forms. The first design employs a shadow mask with approximately 400,000 tiny perforations. A more recent version utilizes an aperture grille that has the form of a vertically-slotted shadow mask. These color picture tubes use three electron guns, as noted previously. However, the arrangement of the three guns is different in the two versions of the shadow-mask arrangement. The chief advantage of the perforated shadow-mask design is that it is adaptable to comparatively large-screen picture tubes. On the other hand, this design has the disadvantage of a rather involved convergence system, which requires skill and experience to adjust or "set up" properly. The chief advantage of the aperture-grille design is that it has a simple convergence system, which is easy to adjust. However, the aperture-grille design has the disadvantage of being impractical for use in large-screen picture tubes.

4.2 CONSTRUCTION OF THE SHADOW-MASK COLOR PICTURE TUBE

As shown in Fig. 4-1, there are three basic components in a shadow-mask picture tube. These are the electron-gun assembly, the shadow mask, and the phosphor-dot screen. All of the guns produce an electron beam that is focused to the same point on the shadow mask so that the beams pass through the same perforation at different angles. Thus, each beam strikes the phosphor-dot screen at a slightly different point. As depicted in Fig. 4-2, the phosphor dots are deposited on the screen so that each beam strikes a single dot. The beam from the red electron gun strikes a red dot, the beam from the green gun strikes a green dot, and the beam from the blue gun strikes a blue dot. This action is the same at any point on the screen so that if, for example, the red gun were operating and the green and blue guns cut off, a red field would be displayed on the screen of the color picture tube. Figure 4-3 shows the order of the phosphor-dot pattern that is deposited on the screen.

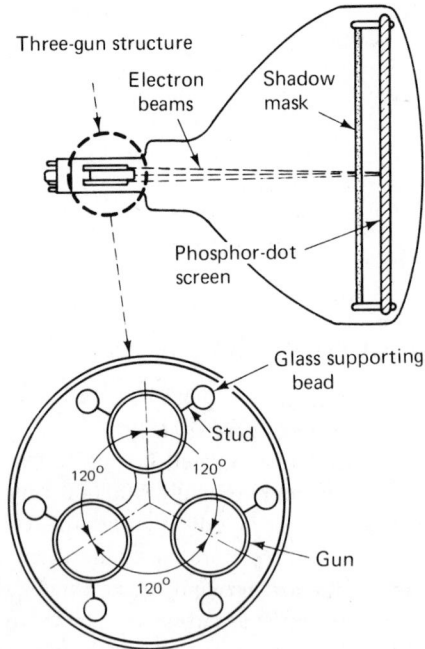

FIGURE 4-1   Basic plan of the shadow-mask color picture tube.

4.3 CONSTRUCTION OF THE APERTURE-GRILLE COLOR PICTURE TUBE

As shown in Fig. 4-4, the fundamental components of the aperture-grille color picture tube are the electron-gun assembly, the aperture grille, and the

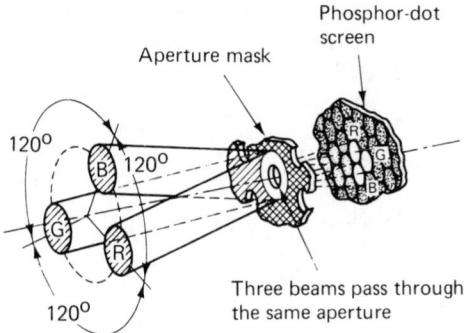

FIGURE 4-2   Each beam from the electron guns is directed to its corresponding phosphor dot.

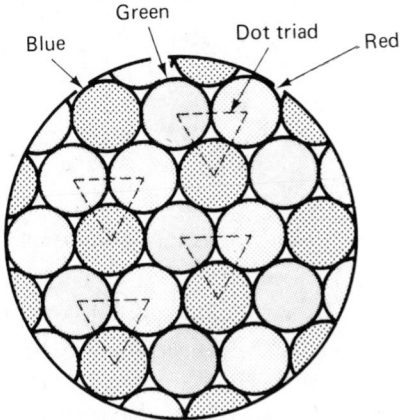

FIGURE 4-3   Phosphor-dot triads form the screen in the shadow-mask type of color picture tube.

phosphor-stripe screen. (*Trinitron* is one trade name for a commercial version of this design.) Note that the three electron guns are placed in line horizontally. All three electron beams pass through the same slot in the grille, but at different angles. Accordingly, the beam from the red gun strikes the red phosphor stripe, the beam from the green gun strikes the green phosphor stripe, and the beam from the blue gun strikes the blue phosphor stripe. This action is the same for any slot at any point on the screen.

It is evident that the aperture-grille picture tube operates from the same signal waveforms as the shadow-mask picture tube. For example, if the red and green guns were energized, and the blue gun cut off, a yellow field would be displayed with either type of picture tube. If all three guns were energized, a white field would be displayed with either type of picture tube. The convergence waveforms and adjustments for the two types of tubes are

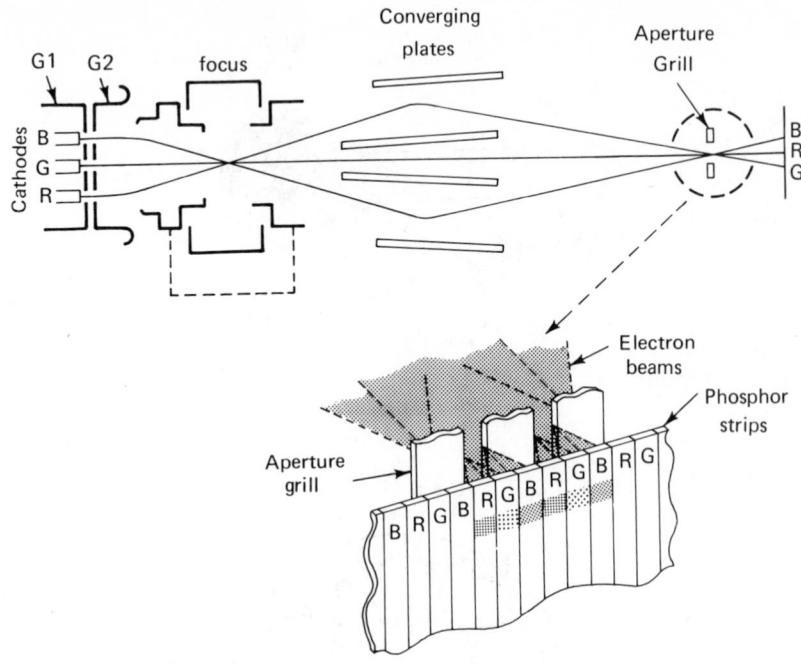

FIGURE 4-4 Construction of an aperture-grille color picture tube.

quite different, the aperture-grille picture tube employing a comparatively simple convergence system. The term *convergence* is defined as the process whereby the three electron beams are maintained in focus at any spot on an aperture grille or shadow mask during the scanning sequence from left-to-right and up-and-down.

In addition to the internal electrodes, external control devices are mounted on a shadow-mask color picture tube, as Fig. 4-5 shows. The magnetic shield prevents disturbance of the electron beams by external magnetic fields, such as the earth's magnetic field. This shield is generally made of mu metal. The deflection yoke is essentially the same as that employed with black-and-white picture tubes. Also, the dynamic convergence assembly comprises electromagnets that are energized by parabolic waveforms. Figure 4-6 depicts the radius of curvature for the locus of the electron beams compared with the radius of the shadow mask and viewing screen. Since the screen has a "flatter" contour (or a greater radius of curvature), corrective forces must be applied by the dynamic-convergence magnets to bring the beams into focus at the edges of the shadow mask.

In addition, the blue-lateral magnet shown in Fig. 4-5 is a static-convergence control. It is an adjustable permanent magnet located near the blue

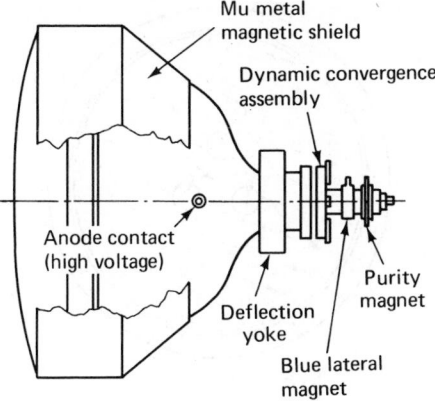

FIGURE 4-5   Components mounted on a typical shadow-mask color picture tube.

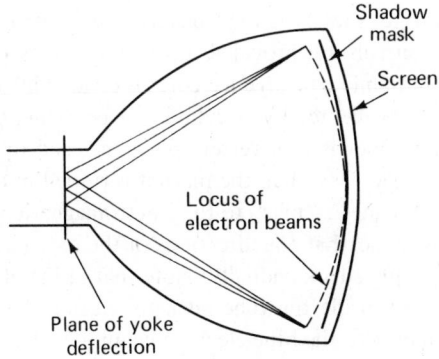

FIGURE 4-6   Picture-tube screen has a greater radius of curvature than that of the electron-beam locus.

electron beam in the neck of the tube. The position of the blue-beam landing on the viewing screen can in turn be adjusted with this control. Note that the purity magnet affects the direction of all three electron beams so that they will enter the field of the deflection yoke precisely at right angles to the magnetic flux lines. Unless the purity adjustment is properly made, the red beam, for example, will shift from the red phosphor dots to blue or green phosphor dots in various screen areas. Figure 4-7 shows the action of the purity magnets. Two ring-shaped permanent magnets are utilized to adjust the field strength. Both rings can be rotated together on the tube neck to adjust the direction of the purity field. The appearances of pure and impure red fields are shown in Fig. 4-8.

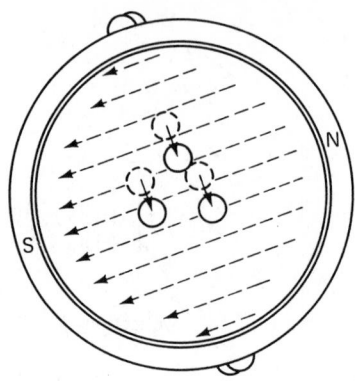

FIGURE 4-7 Action of the purity magnets.

FIGURE 4-8 Correct and incorrect purity adjustments, shown for a red field (see inside back cover for artwork).

In addition to the dynamic convergence controls, supplementary static-convergence controls are provided in many cases, as Fig. 4-9(a) shows. Each of the ferrite dynamic convergence cores is divided into two pole pieces, which are joined at the center by a cylindrical permanent magnet called a *beam magnet*. This magnet is transversely polarized, and operates as a static-convergence control. That is, when the magnet is turned in the ferrite core, it adds a more or less magnetic "bias" to the electromagnetic field. Three beam magnets are employed, so that the directions of the red, green, and blue electron beams can be adjusted individually. Note that in Fig. 4-9(b) the internal pole pieces are utilized inside the tube neck to confine the magnetic field of the blue-lateral corrector to the blue-electron beam.

In Fig. 4-10, note that several permanent magnets are mounted around the picture-tube rim. These are called *rim magnets* and are turned on their axes as required to obtain optimum field purity. The rim magnets supplement purity-magnet action. Some color picture tubes also employ a large coil or electromagnet around or near the magnetic shield. This is called a *degaussing coil*, and it serves to demagnetize the picture-tube mounting brackets and other ferro-magnetic structures since a television receiver chassis tends to become weakly magnetized in the direction of the earth's magnetic field. Nearby ferro-magnetic objects can also affect the magnetic condition of the chassis causing the screen purity to deteriorate. A surge of AC current is automatically passed through the degaussing coil each time the receiver is turned on. This process maintains the color picture tube assembly in a neutral magnetic state.

Figure 4-11 illustrates the effect of convergence-control adjustments. A white-dot or crosshatch pattern is generally used to check the convergence

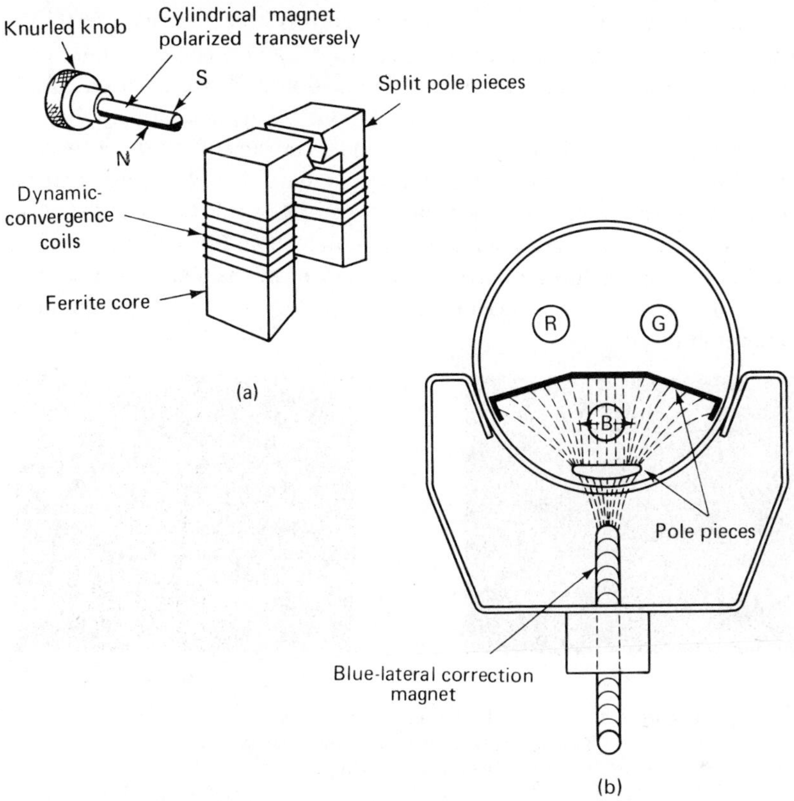

Knurled knob

Cylindrical magnet
polarized transversely

S

N

Split pole pieces

Dynamic-
convergence
coils

Ferrite core

(a)

R   G

+B+

Pole pieces

Blue-lateral correction
magnet

(b)

FIGURE 4-9   Static-convergence controls. (a) Beam-positioning
magnet; (b) blue-lateral corrector magnet.

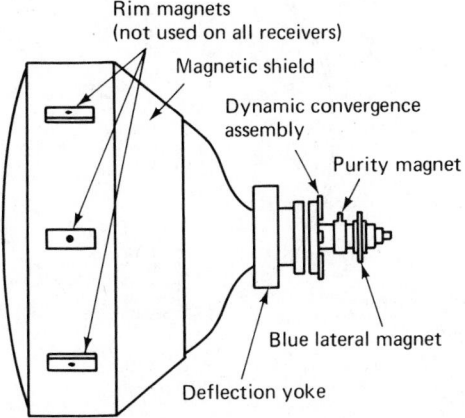

Rim magnets
(not used on all receivers)

Magnetic shield

Dynamic convergence
assembly

Purity magnet

Blue lateral magnet

Deflection yoke

FIGURE 4-10   Rim magnets may be mounted around the picture-
tube shield.

of a picture tube. When the convergence controls have been incorrectly adjusted, white dots are split up into red, green, and blue color dots. Convergence procedures, and description of white-dot and crosshatch generators are reserved for subsequent discussion. At this point, note simply that the static-convergence controls produce an equal movement of the associated color dots at all points on the screen and the dynamic-convergence controls produce a much greater movement of the associated color dots at the edges of the screen than at the center of the screen. Note in Fig. 4-9 that the blue-lateral corrector produces a horizontal movement of the blue dots. However, the beam magnets produce diagonal movements of the associated color dots, as shown in Fig. 4-12.

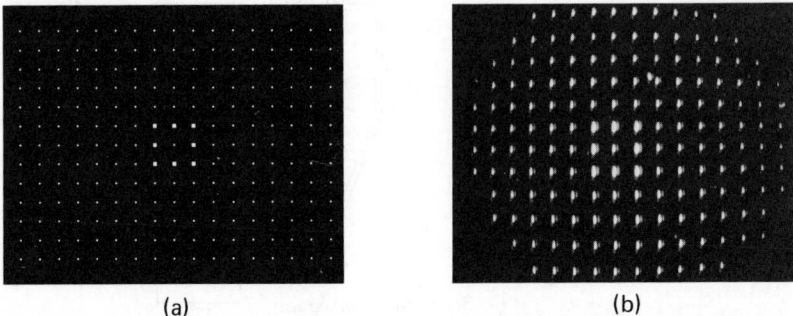

(a)   (b)

FIGURE 4-11 Examples of convergence-control adjustments. (a) Normally converged screen pattern; (b) seriously misconverged screen pattern.

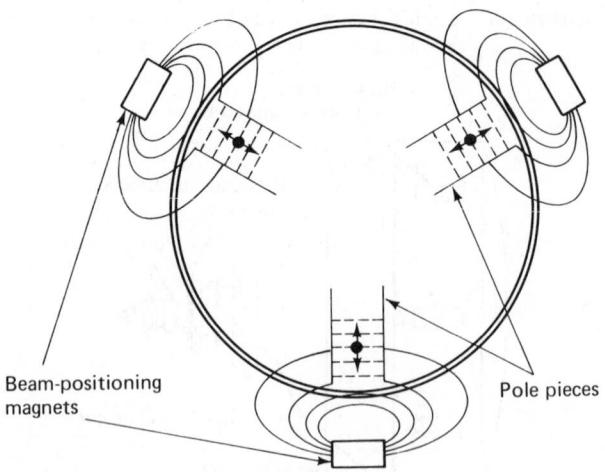

Beam-positioning magnets

Pole pieces

FIGURE 4-12 The beam magnets produce relative diagonal movements of the associated color dots.

4.4   OPERATING FEATURES OF THE APERTURE-GRILLE COLOR PICTURE TUBE

Referring to Fig. 4-4, note that the aperture-grille type of picture tube employs three in-line cathodes. Emitted electrons pass through small holes in the control grid, G1. Next, the electrons are speeded up by the screen grid or accelerating anode, G2. In turn, the electron beams pass through holes in this electrode and proceed at high speed into the focus-electrode region. This is essentially an electrostatic-lens assembly, which brings the electron beams to a point. Only the green electron beam proceeds in a straight line. The red and blue beams come together at a point and then diverge. The focusing action also has the effect of minimizing the diameter of each electron beam. The beams are attracted to the converging plates, which operate at 19,000 volts. This second electron-lens arrangement brings the beams to precise focus on the aperture grille.

Field purity is obtained by proper adjustment of the purity magnet and by correct positioning of the deflection yoke. With reference to Fig. 4-13, the purity magnet is adjusted to obtain optimum purity at the center of the screen. Then, the deflection yoke is moved to a position on the tube neck that provides optimum purity out to the screen edges. A green field is ordinarily used in this procedure. Note that the purity magnet shifts all three beams by the same amount at all points on the screen. On the other hand, the deflection-yoke position shifts the beams much more at the screen edges than at the center of the screen. Another purity control utilized on the aperture-grille tube is called the *neck-twist coil*. It is mounted on the rear of the tube neck. This adjustment affects the red and blue beams only, as depicted in Fig. 4-14. Basically, this arrangement is the same as the purity magnet used with the shadow-mask type of tube. Adjustment of the neck-twist coil is usually made with a red field display.

Convergence at the edges of the screen requires the addition of a parabolic voltage waveform to the converging plates in Fig. 4-4. That is, a parabolic waveform (Fig. 4-15) is superimposed on the DC focusing voltage. This waveform can be tilted one way or the other by means of a sawtooth voltage from the horizontal dynamic control. This tilt is made available to compensate for manufacturing tolerances on replacement picture tubes. The final convergence adjustment control is called the *vertical static control*. This is a DC electromagnetic arrangement that operates as a static convergence control. (See Fig. 4-16.) The coil is mounted so that the vertical beam directions are affected. This is the only vertical convergence control for the aperture-grille type of picture tube, and it is generally adjusted first in the convergence procedure.

Observe in Fig. 4-16 that the cathodes are driven by the outputs of the color video amplifiers. The aperture-grille type of tube is not used as a

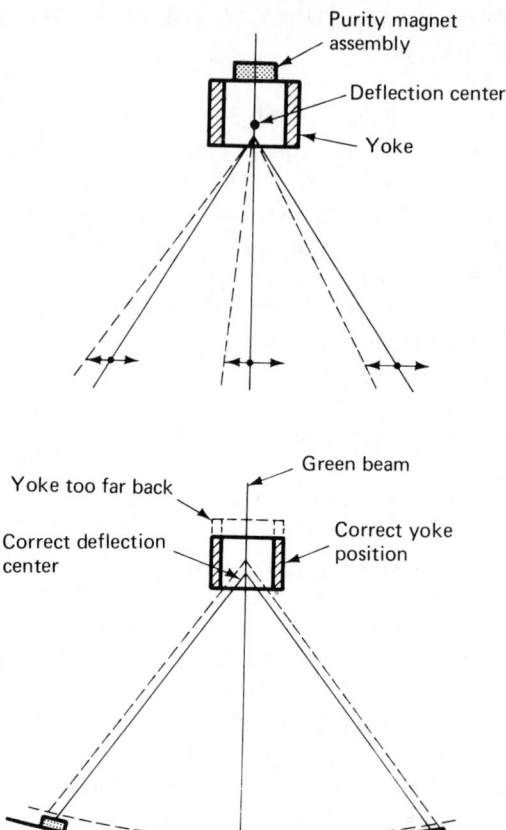

FIGURE 4-13 Field purity requires correct positioning of the deflection yoke.

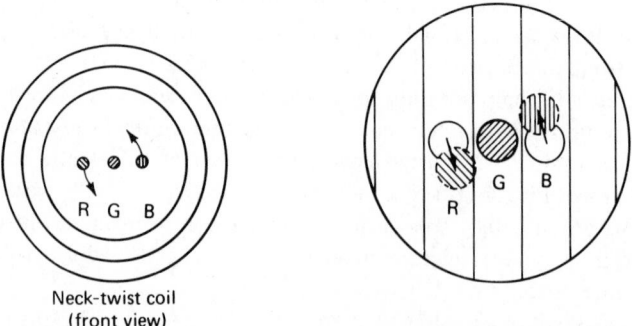

FIGURE 4-14 Outline of a neck-twist coil and its magnetic field action.

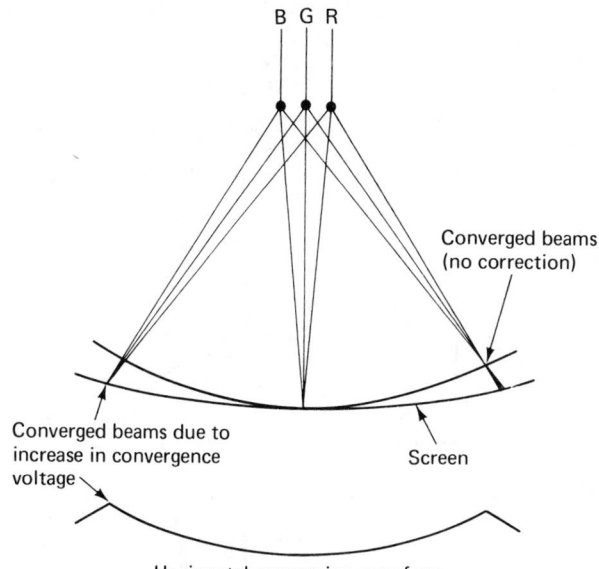

B G R

Converged beams
(no correction)

Converged beams due to
increase in convergence
voltage

Screen

Horizontal converging waveform

FIGURE 4-15   Electron beams are over-converged at the screen
edges unless a parabolic correction voltage is
utilized.

final matrix; all matrixing operations are accomplished in the receiver circuitry. Neon lamps are provided in the cathode circuits to protect the picture
tube against possible excessive cathode potentials. R1 and C1 provide a compensated attenuating pad, which operates as part of the video-amplifier
system to provide correct relative color-signal levels. Thus, the readjusted
chroma values that energize the chroma demodulators must be changed into
unadjusted chroma values before application to the color picture tube. Otherwise, the primary color balance would be incorrect. Blanking pulses are
applied to the control grid of the picture tube to make retrace lines invisible.

Many recent-model color receivers utilize matrix (black-surround)
color picture tubes. This variety of picture tube requires greater yoke positioning accuracy than the older non-matrix tubes in order to eliminate color
fringing. To position the deflection yoke precisely, the following procedure
will be found helpful:

1. Stretch two long rubberbands across a degaussing coil, as depicted
   in Fig. 4-17(a). This forms a crosshair arrangement to indicate
   the precise center of the degaussing coil.

2. Using a grease pencil, draw diagonal lines on the face of the picture
   tube, as those shown in Fig. 4-17(b). This procedure identifies the
   precise center of the screen.

3. Turn the blue and green screen controls off. (Screen controls are
   detailed subsequently.)

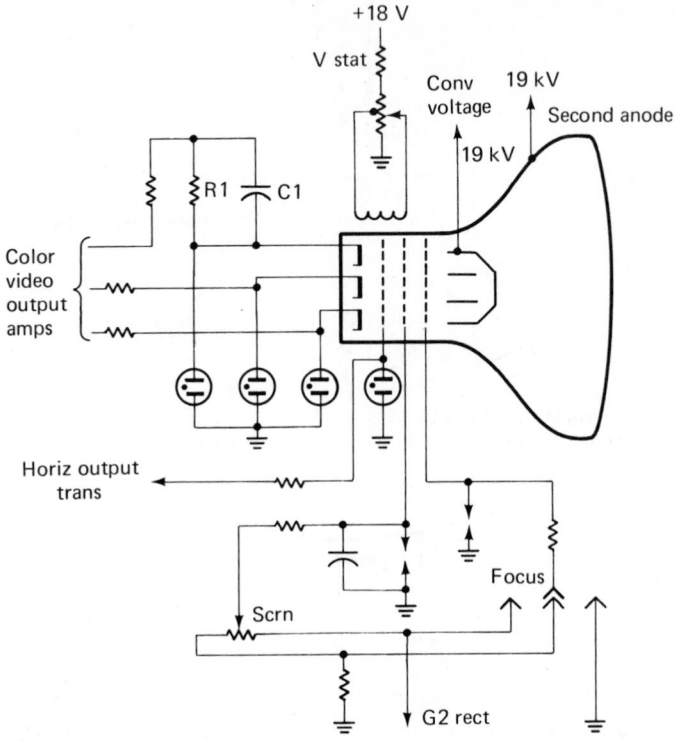

FIGURE 4-16   The vertical static control is an electromagnetic field adjustment.

4. Connect the degaussing coil to a 12-volt DC supply.

5. Center the degaussing coil on the face of the picture tube, as depicted in Fig. 4-17(b).

6. Turn on the 12-volt DC supply.

7. Observe the pattern displayed on the screen. It should appear as a three-bladed green propeller with a red hub on a red background and some blue in the areas between the blades, as depicted in Fig. 4-17(c). The position of the blades is a clue to the correct yoke position. Two of the green blades should be pointing directly into the raster corners (either top or bottom), with the third blade pointing vertically up or down.

8. Center the red hub on the diagonals crossover point with the centering tabs on the picture-tube neck.

9. Loosen the yoke wing bolts and move the yoke forward or backward on the picture-tube neck until the green blades are located correctly in the raster corners; then tighten the wing bolts.

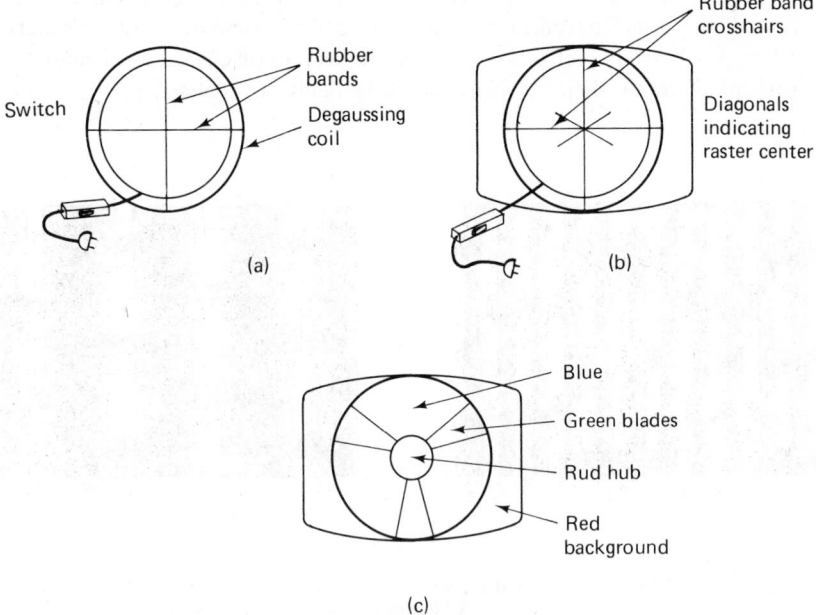

(a)

(b)

(c)

FIGURE 4-17   Procedure for yoke positioning on a matrix color picture tube. (a) Locating the center of the degaussing coil; (b) locating the center of the picture-tube face; (c) pattern displayed when yoke is correctly positioned.

10. Connect the degaussing coil to a 117-volt AC outlet, and demagnetize the picture tube so that a pure red raster can be displayed.

Another design of color picture tube employs a slotted mask and is called the AccuLine system by RCA. These picture tubes have a 90° deflection angle, and the three electron guns are mounted in-line horizontally. The phosphor screen in this color picture tube does not consist of dot triads; instead, it is formed as lines. Figure 4-18(a) shows a photomicrograph of the screen structure, and the slotted apertures are shown in (b). This type of picture tube does not require a dynamic-convergence board, dynamic-convergence yoke, or blue-lateral magnet. Only one screen control is utilized. An unconventional yoke is mounted on the neck of the picture tube, as seen in Fig. 4-19.

In consequence of the precision design of the gun assembly, with the special design of the deflection yoke, dynamic-convergence adjustments are eliminated. To do so, the in-line beams from the electron guns must pass through the center of the deflection yoke in a precisely spaced and precisely horizontal group. The grids in this picture tube have a unit construction with a triple aperture. Only a slight correction is required for adjusting convergence. The purity/convergence device uses a dual assembly of ring magnets mounted behind the deflection yoke on the picture-tube neck. Four rings are provided. Two of them develop four-pole fields, as depicted in Fig. 4-20.

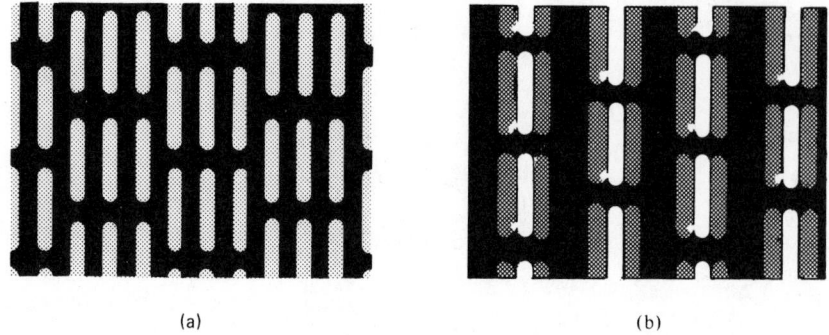

(a)   (b)

FIGURE 4-18   Slotted-mask color picture tube. (a) Photomicrograph of screen; (b) photomicrograph of mask. *(courtesy of RCA)*

FIGURE 4-19   Picture-tube and yoke assembly.

These rings move the outside (B/G) beams equally in opposite directions. The other two rings develop six-pole fields which also move the outer beams equally, but in the same direction.

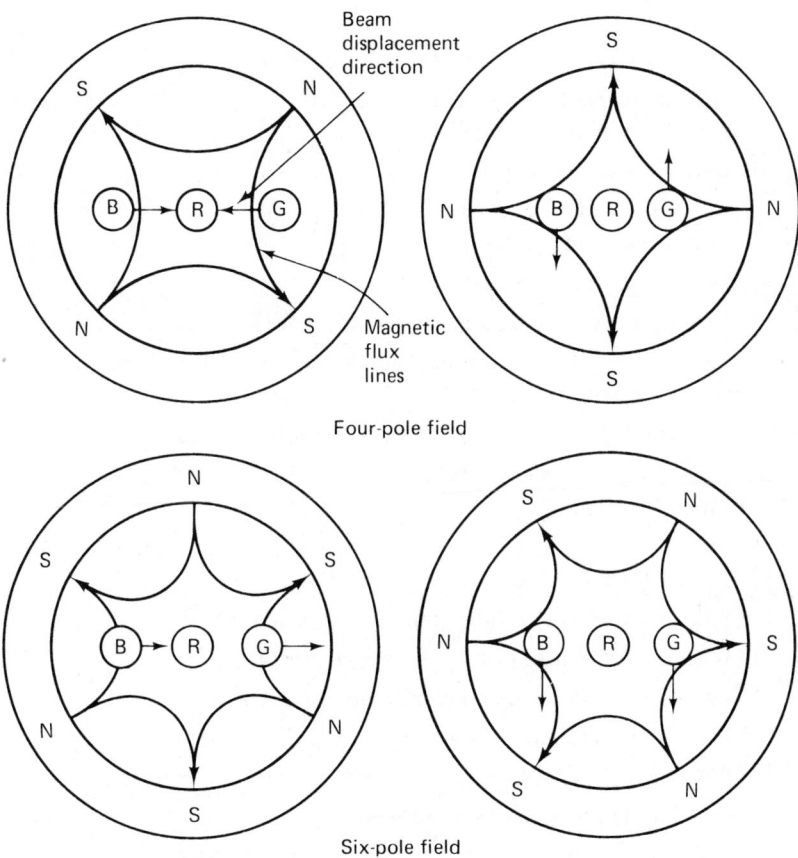

Four-pole field

Six-pole field

FIGURE 4-20   Four ring magnets provide purity and convergence adjustments.

The deflection yoke is called a precision static toroid (PST) yoke, and has a single-layer winding. It is mounted at the factory and cemented to the tube neck. In other words, the yoke cannot be readily replaced. However, it is highly unlikely that the yoke would develop defects during the life of the picture tube. The yoke is wound with comparatively heavy and spaced wire and is heavily insulated. However, if replacement should be necessary, a heat gun can be used to soften the cement between the tube neck and the yoke.

EXERCISES

Questions

1. What are the basic constructional features of a shadow-mask picture tube?

2. How is an aperture-grille picture tube constructed?

3. Do shadow-mask and aperture-grille picture tubes operate from the same form of color signal?

4. Define the term *convergence.*

5. Explain the difference between *static-convergence* and *dynamic-convergence* procedures.

6. Describe a *degaussing coil.*

7. Is an aperture-grille picture tube used as a final matrix?

8. State a typical second-anode accelerating voltage utilized by an aperture-grille picture tube.

9. How many perforations does a shadow mask contain?

10. How many phosphor dots does the screen of a shadow-mask picture tube contain?

11. Explain the function of the converging plates in an aperture-grille picture tube.

12. Why do color picture tubes employ magnetic shields?

13. What is the meaning of the term *screen purity*?

14. Describe the function of a *blue lateral corrector.*

15. Discuss the action of a *beam magnet.*

16. Why are internal pole pieces provided inside a color picture tube?

17. Explain the function of *rim magnets.*

18. What are the basic screen patterns utilized in convergence procedures?

19. Why is the location of the deflection yoke on the neck of a color picture tube critical?

20. Compare the functions of *purity magnets* and *neck twist coils.*

True-False

1. Only one type of color picture tube has been adopted commercially.

2. A shadow-mask picture tube contains an electron-gun assembly, a shadow mask, and a phosphor-dot screen.

3. Single-gun color-TV picture tubes are used in most color receivers.

4. A two-color design of picture tube is found in economy-type color receivers.

5. In normal operation, all three electron beams pass through the same aperture of the shadow mask in a color picture tube.

6. A dot triad is comprised of a red dot, a green dot, and a blue dot.

7. The fundamental components of a Trinitron picture tube are the electron-gun assembly, the aperture grille, and the phosphor-stripe screen.

8. In normal operation all three electron beams pass through the same slot in an aperture grille.

9. An aperture-grille picture tube does not operate from the same signal waveforms that are employed with a shadow-mask tube.

10. Convergence is a process whereby the three electron beams are maintained in focus at any spot on a shadow mask or an aperture grille during the scanning action.

11. Mu metal magnetic shields are used with color picture tube to avoid disturbance from the earth's magnetic field, or other external fields.

12. One of the more recent designs of color picture tube requires no convergence adjustments.

13. A color picture tube that is permanently converged during manufacture has the yoke and convergence devices bonded to the picture-tube neck.

14. Square waveforms are used in the dynamic convergence process.

15. A purity magnet affects the direction of travel of all three electron beams.

16. *Beam magnet* and *purity magnet* are equivalent terms.

17. Rim magnets are used with some color picture tubes to supplement purity-magnet action.

18. A *degaussing coil* is the same device as a *neck twist coil.*

19. Static-convergence controls produce an equal movement of the associated color dots at all points on the screen.

20. Dynamic convergence controls produce a much greater movement of the associated color dots at the edges of the screen than at center-screen.

Multiple Choice

1. An aperture-grille color picture tube _____ be substituted for a shadow-mask color picture tube.
   (a) can
   (b) cannot
   (c) may sometimes
   (d) can (with a deflection-coil change)

2. The most recent design of shadow-mask color picture tube employs _____ electron guns.
   (a) no
   (b) two
   (c) three
   (d) four

3. Aperture-grille color picture tubes are _____ used in large-screen receivers.
   (a) always
   (b) never
   (c) occasionally
   (d) generally

4. Each electron beam in a color picture tube strikes the screen at _____.
   (a) a slightly different point
   (b) the same point
   (c) zero velocity
   (d) variable points

5. When a pure red field is being displayed on the screen of a color picture tube, the _____ gun(s) is/are cut off.
   (a) red
   (b) red, blue, and green
   (c) blue and green
   (b) black-and-white

6. Shadow-mask and aperture-grille color picture tubes operate from _____ signal waveforms.
   (a) different
   (b) partially different
   (c) the same
   (d) reversed

7. Color picture tubes that feature permanent built-in convergence have _____ .
   (a) only one gun
   (b) red, yellow, and blue guns

(c) one gun with two backup guns

(d) red, green, and blue guns

8. A shadow mask has _____ radius of curvature with respect to the picture-tube screen.

(a) a shorter

(b) the same

(c) a longer

(d) variable

9. Screen purity is defined as _____.

(a) physically complete saturation of a hue

(b) physically complete desaturation of a hue

(c) a uniform mixture of a hue with white light

(d) a uniform mixture of a hue with black light

10. Convergence procedures are generally based on a _____ pattern.

(a) station test

(b) white-dot or crosshatch

(c) sine wave

(d) square wave

11. Some color picture-tube assemblies include_____.

(a) automatic degaussing action

(b) reverse color devices

(c) ultraviolet radiators

(d) X-ray intensifiers

12. A blue lateral corrector provides _____.

(a) vertical adjustment of the blue dots

(b) horizontal adjustment of the blue dots

(c) both vertical and horizontal adjustment of the blue dots

(d) diagonal adjustment of the blue dots

13. Only shadow-mask picture tubes contain _____ .

(a) internal pole pieces

(b) three electron guns

(c) red, green, and blue phosphor dots

(d) a shadow mask

14. A_____voltage waveform is employed in dynamic-convergence action.

(a) sinusoidal

(b) parabolic

(c) hyperbolic

(d) elliptical

15. Each cathode in an aperture-grille color picture tube is driven by
    _____.
    (a) the same waveform and peak-to-peak voltage
    (b) different waveforms with the same peak-to-peak voltage
    (c) the same waveform at different peak-to-peak voltages
    (d) none of the above

Problems

1. If a color picture tube operates at 20 kV with a current flow of 1 mA, what is its effective circuit resistance?

2. How much power is dissipated in the resistance that was calculated in problem 1?

3. The regulation of a high-voltage circuit is equal to the difference between its open-circuit voltage and its closed-circuit voltage, divided by its closed-circuit voltage value. If a high-voltage supply provides 20 kV under load and 21 kV under no-load conditions, what is its regulation?

4. If a high-voltage supply provides 18 kV under load and 22 kV under no-load conditions, what is its regulation?

5. It requires 27,500 volts to break down a one-inch air gap between needle points. Breakdown voltage is proportional to gap spacing. How far will a spark "jump" in air from a 20-kV source?

6. A high-voltage probe is to be used with a 20,000 ohms-per-volt multimeter to indicate 20 kV full-scale on the 200-volt range of the meter. What value of multiplier resistor is required in the probe?

7. Another high-voltage probe is to be used with a TVM that has an input resistance of 10 megohms to indicate 20 kV full-scale on the 200-volt range of the meter. What value of multiplier resistor is required in the probe?

8. How much current will be drawn by the high-voltage probe and multimeter arrangement in problem 6?

9. How much current will be drawn by the high-voltage probe and TVM arrangement in problem 8?

10. Calculate the amount of power dissipated by the multiplier resistor in problem 6.

Chapter 5

RF And IF Circuitry And Operation

5.1   RF TUNER REQUIREMENTS

Front-end or *RF-tuner* arrangements for color television receivers are essentially the same as the corresponding arrangements in black-and-white receivers, except that the former provide greater uniformity of frequency response. A front end is required to tune through both the UHF and VHF bands, as Table 5-1 shows. Separate UHF and VHF tuners are utilized, as depicted in Fig. 5-1. When the VHF tuner is in use, the UHF tuner is disabled. When the UHF tuner is in use, its output is fed at IF frequency to the VHF tuner. The VHF amplifier and VHF mixer stages are operated at IF frequency, and the VHF oscillator is disabled. In other words, the IF output from the UHF tuner is amplified through the VHF tuner and then fed to the IF section in the receiver. This additional amplification through the VHF tuner is necessary because the UHF tuner does not provide amplification; it only provides frequency conversion.

Various types of transistors are utilized in front-end configurations. For example, *bipolar, junction field-effect* (JFET), *insulated-gate field-effect* (IGFET or MOSFET), *dual-gate MOSFET*, and *dual-gate with Zener*

TABLE 5-1

FCC TELEVISION CHANNEL ALLOCATIONS

P = Picture carrier freq. (MHz)
S = Sound carrier freq. (MHz)
Freq. limits of channel (MHz)

P/S freq. (MHz)	Channel number	Freq. limits (MHz)
		54
P 55.25 / S 59.75	2	
		60
P 61.25 / S 65.75	3	
		66
P 67.25 / S 71.75	4	
		72
		76
P 77.25 / S 81.75	5	
		82
P 83.25 / S 87.75	6	
		88
		174
P 175.25 / S 179.75	7	
		180
P 181.25 / S 185.75	8	
		186
P 187.25 / S 191.75	9	
		192
P 193.25 / S 197.75	10	
		198
P 199.25 / S 203.75	11	
		204
P 205.25 / S 209.75	12	
		210
P 211.25 / S 215.75	13	
		216
		470
P 471.25 / S 475.75	14	
		476
P 477.25 / S 481.75	15	
		482
P 483.25 / S 487.75	16	
		488
P 489.25 / S 493.75	17	
		494
P 495.25 / S 499.75	18	
		500
P 501.25 / S 505.75	19	
		506
P 507.25 / S 511.75	20	
		512
P 513.25 / S 517.75	21	
		518
P 519.25 / S 523.75	22	
		524
P 525.25 / S 529.75	23	
		530
P 531.25 / S 535.75	24	
		536
P 537.25 / S 541.75	25	
		542

P/S freq. (MHz)	Channel number	Freq. limits (MHz)
		542
P 543.25 / S 547.75	26	
		548
P 549.25 / S 553.75	27	
		554
P 555.25 / S 559.75	28	
		560
P 561.25 / S 565.75	29	
		566
P 567.25 / S 571.75	30	
		572
P 573.25 / S 577.75	31	
		578
P 579.25 / S 583.75	32	
		584
P 585.25 / S 589.75	33	
		590
P 591.25 / S 595.75	34	
		596
P 597.25 / S 601.75	35	
		602
P 603.25 / S 607.75	36	
		608
P 609.25 / S 613.75	37	
		614
P 615.25 / S 619.75	38	
		620
P 621.25 / S 625.75	39	
		626
P 627.25 / S 631.75	40	
		632
P 633.25 / S 637.75	41	
		638
P 639.25 / S 643.75	42	
		644
P 645.25 / S 649.75	43	
		650
P 651.25 / S 655.75	44	
		656
P 657.25 / S 661.75	45	
		662
P 663.25 / S 667.75	46	
		668
P 669.25 / S 673.75	47	
		674
P 675.25 / S 679.75	48	
		680
P 681.25 / S 685.75	49	
		686
P 687.25 / S 691.75	50	
		692
P 693.25 / S 697.75	51	
		698
P 699.25 / S 703.75	52	
		704
P 705.25 / S 709.75	53	
		710
P 711.25 / S 715.75	54	
		716

P/S freq. (MHz)	Channel number	Freq. limits (MHz)
		716
P 717.25 / S 721.75	55	
		722
P 723.25 / S 727.75	56	
		728
P 729.25 / S 733.75	57	
		734
P 735.25 / S 739.75	58	
		740
P 741.25 / S 745.75	59	
		746
P 747.25 / S 751.75	60	
		752
P 753.25 / S 757.75	61	
		758
P 759.25 / S 763.75	62	
		764
P 765.25 / S 769.75	63	
		770
P 771.25 / S 775.75	64	
		776
P 777.25 / S 781.75	65	
		782
P 783.25 / S 787.75	66	
		788
P 789.25 / S 793.75	67	
		794
P 795.25 / S 799.75	68	
		800
P 801.25 / S 805.75	69	
		806
P 807.25 / S 811.75	70	
		812
P 813.25 / S 817.75	71	
		818
P 819.25 / S 823.75	72	
		824
P 825.25 / S 829.75	73	
		830
P 831.25 / S 835.75	74	
		836
P 837.25 / S 841.75	75	
		842
P 843.25 / S 847.75	76	
		848
P 849.25 / S 853.75	77	
		854
P 855.25 / S 859.75	78	
		860
P 861.25 / S 865.75	79	
		866
P 867.25 / S 871.75	80	
		872
P 873.25 / S 877.75	81	
		878
P 879.25 / S 883.75	82	
		884
P 885.25 / S 889.75	83	
		900

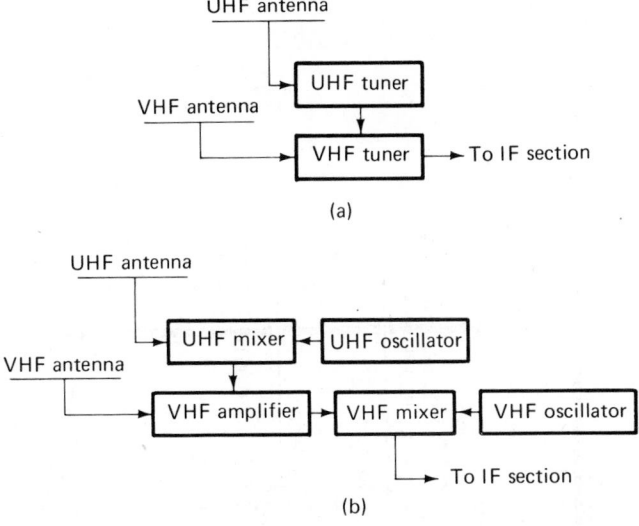

FIGURE 5-1   Front end of a color-television receiver. (a) Basic sections; (b) subsectional arrangement.

MOSFET types are used by different manufacturers. Silicon bipolar transistors are generally used as UHF oscillators. This design is operable up to 500 MHz in the common-emitter (CE) configuration, and up to 1,200 MHz in the common-base (CB) configuration. Figure 5-2 shows the configuration for a typical UHF tuner. Resonant lines are employed instead of lumped inductance and capacitance. The incoming UHF signal is fed into a mixer stage where the oscillator output *heterodynes* (combines frequencies) with the UHF signal in a semiconductor diode. In turn, an IF output is obtained in the 40-MHz range. Note that an automatic fine-tuning (AFT) diode is provided. This is a *varactor diode* that operates to prevent the UHF oscillator from drifting off-frequency.

A typical tube-type UHF tuner configuration is shown in Fig. 5-3. Its operation is essentially the same as explained previously for a solid-state UHF tuner. Servicing of a UHF tuner is usually limited to tube replacement and, sometimes, to diode replacement. Most technicians are inexperienced with UHF circuit repair, and do not attempt to troubleshoot the assembly. A defective UHF tuner may be sent to a specialized tuner service shop, or it may be discarded and replaced with a new UHF tuner.

Figure 5-4 depicts a VHF tuner configuration. A dual-gate MOSFET of the N-channel depletion type is used in the RF stage. Note that the signal voltage is applied to one gate, and the AGC voltage is applied to the other gate. The AGC voltage normally ranges from −5 to +6.7 volts. A dual-gate amplifier stage can be compared with a cascode configuration; it has good

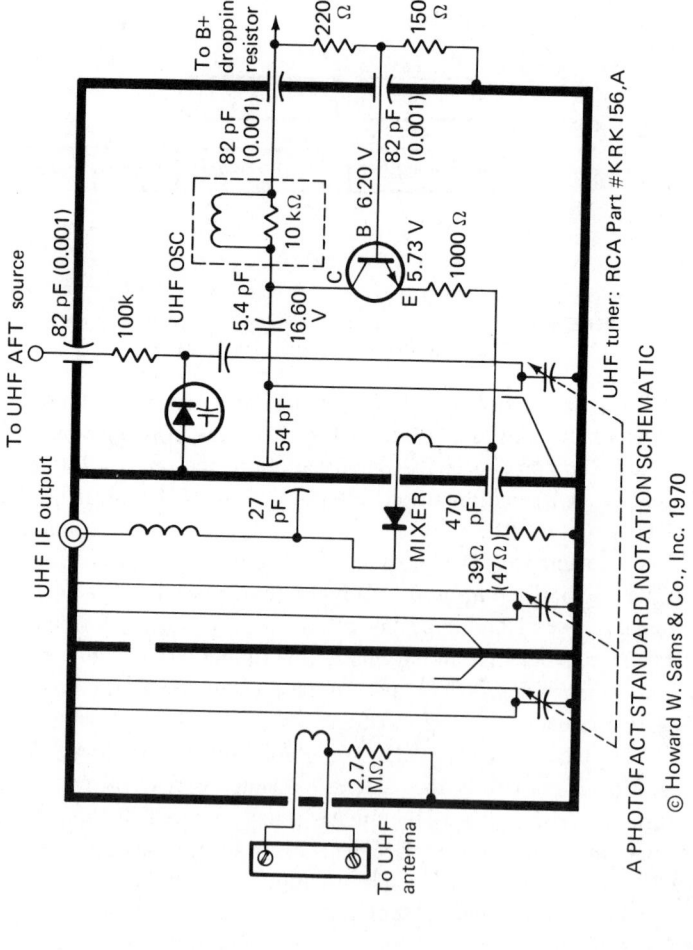

FIGURE 5-2 A UHF tuner configuration. *(Courtesy of Howard W. Sams & Co., Inc.)*

84

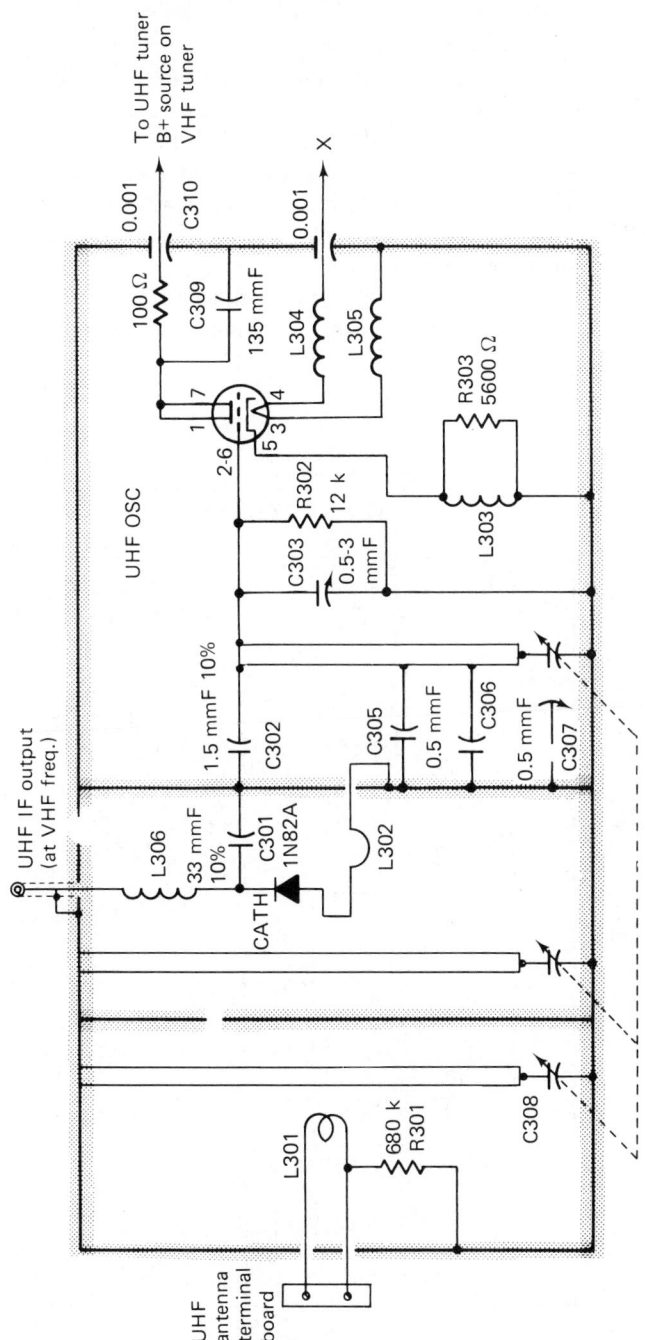

FIGURE 5-3 A typical tube-type UHF tuner configuration.

85

stability and does not require neutralization. The mixer section in Fig. 5-4 employs a pair of bipolar transistors in a standard cascode configuration. Observe that Q1 operates in the CE mode and drives Q2, which operates in the CB mode. The oscillator stage utilizes a bipolar transistor in a Colpitts configuration. Q4 operates as a varactor diode in an AFC circuit to keep the VHF oscillator from drifting off-frequency. The oscillator operates at 41.25 MHz above the sound-carrier frequency.

Next, consider the tube-type turret tuner diagrammed in Fig. 5-5. This configuration is typical of many older-model tube-type receivers. A

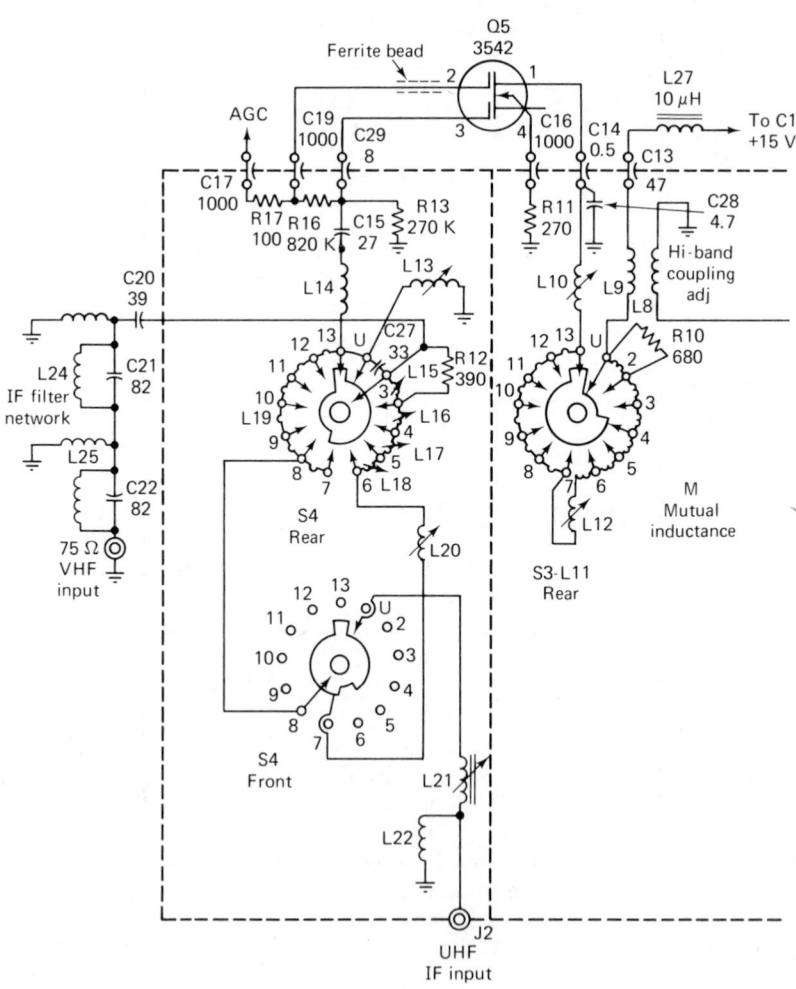

FIGURE 5-4   A VHF tuner configuration. *(courtesy of RCA)*

triode RF amplifier and triode oscillator are utilized, with a pentode mixer tube. In case of malfunction, the tubes are checked or replaced first. That is, tubes usually fail before other components. In case tube replacement does not eliminate the trouble symptom, the technician generally inspects the switch contacts. If spring pressure is weak, this condition is corrected. In case the contacts are dirty or corroded, the contacts are sprayed with a standard cleaner. In many cases, this procedure will restore the tuner to normal opera-

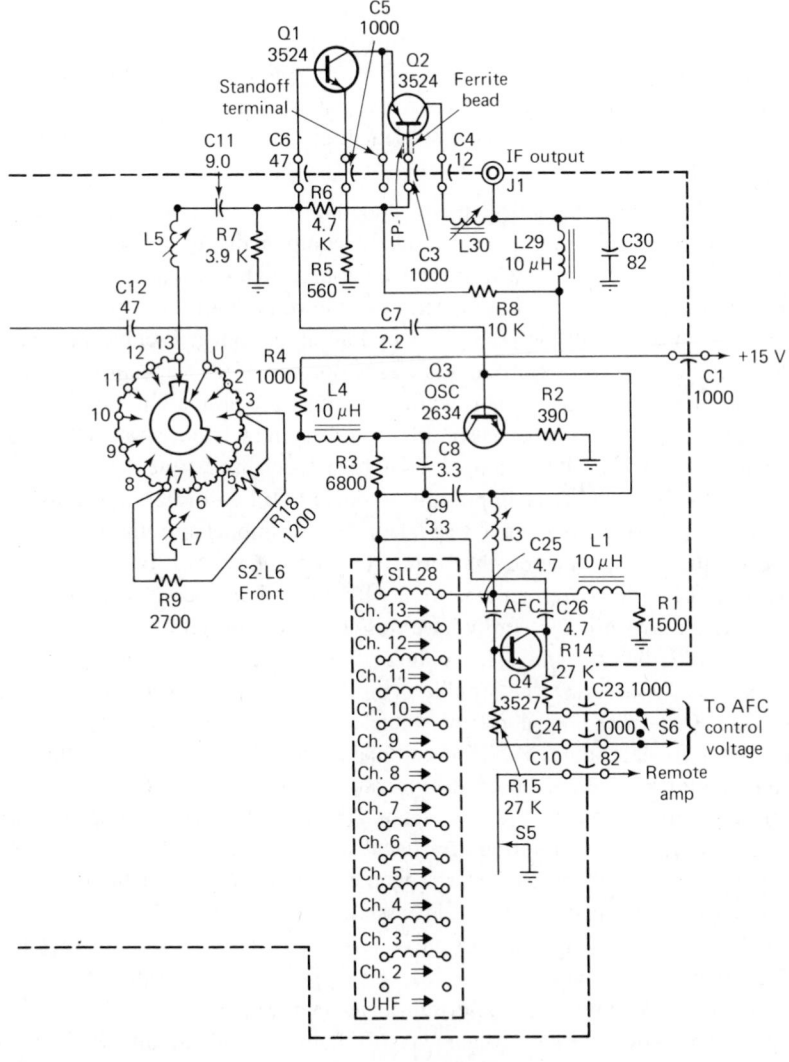

FIGURE 5-4 (continued)

tion. However, in case the malfunction persists, the technician may choose to troubleshoot the circuitry or to send the tuner to a specialized repair shop. Sometimes, a defective tuner is simply discarded and replaced with a new tuner.

Tuner troubleshooting is a comparatively demanding procedure because lead lengths in the VHF circuitry are critical, as is component placement. In other words, when a VHF component is replaced, it is often important to put it in precisely the same position as the defective component and to use precisely the same length of connecting leads. Of course, an exact replacement should be utilized, or an equivalent that is recommended in the receiver service data. Preliminary trouble analysis is usually made by means of DC-voltage and resistance measurements. Note the DC voltages specified at the tube terminals in Fig. 5-5. Receiver service data may include resistance charts, which tabulate specified resistance values from each tube terminal to chassis ground. Such resistance charts facilitate in-circuit resistance measurements and evaluation.

Excluding defective tubes and faulty contacts, leaky or open-circuited capacitors are the most likely components to cause trouble in tuners. A leaky capacitor will usually show up on DC-voltage measurements, and often on resistance measurement. On the other hand, an open-circuited capacitor must usually be pinpointed by logical reasoning. Technicians generally make substitution tests of suspected open-circuited capacitors. One of the problems in tuner repair work is frequent unavailability of suitable replacement parts. For this reason, the technician may choose not to "tackle" a malfunctioning tuner, but to send it to a specialized tuner repair depot. When this is done, it is important to include all information about the receiver from which the tuner was taken—chassis, model and run numbers. It is good practice to include the correct type of good tubes. In other words, many similar tuners are manufactured with parallel and series heaters, and a tuner depot does not necessarily have this detailed information available.

It is helpful and often reduces the service charge if the technician takes time to describe the trouble symptoms (or the defect) as clearly as possible. Otherwise, the depot personnel must spend time running tests on the tuner to determine what the complaint may be. Note well if a tuner is from a color receiver because the tuner alignment procedure is comparatively different with respect to the mixer-plate IF coil and a black-and-white tuner alignment. It is also important to install all shields that are normally used with a tuner, inasmuch as correct alignment or adjustment of the oscillator section cannot be made if a shield is missing.

A typical RF response curve is shown in Fig. 5-6. Note that the picture-carrier and sound-carrier markers appear on the peaks of the frequency-response curve. The color-subcarrier marker is located 920 kHz from the sound-carrier marker. Alignment of the front-end circuits is directed toward obtaining equal peak amplitudes with minimum sag between the

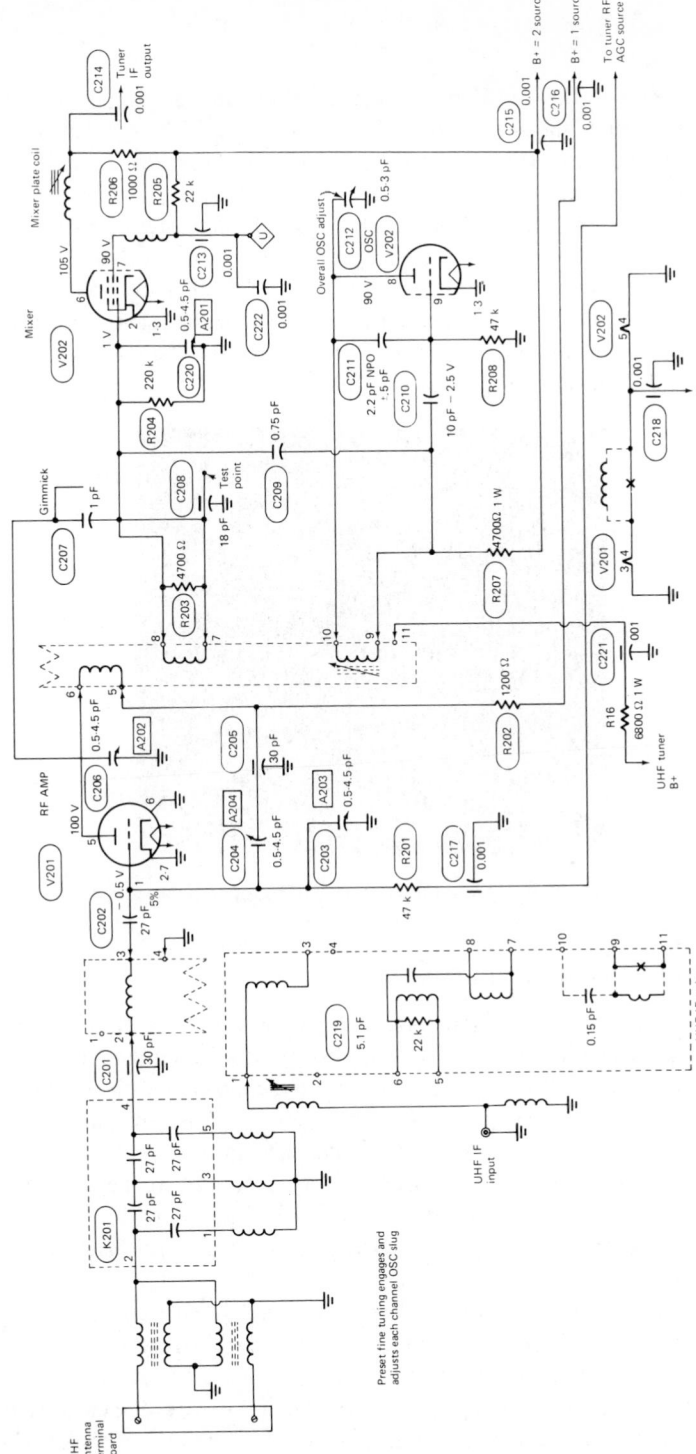

FIGURE 5-5 A tube-type turret tuner configuration.

peaks. Observe in Fig. 5-4 that separate tuned circuits are employed to process the IF output signal from the UHF tuner. Thus, L21, supplemented by series arrangements of the VHF coils, is employed in UHF operation. In alignment procedures, L21 is the only inductor in the VHF tuner that is adjusted for UHF reception. This indicates that the other inductors, L13, L20, L10, L212, L7, and L5, etc., are adjusted for VHF reception. Note that L30 operates at IF frequency during either VHF or UHF reception. Troubleshooting the RF tuner usually involves cleaning the contacts and adjusting the contact-spring pressure, if required. Figure 5-7 illustrates the procedure that is followed for a widely used type of VHF tuner.

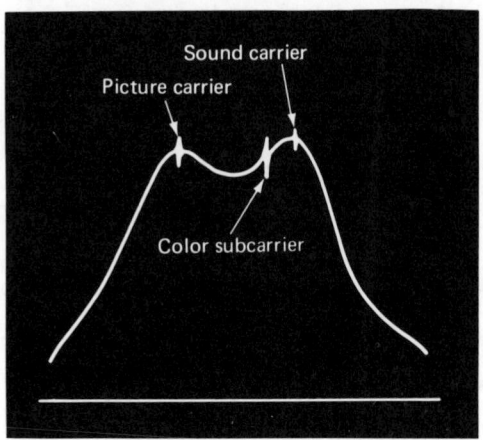

FIGURE 5-6   A typical RF tuner frequency-response curve.

5.2   ALL-ELECTRONIC TUNER

As illustrated in Fig. 5-8, a standard tuner employs an RF amplifier, a mixer, and a local oscillator, with four tunable circuits. These are the input and output circuits of the RF amplifier, the input of the mixer, and the tank circuit of the oscillator. Tunable circuits have two general forms. In some tuners, a separate resonant circuit is used on each channel. In other tuners, an inductance is tapped at 12 points to make it resonant at each of the channel frequencies. A total of 13 resonant frequencies is required, with one for each channel and one for the output of the UHF tuner, which is normally the same as the IF frequency.

In an all-electronic tuner, electronic devices perform the functions of the four switches which select the tap on the tuned circuits. The arrangement depicted in Fig. 5-9 has been simplified for clarity; it shows the elements of an electronic tuning system for selection of two channels. Similar banks of diode switches are utilized for each of the remaining channels. These diodes perform all switching operations, with the exception of the channel

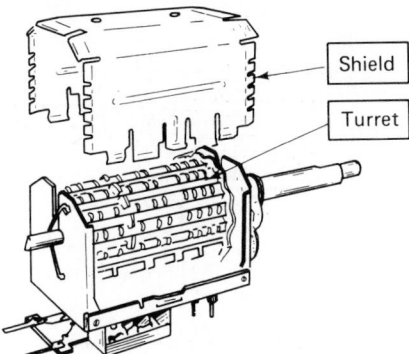

(a)   Remove the shield, exposing the turret.

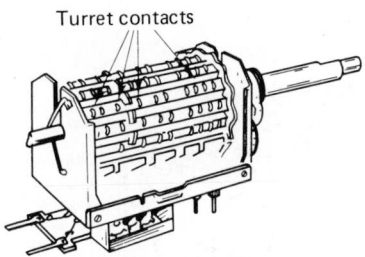

(b)   Spray the turret contacts and then wipe them with
      a soft cloth. Rotate the channel selector several
      times and respray the contacts, leaving a thin film.

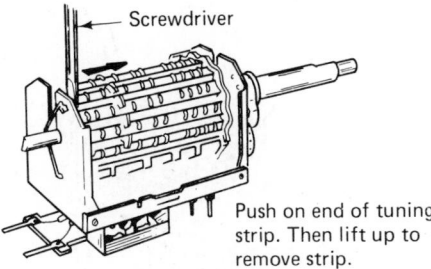

(c)   Remove any three adjacent tuning strips.

(d)   Rotate the channel selector until the stationary contacts
      can be seen.

(e)   Spray the stationary contacts and then wipe them with a
      soft cloth.

FIGURE 5-7   VHF tuner contact-cleaning procedure. *(courtesy
                  of Heath Company)*

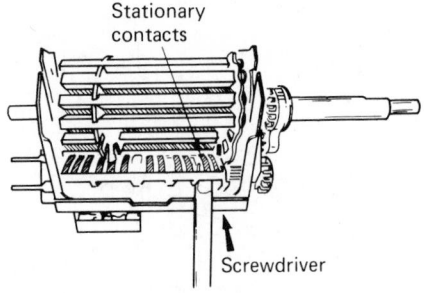

(f) Check the contacts to make sure they are all protruding an equal amount toward the turret. Bend the contacts toward the turret slightly by carefully applying pressure underneath the contacts with a small screwdriver.

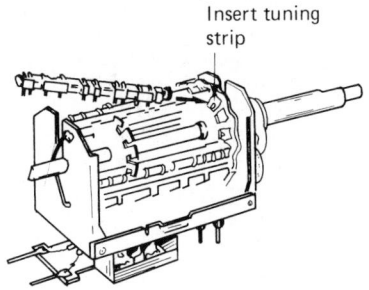

(g) Reinstall the tuning strips, being careful to note the number sequence of the strips.

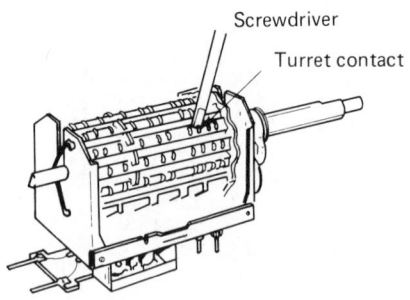

(h) Rotate the channel selector while viewing both the turret and stationary contacts simultaneously. If any individual turret contact does not touch the stationary contact, bend the turret contact out slightly with a screwdriver.

(j) Install the tuner shield.

This completes the contact cleaning procedure.

FIGURE 5-7   VHF tuner contact-cleaning procedure (continued).

indicator control. In other words, the operator sets control S1 to the desired channel, thereby applying B+ voltage to selected banks of diodes, which in turn automatically switch the inductance values as required.

In Fig. 5-9, positive voltage is fed to the anodes of D1301, D1302, D1303, and D1304, thereby biasing them into conduction. In turn, the resistance of these diodes is made very low, which effectively places the lower ends of coils L1301, L1302, L1303, and L1304 at ground potential, through capacitors C1301, C1302, C1303, and C1304. The result is the same as if a mechanical switch were used instead of each of the diodes. Next, since the foregoing diode is forward-biased, whatever voltage is applied to its anode also appears at its cathodes (less the 0.7 volt forward-drop). Thus the positive voltage is conducted to the cathodes of all the other diodes, D1201, D1202, D1203, D1204, etc., and holds these diodes in cutoff.

Fine tuning is accomplished electronically also, by means of an automatic fine tuning (AFT) configuration. It employs a variable capacitance across the oscillator tank circuit to maintain the correct frequency of oscillation. A simplified AFT circuit is shown in Fig. 5-10. If the local oscillator frequency tends to drift, the frequency of the video-IF carrier from the mixer tends to shift accordingly. This tendency causes an error voltage to be produced by the discriminator. In turn, the error voltage changes the effective capacitance of a varactor (tuning) diode and pulls the oscillator back to correct frequency. In the all-electronic tuner, separate oscillators are provided for the low-band and the high-band channels. This permits the use of optimum varactor diodes for each band. In other words, better control action is obtained if a comparatively low-capacitance varactor diode is used on the high band.

5.3   IF AMPLIFIER REQUIREMENTS

As depicted in Fig. 5-11, most of the gain in the picture channel is contributed by the IF section; thus, the IF gain in this example is 58 dB. Whereas the RF-amplifier gain is less than one-half as much, the video-detector gain is approximately one-sixth as much, and the video amplifier gain is approximately one-half as much as that of the IF section. Most of the selectivity in the picture channel is also contributed by the IF section. The most important contribution of the RF amplifier is *preselection*, which minimizes the possibility of interference from other TV stations. Otherwise, the antenna-input signal could be applied directly to the mixer. Since the IF amplifier operates in the 40-MHz region, greater stage gain can be realized than from the RF amplifier, which operates at frequencies of up to 216 MHz. Moreover, since the IF amplifier operates over a fixed frequency band, it is much easier to obtain a frequency-response characteristic that provides maximum selectivity.

With reference to Fig. 5-12, the incoming IF signal is coupled via capacitor C101 to the primary of T101. Two traps are connected in series

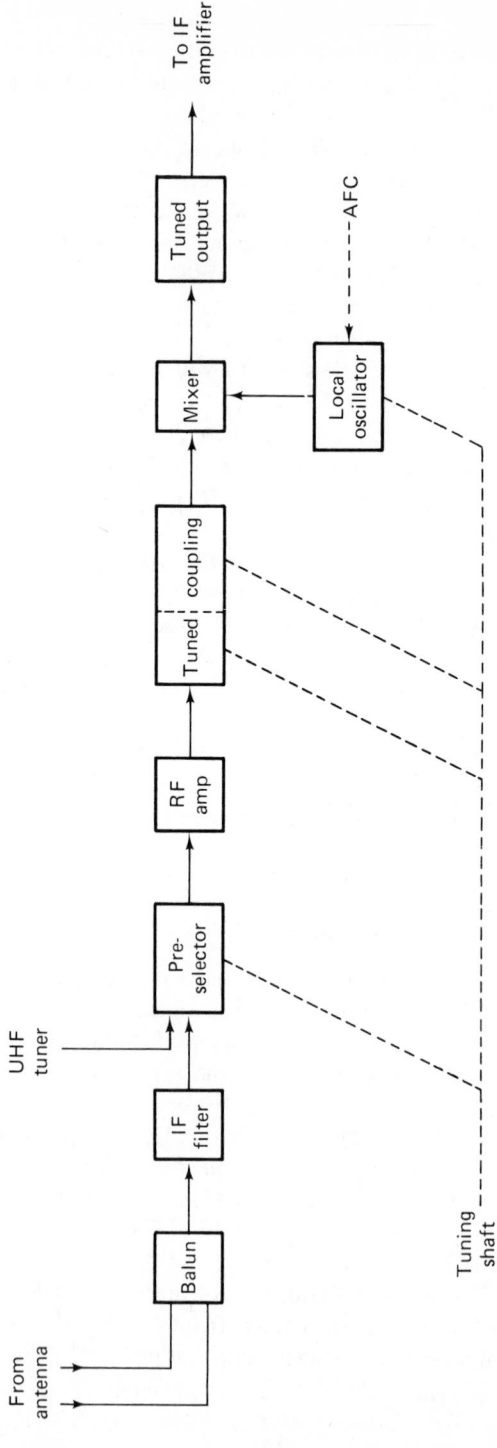

FIGURE 5-8 Locations of the tuned circuits in a standard tuner.

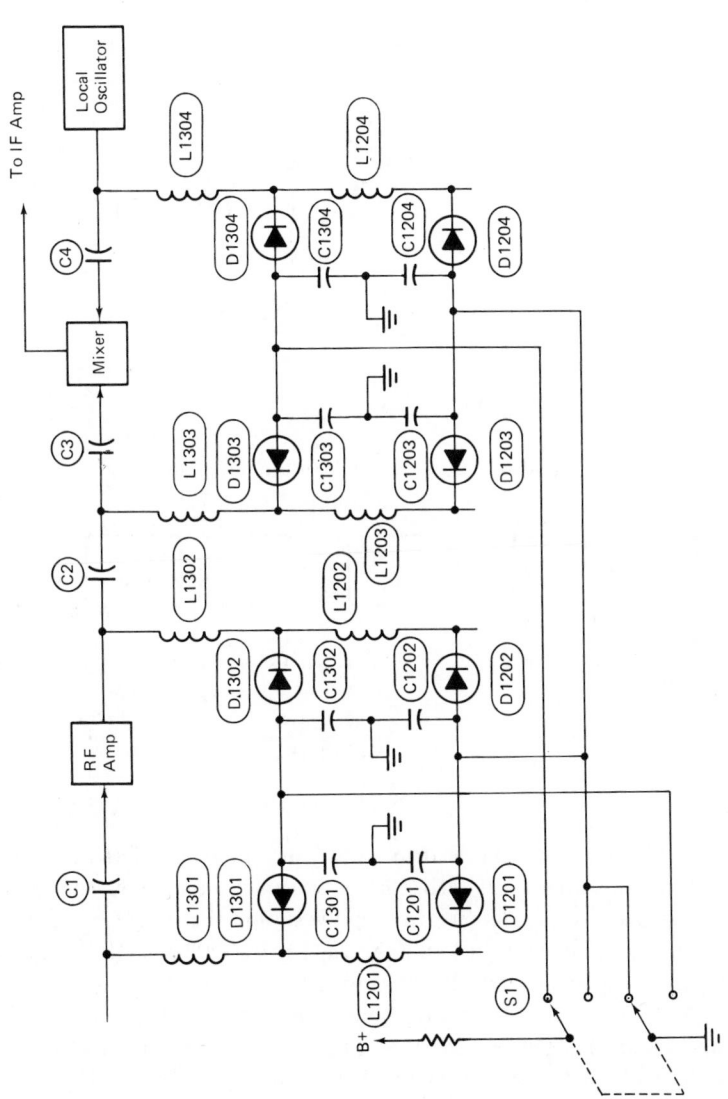

FIGURE 5-9 Partial schematic for two channels of an all-electronic tuner.

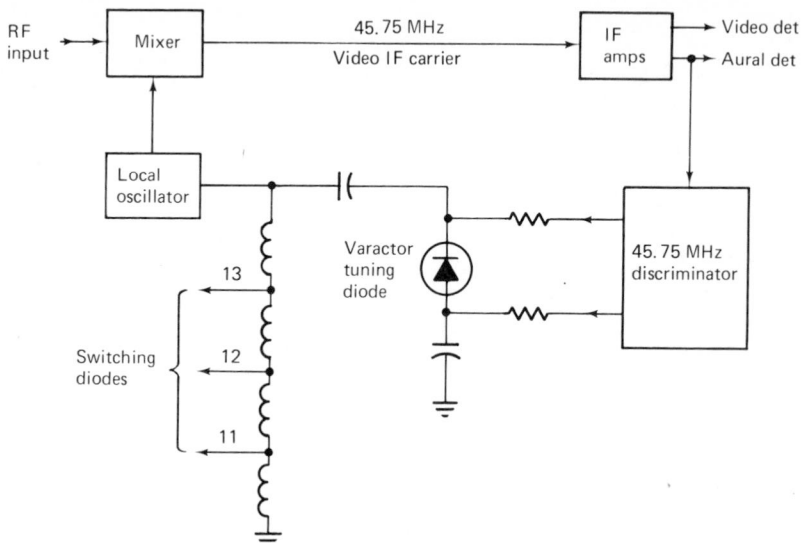

FIGURE 5-10   Basic automatic fine-tuning arrangement.

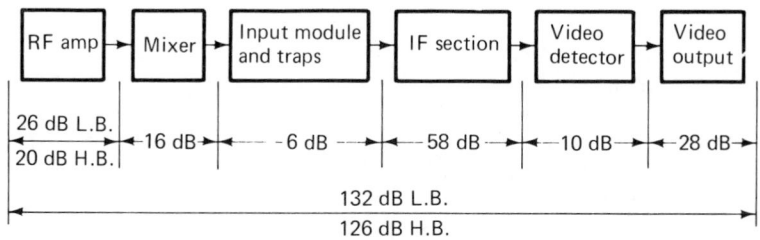

FIGURE 5-11   Sectional dB gain values for a typical receiver.

with the common ground lead to transformer T101. Coil L101 and capacitor C104 comprise the 47.25-MHz adjacent-channel *sound trap*. The sound trap eliminates sound interference from a TV station that might be operating on the adjacent channel. Coil L102 and capacitors C103 and C105 comprise the accompanying sound trap. It serves to reduce the accompanying sound signal to a level sufficiently low so that interference with the chroma signal is not visible. Note that the secondary of T101 is coupled by an impedance-matching network to the base of Q101. This network comprises C102 and R102. AGC voltage is applied to the base of Q101 via R103. Details are reserved for subsequent discussion.

Next, the amplified IF signal from the collector of Q101 drops across the L103-C108 load and is coupled through C107 and the impedance-matching resistor R108 to the base of Q102. AGC voltage is applied to the base of Q102 through R109. From the collector of Q102, the IF signal then

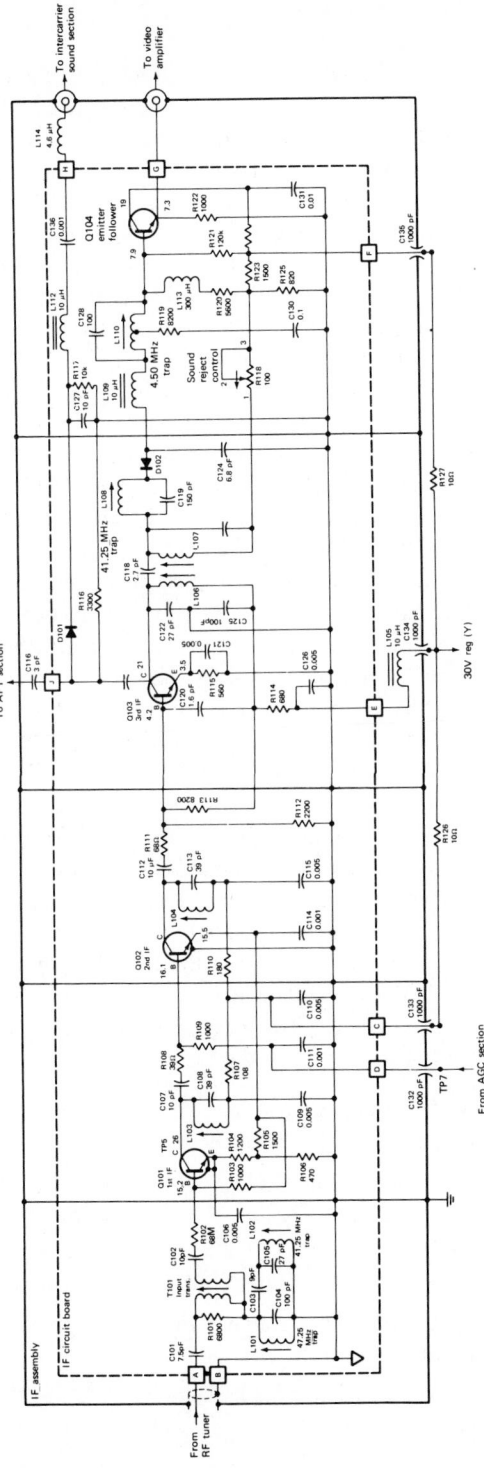

FIGURE 5-12 Configuration of a typical IF amplifier.

drops across the L104-C113 load and is coupled via C112 and R111 to the base of Q103. Observe that the DC bias on the base of Q103 is produced by a voltage divider comprising R112, R113, and R114. Stability for the third IF stage is provided by current feedback across R114. Note that the signal at the collector of Q103 is coupled through C117 to the 4.5-MHz sound-detector diode D101. In turn, the 4.5-MHz sound signal drops across R117 and is coupled via L112, C136, and L114 to the sound-output jack. C127 bypasses the undesired 45.75-MHz signal to ground. Note that C116 couples the signal at the anode end of D101 to the AFT circuit.

A double-tuned output circuit for Q103 is provided by L106 and L107 with capacitors C122, C125, and C123. In addition to providing inductive coupling, C118 also provides capacitive coupling to develop a properly shaped frequency-response curve. Note that a neutralizing voltage is developed across C125 and is fed back to the base of Q103 through C120. An additional sound trap consisting of L108 and C119 rejects the residual 41.25-MHz signal, which would otherwise have entered the video detector. Observe that this sound trap is loaded both by the video-detector load circuit and by L107-C123. The trap's tuning can be trimmed slightly by adjustment of R118. This is a maintenance adjustment control called the *sound-reject control*. From a technical viewpoint, the first half of the video-detector stage is an IF circuit, and the latter half is a video-frequency circuit. We will briefly consider the complete video-detector circuit as a single unit at this time.

It is instructive to observe the basic tube-type IF-amplifier configuration depicted in Fig. 5-13. This is a three-stage arrangement that employs pentode tubes. AGC voltage is applied to the first two stages. Tube-type and solid-state IF strips function in the same general manner. When malfunction occurs in a tube-type IF amplifier, the tubes are checked or replaced at the outset, because a tube is much more likely to fail than are other components. However, in case a tube is not causing the trouble symptom, the technician must "dig deeper." Many technicians make signal-substitution tests at points A through H in Fig. 5-13. A TV analyzer is often used for this purpose, as shown in Fig. 5-14. The basic procedure is as follows:

1. Disconnect the coaxial test lead from the UHF antenna terminals and unplug the power cord of the receiver.

2. Remove the back and cabinet of the receiver, as required, for access to the chassis. Disable the tuner by removing the IF-input plug from the IF amplifier, or equivalent.

3. Reapply power with a cheater cord.

4. Referring to Fig. 5-14, connect the black test lead from the analyzer's ground jack to the chassis of the receiver.

5. Connect the coaxial test lead to the RF jack on the analyzer. Connect the black clip of the coaxial test lead to the chassis of the

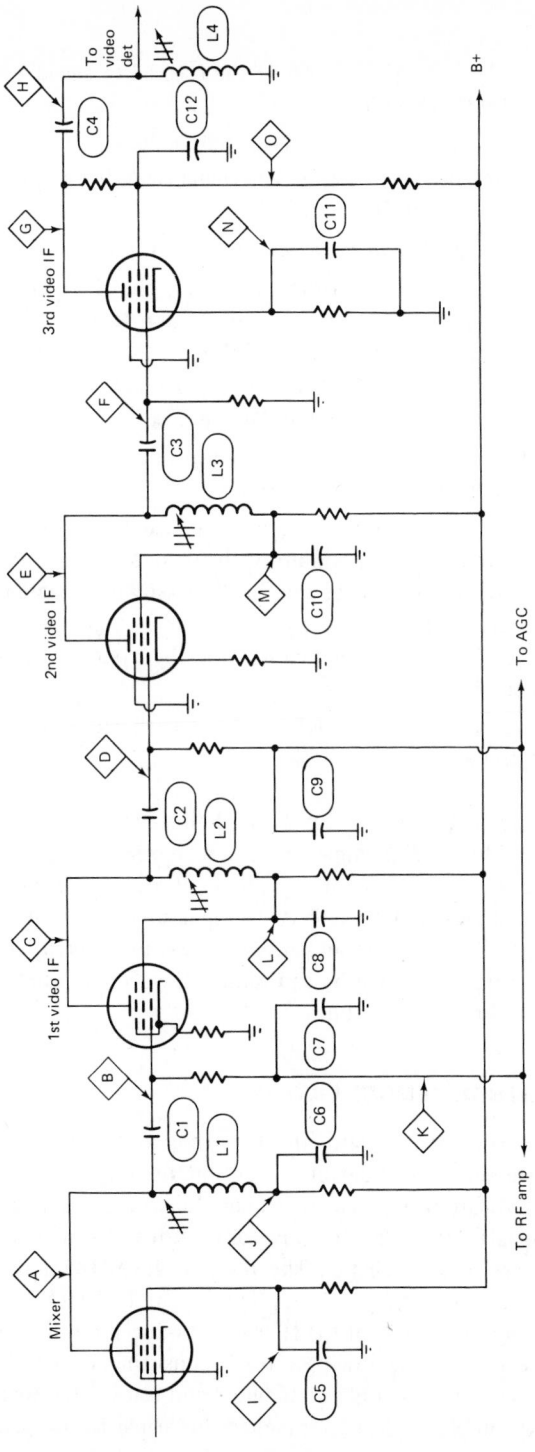

FIGURE 5-13 A tube-type IF amplifier configuration, with test points indicated.

receiver. Connect the red clip of the coaxial test lead to the input of the third IF amplifier.

6. Set the RF-attenuator control on the analyzer to maximum. This setting is required because the test signal is neither amplified by the first nor the second IF amplifiers.

7. Set the RF selector on the analyzer to its IF position.

8. Adjust the IF control on the analyzer to the IF frequency of the receiver under test. Most receivers have a 45-MHz IF, but some older receivers might have a 25-MHz IF system. When the frequency is correctly adjusted, a test pattern normally appears on the picture-tube screen, as illustrated in Fig. 5-15.

9. Reduce the setting of the RF attenuator on the analyzer until snow becomes visible in the screen display.

10. Move the red clip of the coaxial test lead from the input of the third IF amplifier to the input of the second IF amplifier. Normally, the snow will disappear and the contrast will increase because of the amplification provided by the second IF stage. If this response is not obtained, it is an indication that the trouble will be found in the second IF stage.

11. The foregoing procedure is repeated for the first IF stage, if required, to localize the defective stage.

Referring to Fig. 5-13, test points I through O indicate useful DC-voltage check points. For example, in case a bypass capacitor is short-circuited, zero DC voltage will be measured. Or, if a bypass capacitor is quite leaky, a subnormal DC-voltage value will be measured. Note that if a bypass capacitor is "open," DC-voltage measurements are unaffected. However, the gain of the associated stage will be subnormal. Suspected "open" capacitors are usually checked by substitution.

5.4   AMPLITUDE DEMODULATION

Diode detector D102 demodulates the IF signal so that the Y and chroma signals are developed in the detector load circuit. Figure 5-16 depicts the process of demodulation. Observe that demodulation develops the *envelope* (modulating signal) from the IF waveform. Thus, the combined Y and chroma signals are made available. The residual 4.5-MHz intercarrier sound signal is eliminated by a bridged-T trap comprising L110, C128, and R119. Note also that any harmonics of the IF signal produced by the demodulation process are rejected by the low-pass filter consisting of L109 and C124. Otherwise, these harmonics could be radiated and cause interference with the signal in the RF amplifier. The complete color signal is dropped across the

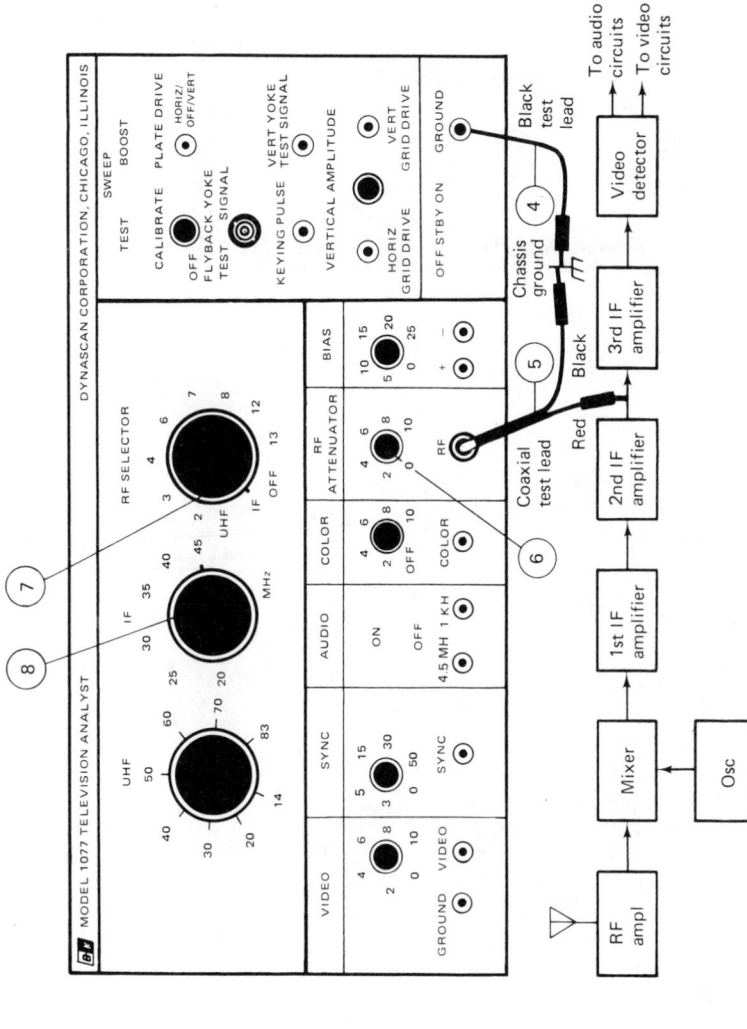

FIGURE 5-14 Analyzer connections and control settings for IF signal-substitution tests. *(courtesy of B&K Precision Mfg. Co., Division of Dynascan Corp.)*

101

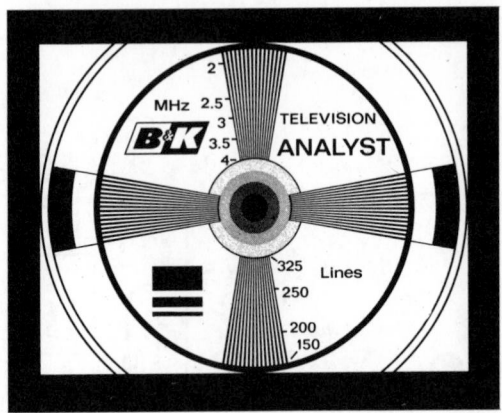

FIGURE 5-15   Screen pattern produced by a TV analyzer. *(courtesy of B&K Precision Mfg. Co., Division of Dynascan Corp.)*

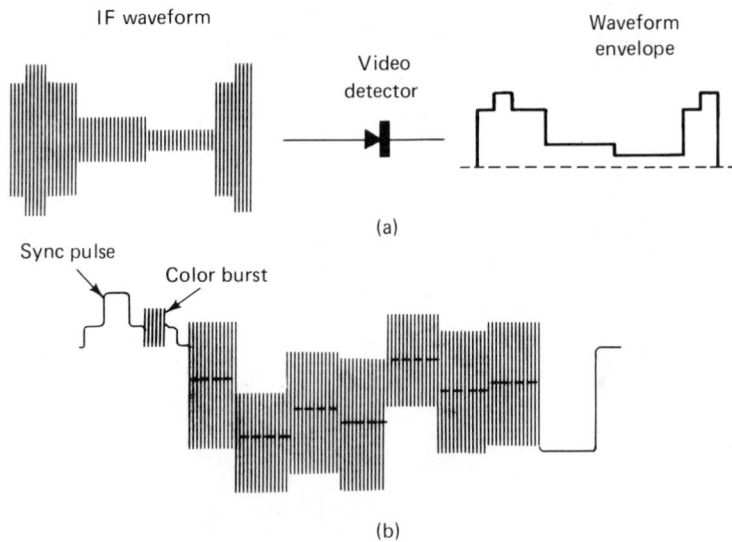

FIGURE 5-16   Video-detector action. (a) Development of the envelope from the IF waveform; (b) detector output waveform for a color-bar signal.

load circuit comprising L113 and R120. L113 is a shunt peaking coil, which compensates for the bypassing action of stray circuit capacitances. Thus, a uniform frequency response is maintained up to 4 MHz.

Figure 5-12 shows Q104 operating in an emitter-follower circuit. It serves to match the comparatively high output impedance of the video-detector stage to the low input impedance of the following video-amplifier stage. Note that the video detector is DC-coupled to the emitter follower, which in turn is DC-coupled to the video amplifier. In this manner, the DC component of the complete color signal is retained. As will be explained subsequently in greater detail, the DC component corresponds to the background illumination of a televised scene. For example, a daylight scene has a bright background and a corresponding high-level DC component. On the other hand, a moonlight or evening scene has a dim background and a corresponding low-level DC component. It is evident that if AC coupling were employed in the video-detector output circuitry, the DC component would be blocked. In turn, daytime and evening scenes would be reproduced with the same background illumination on the picture-tube screen.

Observe in Fig. 5-12 that three *test points* are called out. These are special terminals provided on the IF circuit board for convenient measurement of operating voltages and waveforms. When trouble symptoms that indicate a defect in the IF section appear, these key check points facilitate preliminary troubleshooting tests. For example, measurement of the AGC voltage will indicate whether a no-signal, weak-signal, or overloading-signal condition is being caused by a fault in the IF section or in the AGC section. In case the AGC voltage is normal, a demodulator probe can be applied at the collector of Q101, and the IF signal (if any) displayed on a scope screen. If the IF signal is normal at this point, the next signal-tracing test can be made at the input of the video detector. Thus, preliminary trouble localization can be made without extensive disassembly procedures.

Various types of IF response curves are utilized in different color-TV receivers, and the receiver service data should be consulted in case of doubt. The tuned coupling and trap circuits are adjusted to provide a specified frequency-response curve. As Fig. 5-17 shows, a trap cuts steeply into the skirt response of the coupling circuits providing extensive attenuation at a given frequency, thereby also providing the required IF selectivity. Figure 5-18 shows a widely-used contour for the IF response curve. This example employs a 6-dB bandwidth of 3.58 MHz. Note that the picture-carrier and chroma-subcarrier frequencies fall halfway down on either side of the response curve. Adjacent- and accompanying-sound trapping is practically complete.

It is instructive to consider some of the signal-processing principles that are involved in this IF response characteristic. As in conventional black-and-white receivers, the picture-carrier frequency is placed halfway down the side of the IF response curve to provide optimum vestigial-sideband response. This placement of the chroma subcarrier entails other considerations. The chroma signal in Fig. 5-18 occupies the frequency interval from 3.08 to 4.08 MHz, and is a double-sideband signal. Therefore, it would seem that the IF response curve should be flat from 42.67 MHz to 41.67 MHz. Note in passing that some receivers have been designed with essentially flat response through

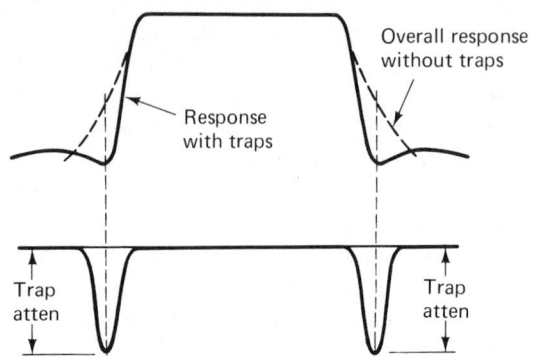

FIGURE 5-17   Trap action on coupling-circuit response.

the chroma-signal interval, as shown in Fig. 5-19. However, the design trend has been toward the curve contour depicted in Fig. 5-18.

There are two chief reasons for utilizing the curve contour shown in Fig. 5-18. The first is a manufacturing economy; that is, fewer IF components are required to provide a bandwidth of 3.58 MHz than a bandwidth of 4.1 MHz. The second reason is realization of a more linear phase characteristic toward the high end of the chroma interval. Observe in Fig. 5-18 that the IF curve slopes gradually from 41.67 MHz to 41.25 MHz, whereas in Fig. 5-19 the IF curve falls off abruptly from 41.67 to 41.25 MHz. An abrupt change in frequency response is inevitably associated with even greater change in the phase characteristic. Since a nonlinear phase characteristic distorts pulse waveforms and other complex waveforms, the IF contour shown in Fig. 5-18 entails less transient distortion than the comparatively flat frequency response of Fig. 5-19.

A trade-off is involved concerning signal-to-noise ratio and resulting impairment of weak incoming signals; that is, the color subcarrier is attenuated 50%, and the high chroma frequencies are attenuated 75% by the IF response shown in Fig. 5-18. Although this processing characteristic goes unnoticed in strong-signal reception, it becomes apparent in fringe-area reception. Thus, when the signal-to-noise ratio is poor, attenuation of the chroma signal effectively produces an even poorer signal-to-noise ratio. The practical effect is that more snow is displayed on the picture-tube screen. Under marginal reception conditions color sync is more likely to be lost. Thus, in summary, any design trade-off involves a choice between evils.

To anticipate subsequent discussion, it should be noted that the progressive attenuation of the chroma signal seen in Fig. 5-18 is compensated for in the following bandpass-amplifier circuit. That is, the bandpass amplifier is aligned with a rising response to compensate for the falling response through the IF chroma interval. The overall chroma-system response is uniform. Note that this does not improve the signal-to-noise ratio that exists at

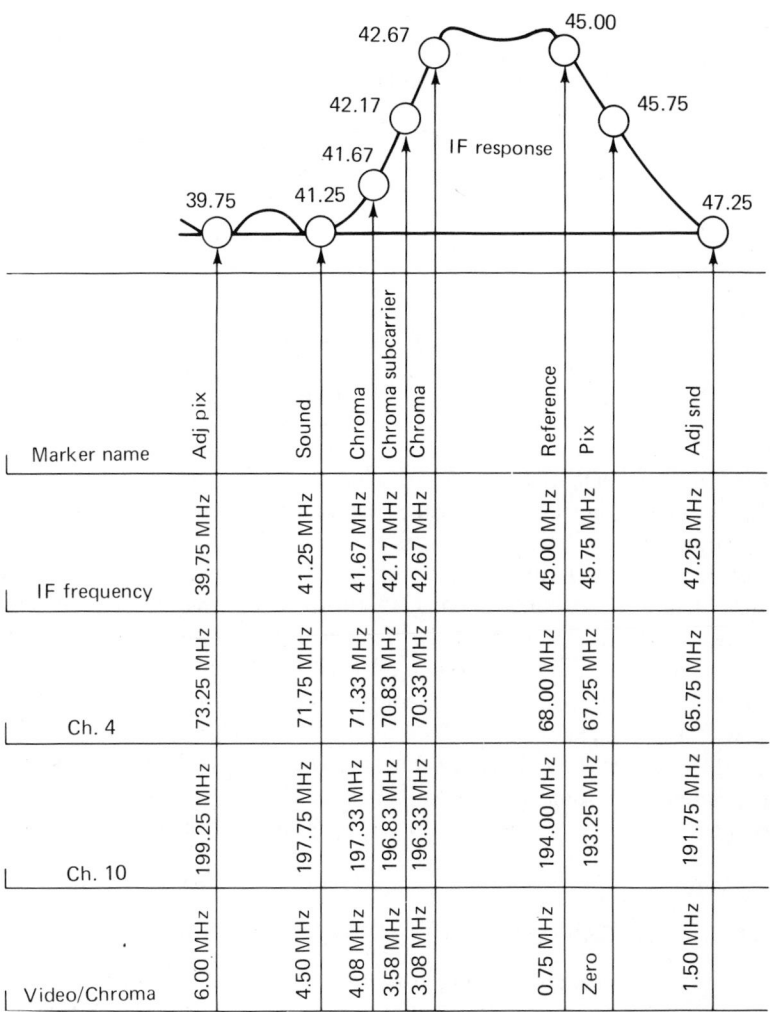

Marker name	Adj pix		Sound	Chroma	Chroma subcarrier	Chroma		Reference	Pix	Adj snd
IF frequency	39.75 MHz		41.25 MHz	41.67 MHz	42.17 MHz	42.67 MHz		45.00 MHz	45.75 MHz	47.25 MHz
Ch. 4	73.25 MHz		71.75 MHz	71.33 MHz	70.83 MHz	70.33 MHz		68.00 MHz	67.25 MHz	65.75 MHz
Ch. 10	199.25 MHz		197.75 MHz	197.33 MHz	196.83 MHz	196.33 MHz		194.00 MHz	193.25 MHz	191.75 MHz
Video/Chroma	6.00 MHz		4.50 MHz	4.08 MHz	3.58 MHz	3.08 MHz		0.75 MHz	Zero	1.50 MHz

FIGURE 5-18   A widely used contour for the IF response curve.
(courtesy of B & K Manufacturing Company)

the output of the IF amplifier. Once a given noise level is established or introduced into a system, the only method of noise reduction available is reduction of the channel bandwidth. However, since this approach entails intolerable signal distortion, the signal-to-noise ratio at the output of the IF amplifier cannot be improved in practice.

Note in Fig. 5-16 that the picture-carrier frequency appears on the left-hand side of the response curve, and the sound-carrier frequency appears on the right-hand side. In Fig. 5-19, the picture-carrier frequency appears on

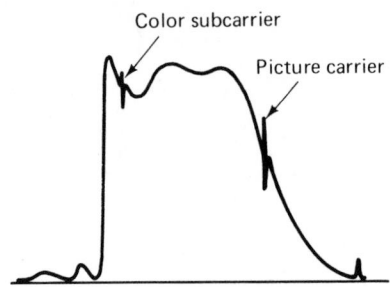

FIGURE 5-19   An IF response curve with essentially uniform
response through the chroma-signal interval.

the right-hand side of the curve, and the sound-carrier frequency appears on
the left-hand side. In other words, the frequency progression of the signal
spectrum becomes reversed through the mixer stage. This reversal is a consequence of the fact that the local oscillator operates at a frequency higher
than that of the incoming signal. In turn, the *heterodyne*, beat output, involves an effective reversal of the frequency progression. Of course, there is
no theoretical reason why the local oscillator could not operate at a frequency lower than that of the incoming signal. On the other hand, there *are*
practical considerations that favor high-side oscillator operation. The chief
consideration is the reduction of certain types of interference that could be
encountered in some locations. Also, high-side operation ensures that the
oscillator frequency cannot fall within any VHF TV channel from 2 to 13.

5.5   INTEGRATED-CIRCUIT LC-FILTER IF STRIP

Selectivity is generally considered to be the most important characteristic of
the IF amplifier in a color-TV receiver. Figure 5-20(a) shows the IF frequency-response curve for a typical modern high-quality receiver. This
frequency response is obtained by means of transistors and conventional
tuned circuits. On the other hand, there is a trend to the use of integrated-
circuit LC-filter IF amplifiers. This design is unconventional, inasmuch as it
utilizes a single bandpass filter as its tuning system. Figure 5-20(b) shows the
frequency-response curve for the LC-filter type of IF amplifier, and the schematic diagram is seen in Fig. 5-21. Note that the tuner output is coupled to
the LC filter through transistor Q325, which operates in the common-base
mode. This transistor provides some gain, and presents a constant impedance
to the tuner output circuit and to the filter input circuit; it also serves as a
buffer stage.

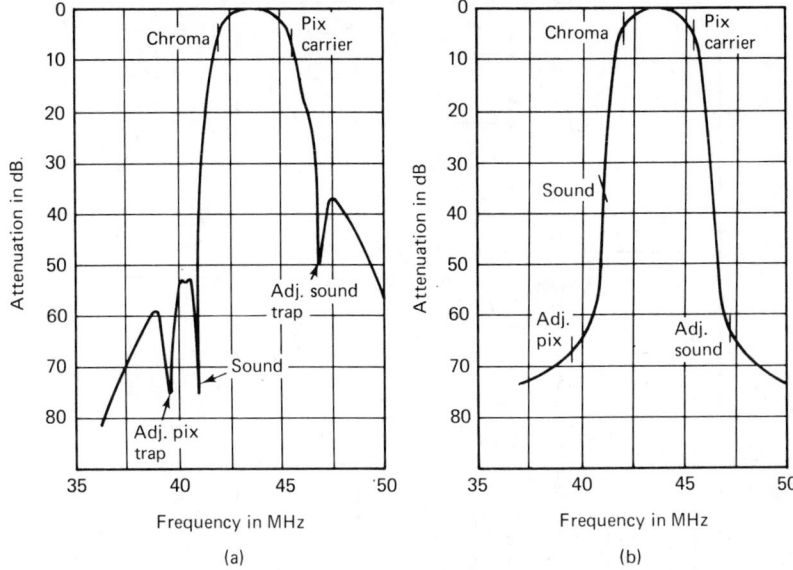

FIGURE 5-20   Comparative IF response curves. (a) Conventional color-IF response; (b) IC LC-filter color-IF response. *(courtesy of Heath Company)*

Observe in Fig. 5-21 that the LC filter is positioned ahead of the two-stage IC amplifier. Ample gain is provided by the IC's, including detection, an AFC output, and a choice of high- or low-impedance video outputs. These two IC stages are coupled by a broadly tuned transformer. Accordingly, the transformer is easily adjusted, and no special alignment equipment is required. Note that the 4.5-MHz sound-output signal is taken from the low-impedance output of IC326. Branching from the same output is a 4.5-MHz trap that is coupled to Q326, an emitter-follower stage that provides low-impedance output from driving the AGC and chroma sections. Two advantages of an LC-filter IF strip are an improved selectivity characteristic, and simplified alignment procedure. Long-term frequency stability of the LC-filter system is also much better than that of old-style IF-amplifier configurations.

EXERCISES

Questions

1. What is the VHF frequency range? The UHF frequency range?
2. How does a UHF tuner function in combination with a VHF tuner?

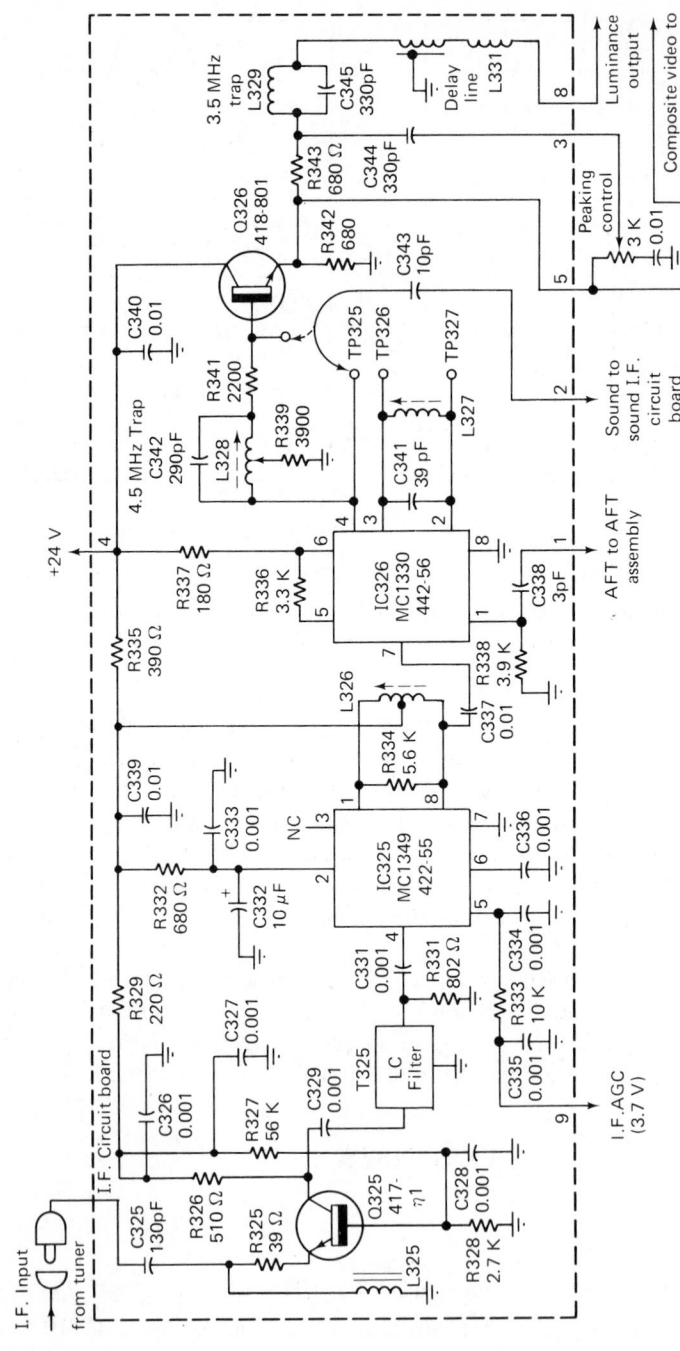

FIGURE 5-21 Configuration of an IC LC-filter color-IF amplifier. *(courtesy of Heath Company)*

3. Does a UHF tuner provide substantial amplification?

4. Briefly describe automatic fine-tuning action in a tuner.

5. How many stages of RF amplification does a VHF tuner usually provide?

6. Explain the basic function of a tuner.

7. Where are the sound and picture carriers normally located along the RF-tuner frequency-response curve?

8. Discuss the function of an adjacent-channel sound trap.

9. Describe the chief differences between a tube-type tuner and a solid-state tuner.

10. What are the basic functions of an IF amplifier?

11. How can the picture detector produce interference in the picture?

12. Are all stages in an IF amplifier AGC controlled?

13. Define the term *6-dB bandwidth*.

14. Does the sound signal proceed through the IF amplifier at the same level as the picture signal?

15. How does the bandwidth of an RF tuner compare with the bandwidth of an IF amplifier?

16. Where is the color subcarrier generally located along the IF frequency-response curve?

17. Why is the local oscillator operated on the high side of the incoming signal in most tuners?

18. Explain the reversal of picture/sound frequency progression that occurs in most VHF tuners.

19. Can noise be trapped out of an IF signal?

20. What is the typical gain in dB of an RF amplifier and mixer section? Of an IF amplifier section?

True-False

1. Front ends for color-TV receivers are fundamentally different from front ends for black-and-white receivers.

2. Separate tuners are employed for the VHF range and for the UHF range.

3. A UHF tuner provides only heterodyne action, without amplification.

4. During UHF reception, the VHF tuner is operated as a two-stage IF amplifier.

5. Audio-frequency bipolar transistors are generally used as UHF oscillators.

6. Resonant lines are employed instead of lumped inductance and capacitance in UHF tuning arrangements.

7. Diodes are generally utilized as mixers in UHF tuners.

8. Transistors are generally employed as mixers in VHF tuners.

9. When a malfunction occurs in a tuner, transistors are checked first.

10. An automatic fine-tuning system keeps the horizontal oscillator on-frequency.

11. MOSFET's require a typical AGC voltage range from –5 to +6.7 volts.

12. Transistors operate in series in a cascode configuration.

13. The first transistor in a cascode circuit operates in the CE mode, and the second transistor operates in the CB mode.

14. A local oscillator generally operates at 4.25 MHz above the sound-carrier frequency in a VHF tuner.

15. Although a local oscillator can be operated on either the high side or the low side of the incoming signal, the output from the mixer always has a lower frequency than the input.

16. Most of the selectivity and gain in a TV receiver is provided by the IF amplifier.

17. A video detector typically provides a 10-dB gain, whether a transistor or a diode is utilized.

18. A VHF tuner functions primarily as a preselector.

19. Mixer output frequencies are always harmonics of the input frequencies.

20. Local oscillators function to reinsert the color subcarrier into the chroma signal.

21. Tube-type VHF tuners operate on a basically different principle, compared with solid-state tuners.

22. When malfunction occurs in a tube-type tuner, tubes are always checked first.

23. DC operating voltages are much higher in solid-state tuners than in tube-type tuners.

24. A typical IF frequency response curve in a color receiver has a 6-dB bandwidth of 3.58 MHz.

25. High-side operation of the local oscillator ensures that the oscillator frequency does not fall in any VHF TV channel from 2 to 13.

Multiple Choice

1. A dual-gate MOSFET stage can be compared with a _____ .
   (a) cascode configuration
   (b) differential stage
   (c) complementary symmetry arrangement
   (d) bootstrap circuit

2. Heterodyne action involves _____ .
   (a) trap action
   (b) generation of sum and difference frequencies
   (c) generation of subharmonics
   (d) reinsertion of the color subcarrier

3. A MOSFET is the same as a/an _____ .
   (a) bipolar transistor
   (b) varactor diode
   (c) IGFET
   (d) thermistor

4. Bipolar transistors have the highest alpha cutoff frequency when operated in the _____ configuration.
   (a) common-collector
   (b) common-base
   (č) common-emitter
   (d) emitter-follower

5. A diode mixer operates with a/an _____ .
   (a) high gain
   (b) low gain
   (c) no gain
   (d) appreciable loss

6. Resonant lines are used in UHF tuner as_____.
   (a) tuned circuits
   (b) untuned circuits
   (c) high-frequency rectifiers
   (d) high-frequency demodulators

7. A varactor diode operates as a/an _____ .
   (a) detector
   (b) variable inductor
   (c) variable capacitor
   (d) variable demodulator

8. A VHF tuner has an RF response curve with a −6 dB bandwidth of approximately _____.
   (a) 4.5 MHz
   (b) 6 MHz

   (c) 9 MHz

   (d) 455 kHz

9. Preselection is the _____characteristic of an RF tuner.

   (a) least important

   (b) most important

   (c) least desirable

   (d) least avoidable

10. Picture interference could be caused by harmonics from the _____.

   (a) antenna lead-in

   (b) channel-selector switch

   (c) varactor diode

   (d) picture-detector diode

11. A typical IF section provides a gain of _____ dB.

   (a) 58

   (b) 126

   (c) 132

   (d) −6

12. Decibel gain values _____ .

   (a) add

   (b) multiply

   (c) divide

   (d) cancel

13. AGC voltage is applied to _____ IF stages in a color receiver.

   (a) all of the

   (b) none of the

   (c) all but the first of the

   (d) all but the last of the

14. IF stages may require _____ .

   (a) neutralization

   (b) bootstrapping

   (c) reflexing

   (d) cancellation

15. Tube-type IF amplifiers usually have _____ stages than/as solid-state IF amplifiers.

   (a) more

   (b) less

   (c) the same number of

   (d) higher-gain

16. A video detector operates as a _____.
    (a) rectifier followed by incomplete filtering
    (b) rectifier followed by complete filtering
    (c) rectifier followed by no filtering
    (d) linear amplifier

17. Background illumination in a televised scene corresponds to the _____.
    (a) AC component in the video signal
    (b) DC component in the video signal
    (c) sync pulses in the composite video signal
    (d) subcarrier amplitude in the complete color signal

18. When malfunction occurs in a tube-type IF amplifier, _____ .
    (a) DC voltages are measured first
    (b) capacitors are checked first
    (c) resistors are checked first
    (d) tubes are checked first

19. Stage gains in a tube-type IF amplifier are_____ solid-state IF amplifiers.
    (a) much higher than in
    (b) much less than in
    (c) about the same as in
    (d) negative, compared to

20. High-side local-oscillator operation results in _____ .
    (a) reversal of the IF frequency progression
    (b) zero IF frequency progression
    (c) loss of horizontal sync
    (d) color distortion

Problems

1. An IF section comprises three stages; the first stage has a gain of 15 times, the second stage has a gain of 20 times, and the third stage has a gain of 25 times. What is the total gain of the IF section?

2. What is the dB voltage gain and the dB power gain of the amplifier in problem 1?

3. An amplifier has a dynamic range of 40 dB; what is its corresponding signal-voltage range?

4. A 15-pF capacitor has approximately 2000 ohms of reactance at 4.5 MHz; how much reactance will the capacitor have at 45 MHz?

5. If a video signal has a sine waveform and a frequency of 300 Hz, how many bars will be displayed on the picture-tube screen?

6. How many bars would be displayed in the preceding problem if the video frequency is increased to 78,750 Hz? Explain your answer.

7. What is the repetition rate of a sync-buzz signal? Explain your answer.

8. A transistor-IF stage provides a voltage gain of 10 times and a power gain of 20 dB. What is the current gain of the stage?

9. A transistor operating in the CE mode has an input resistance of 1300 ohms. If the base-emitter voltage changes from 0.25 volt to 0.26 volt, how much does the input current change?

10. If the base-emitter voltage in problem 9 should change from 0.25 volt to 0.1 volt, how much would the input current change?

Chapter 6

Video Amplifier, Synchronizing And AGC Sections

6.1   VIDEO AMPLIFIER REQUIREMENTS

As noted in Chap. 5, the output circuit of the video detector is the input circuit of the video amplifier. Although the video amplifier has full response up to 4.1 MHz, the following Y amplifier has attenuated frequency response from 3 MHz to 4.5 MHz, as depicted in Fig. 6-1. Note that a color-subcarrier trap is placed between the video amplifier and the Y amplifier, and that the chroma section is energized from the output of the video amplifier. Observe that the video amplifier processes the complete chroma signal, whereas the Y amplifier processes only the Y-signal component. Although a small portion of the chroma signal proceeds through the Y amplifier, that portion is reduced extensively in amplitude by the color-subcarrier trap. The residual chroma signal that passes through the Y amplifier does not produce visible interference because of the frequency-interleaving process.

The frequency-response characteristic of the Y amplifier involves a trade-off. That is, fine detail in the black-and-white image is degraded to the extent that the high video frequencies are attenuated. It is impractical in the present state of the art to employ full frequency response up to 4.1 MHz in

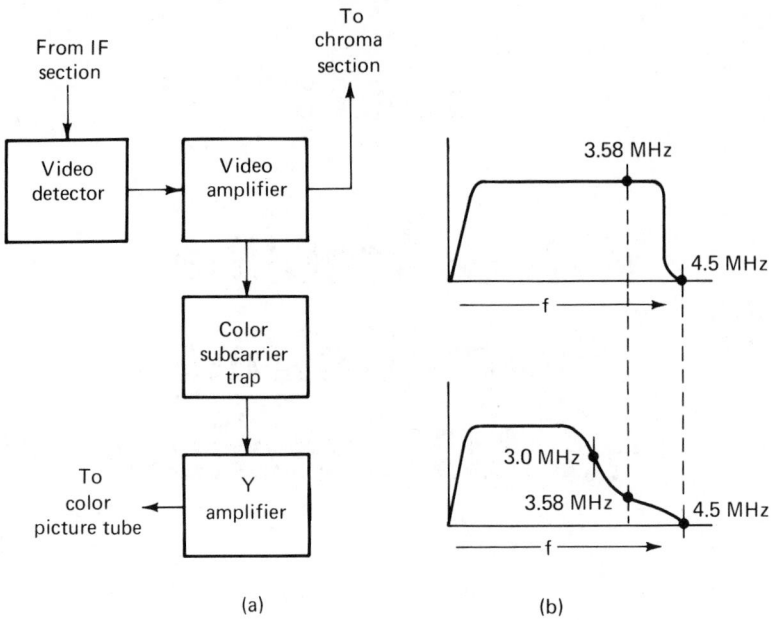

(a)   (b)

FIGURE 6-1   Video-frequency section of a color-television receiver. (a) Block diagram; (b) frequency response of the video-amplifier and Y-amplifier sections.

the Y amplifier. Inherent nonlinearities in the output system, particularly in the color picture tube, result in objectionable interference of the chroma signal with the Y signal if the high video frequencies are not attenuated. Therefore, the Y-amplifier response curve is contoured to optimize picture reproduction under the prevailing limitations.

Figure 6-2 shows the signal distribution in the video and Y amplifiers. An NTSC color-bar waveform is used in this example. The chroma-signal component is rejected by the color-subcarrier trap of the Y amplifier. Then the chroma bandpass amplifier processes the chroma-signal component only. This amplifier has a frequency-response characteristic that largely rejects the Y-signal component. Although the chroma-bandpass amplifier is a part of the video-frequency system from a technical point of view, it is more instructive to consider the chroma-signal processing system separately. Similarly, although the intercarrier-sound IF section is a part of the video-frequency system, it is preferable to consider this subsystem separately.

6.2   VIDEO AND Y-AMPLIFIER CIRCUITRY

Referring to Fig. 6-3, note that the complete color signal from the video detector is coupled through resistor R816 to the sync, AGC, chroma, and Y-amplifier sections. L807 and C806 function as a color-subcarrier trap. The Y

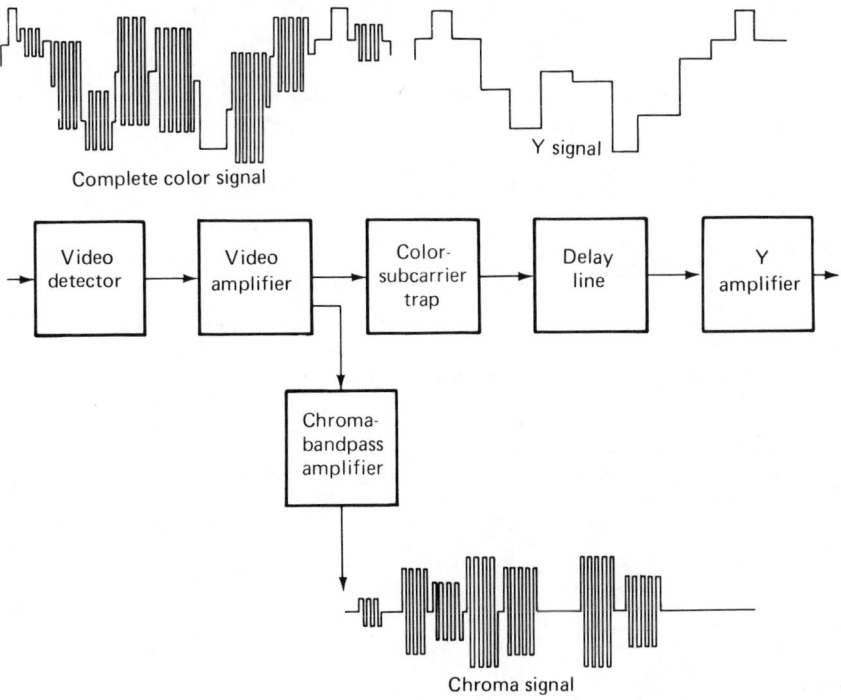

FIGURE 6-2   Signal distribution in the video and Y amplifier.

signal is then delayed by almost 1 microsecond in passage through the delay line L808. The first Y, or *luminance amplifier stage*, consists of transistor Q251 operating in the CB mode. Note that the Y signal is applied to the emitter of Q251 with output taken from the collector. This collector circuit has an adjustable impedance, determined by the setting of the contrast control (not shown). Thus, the output signal level increases when the collector load impedance is increased. Next, the Y signal is fed through the parallel combination of R257 and C253 to the Dots-Normal switch. This RC circuit provides a high-frequency boost to compensate for the bypassing action of stray capacitance in the contrast-control branch.

Note that the base of Q251 in Fig. 6-3 is biased through a maintenance control called the *brightness minimum control*. This is a range-setting adjustment for the brightness control (not shown). Manual control of the brightness level is supplemented by an automatic brightness-control subsection, whose function will be explained in greater detail subsequently. The Dots-Normal switch is a maintenance switch, which is operated only during servicing procedures. Its action is discussed in a following chapter, and simply note at this point that R260 maintains the base bias voltage on Q252 when the switch is in its Dots position. From the collector of Q251, the Y signal is

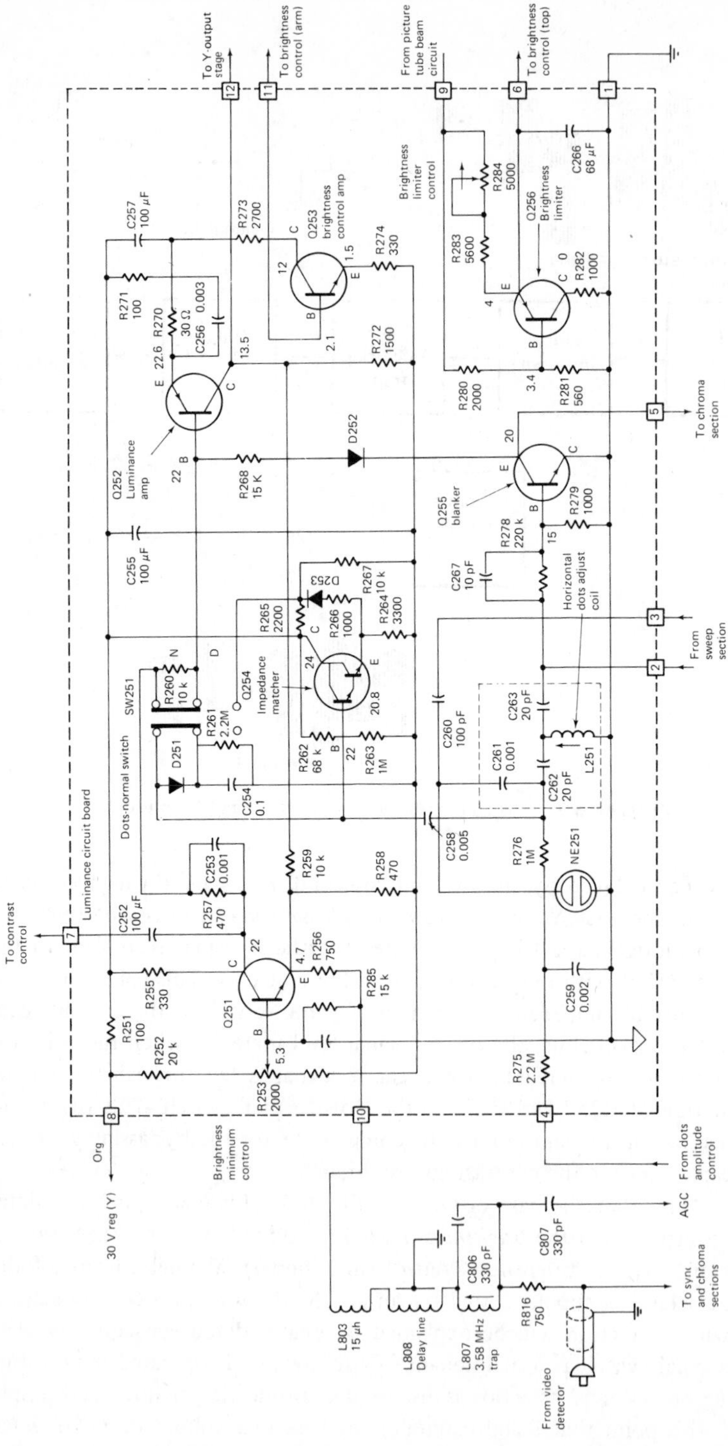

FIGURE 6-3 A Y-amplifier configuration.

normally fed to the base of Q252. DC coupling is employed throughout in order to retain the DC component of the Y signal.

The amplified signal from Q252 in Fig. 6-3 then drops across the collector load resistor R272. Observe that a feedback loop is provided from the collector of Q252 to the emitter of Q251. This negative feedback provides bias stability over a considerable range of temperature variation. Finally, the output signal from Q252 is fed to the Y output stage. Since this stage is included on another circuit board in this example, it will be discussed subsequently.

6.3   BRIGHTNESS CONTROL

Brightness control in the example under discussion is accomplished by varying the DC current through Q252 in Fig. 6-3. Note that the setting of the brightness control (not shown) determines the base bias on Q253, the brightness-control amplifier transistor. The collector current from Q253 flows in turn through R271, the emitter resistor for Q252. An increase in current flow through Q253 results in a decrease of current flow through Q252. A current decrease causes a collector-voltage increase, which is DC-coupled through the following circuitry to the picture tube, and, in turn, results in a decrease of screen brightness. On the other hand, a decrease in current flow through Q253 has the final result of increasing the screen brightness. The screen brightness cannot be increased manually beyond a preset limit; this automatic brightness control action prevents accidental damage to the color picture tube.

The brightness-limiter control R284 in Fig. 6-3 establishes a preset limit to the screen brightness that can be obtained by advancing the manual brightness control. When the manual brightness control is advanced past this preset limit, the forward base bias on Q253 starts to decrease, and automatic brightness-limiting action is initiated. Note that the beam current of the picture tube flows through R283 and R284. In turn, the current flow through Q256 is varied. However, this current variation starts abruptly (much like Zener action) because the bias on Q256 is a combination of fixed and varying bias voltages. Note that the current through R284 flows in branched paths through Q256 and the beam circuit of the picture tube. Thus, if the brightness control is turned to minimum, all of the current through R284 flows through Q256. Then when the brightness control is advanced, part of the beam current flows through the limiter control, and part through Q256. Next, observe how the limiting action takes place.

Observe in Fig. 6-3 that the "Zener" voltage on the emitter of Q256 is approximately 4 volts; this voltage is applied to the top of the brightness control. When this control is advanced to a point where the picture tube draws the preset limit of current, all of the current flowing through the limiter control will be beam current. At this point, there is no current remaining to sustain conduction in Q256, and its "Zener" action stops. That is, the emitter voltage no longer remains at 4 volts, but falls to a lower value. In turn, the voltage at the top of the brightness control decreases, and the voltage at the arm also decreases. Consequently, the screen brightness can no longer be increased by advancing the brightness control.

The blanker-amplifier transistor Q255 in Fig. 6-3 is driven by retrace pulses from the sweep section. Blanking pulses are coupled via diode D252 to the base of Q252. In this manner, the picture-tube beam current is cut off during retrace intervals so that retrace lines are eliminated. Diode D252 operates as a rectifier so that only negative-polarity pulses are applied to the base of Q252. Note that the output from Q255 is also fed to the chroma section of the receiver to disable the chroma channel during retrace. Retrace pulses from the sweep section are also utilized to energize the dot generator, which is employed only during convergence procedures. It is instructive to note the circuit action in the dot-generator section.

Observe in Fig. 6-3 that retrace pulses are applied to L251 and its associated capacitors. This is a ringing-circuit arrangement, which is shock-excited into oscillation by the horizontal-flyback pulses. In turn, a sine wave is generated, the peaks of which produce a series of horizontal dots on the picture-tube screen. At the same time, neon bulb NE-251 is operating as a relaxation oscillator at a frequency that produces a series of vertical dots on the picture-tube screen. It is evident that both series of dot displays will be synchronized with the sweep rates because the ringing coil is energized by the horizontal flyback pulses, and the neon-bulb oscillator is synchronized by the vertical-retrace pulses. Transistor Q254 comprises a Darlington pair, which provides high current gain and impedance matching between the dot generator and the Y amplifier. Diode D253 operates as an amplitude limiter so that all dots are clipped to a uniform height.

It is helpful to consider the typical tube-type Y-amplifier configuration shown in Fig. 6-4. A pentode tube is employed. In case of malfunction, tube replacement is observed. However, if this does not restore normal operation, circuit tests must be made. Signal-tracing with an oscilloscope is usually informative; the waveform displays will indicate whether the trouble is in the input or the output section of the Y amplifier. Open coupling capacitors are also pinpointed by signal-tracing tests. As an illustration, if C74 were "open" in Fig. 6-4, the incoming signal would be found at the input terminal of the capacitor, but no signal would be displayed at the output terminal of the capacitor. Some technicians prefer to make signal-injection tests in Y amplifiers. A TV analyzer can be utilized, or a conventional signal generator may be operated on its AM function. The generator frequency is set to 3.58 MHz for the test. In turn, a horizontal bar pattern is normally displayed on the picture-tube screen.

After the trouble area has been localized, DC-voltage and resistance measurements are usually made to pinpoint the defective component. Capacitors are the most probable troublemakers; note that electrolytic capacitors

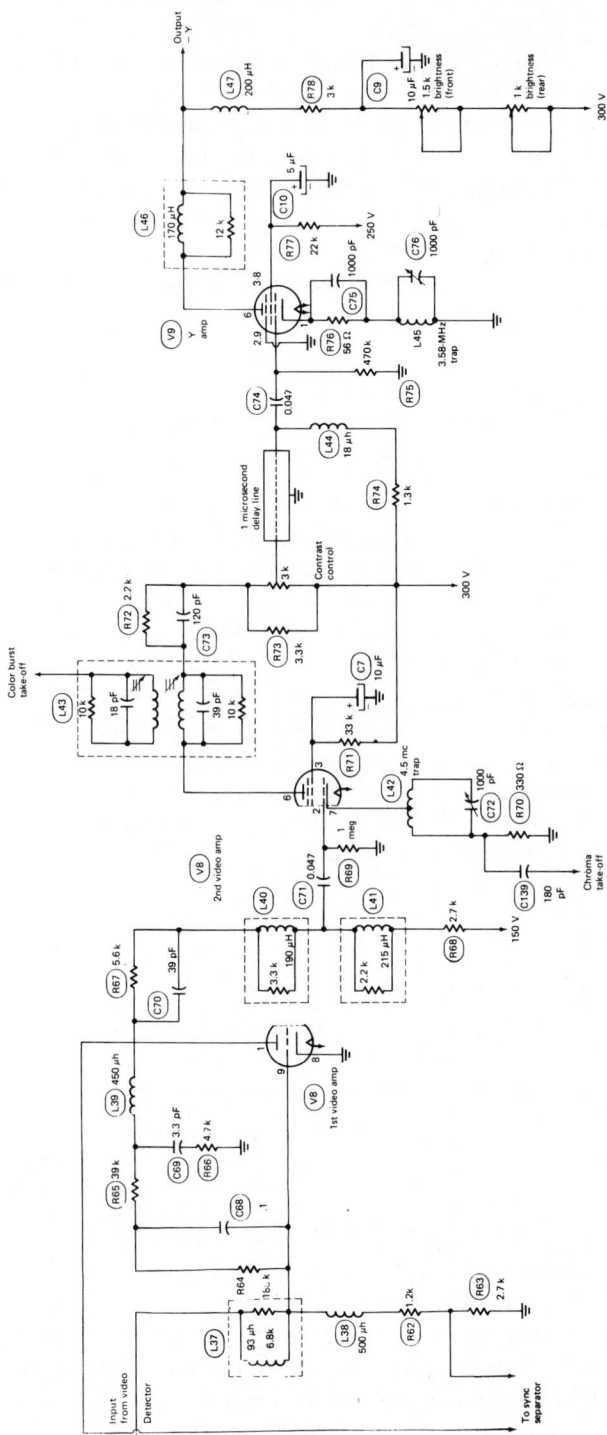

FIGURE 6-4   Typical tube-type Y amplifier configuration.

such as C9 are more likely to fail than paper capacitors such as C74 (Fig. 6-4). Again, paper capacitors are more likely to fail than mica or ceramic capacitors. It is helpful to keep these statistical facts in mind because a troubleshooting job sometimes "boils down" to making systematic substitution tests. As a practical note, certain kinds of failures occur in particular receivers more frequently than in other receivers. Such failures are published in servicing magazines and provide useful reference. For example, "Troubleshooting Tips" appears in *Electronic Servicing Magazine*. This feature covers reported failures in particular receivers.

6.4 HORIZONTAL AND VERTICAL SYNCHRONIZING SECTIONS

Synchronizing pulses are included in the complete color signal to keep the horizontal and vertical deflection oscillators in step with the field and frame rates of the transmitted signal. (See Fig. 6-5.) The first function of the synchronizing section is to clip the sync tips from the blanking pedestals, as depicted in Fig. 6-6. Thus, the pedestals and the camera signal are rejected, and the clipped sync pulses are passed into the sync section. Basic *sync-clipper* arrangements are shown in Fig. 6-7. These are clipper circuits that operate with signal-developed bias. The clipping level tends to "follow" the incoming signal amplitude and provides a substantial dynamic range to accommodate both strong-signal and weak-signal reception. Note that capacitor C becomes charged on positive peaks of the applied signal, and the clipping device operates in Class C. An advantage of the transistor circuit is the gain that is provided from base to collector.

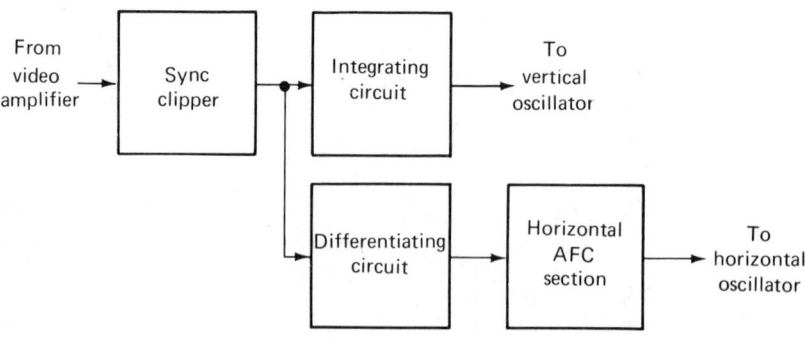

FIGURE 6-5   Plan of the synchronizing system.

The horizontal-sync pulses are then separated from the vertical-sync pulses by means of a differentiating circuit shown in Fig. 6-8(a). Horizontal-sync pulses are transformed into positive and negative "spikes," and the ser-

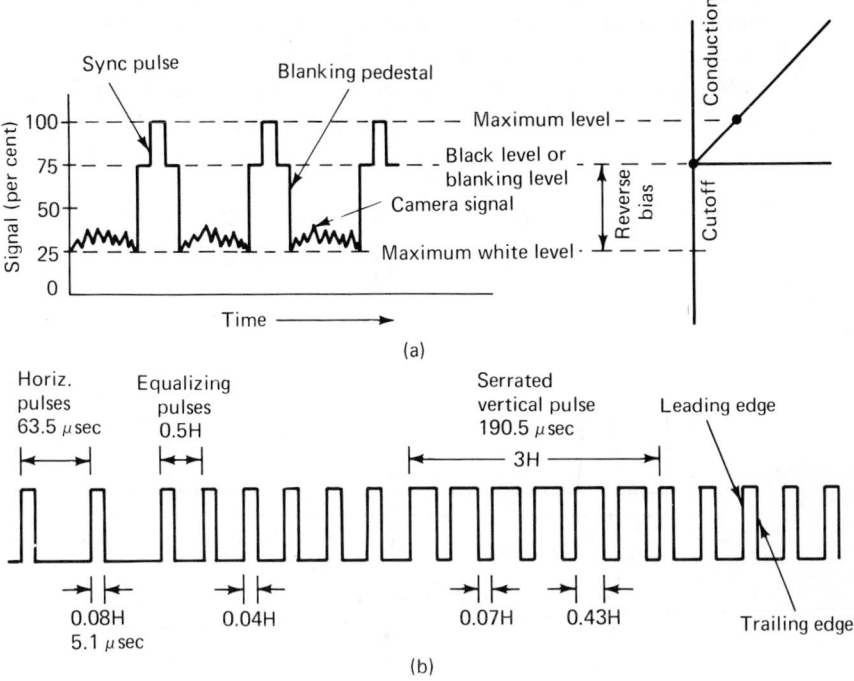

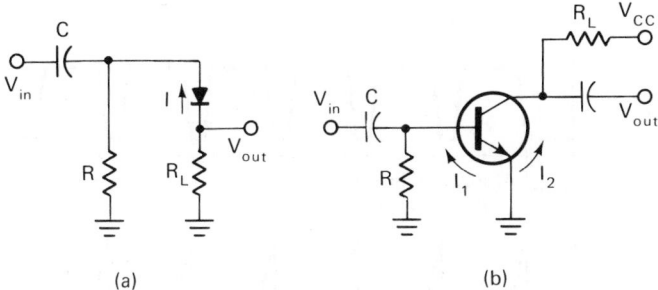

FIGURE 6-6   Sync-separator action. (a) Sync-clipper section; (b) clipped sync pulses.

FIGURE 6-7   Basic sync-clipper arrangements. (a) Diode-clipper circuit; (b) transistor clipper circuit.

rations in the vertical-sync pulses are also changed into positive and negative "spikes." Note that a vertical-sync pulse produces differentiated spikes at twice the repetition rate of those produced by horizontal-sync pulses. This is an aspect of interlaced scanning, whereby the horizontal oscillator is prevented from drifting off-frequency during the passage of vertical-sync pulses;

although there is a half-line difference in the occurrence of successive vertical-sync pulses, horizontal-oscillator operation is unaffected because the oscillator is locked by every other spike during the passage of a vertical-sync pulse.

Vertical-sync pulses are separated from horizontal-sync pulses by means of an integrating circuit shown in Fig. 6-9(a). A horizontal-sync pulse

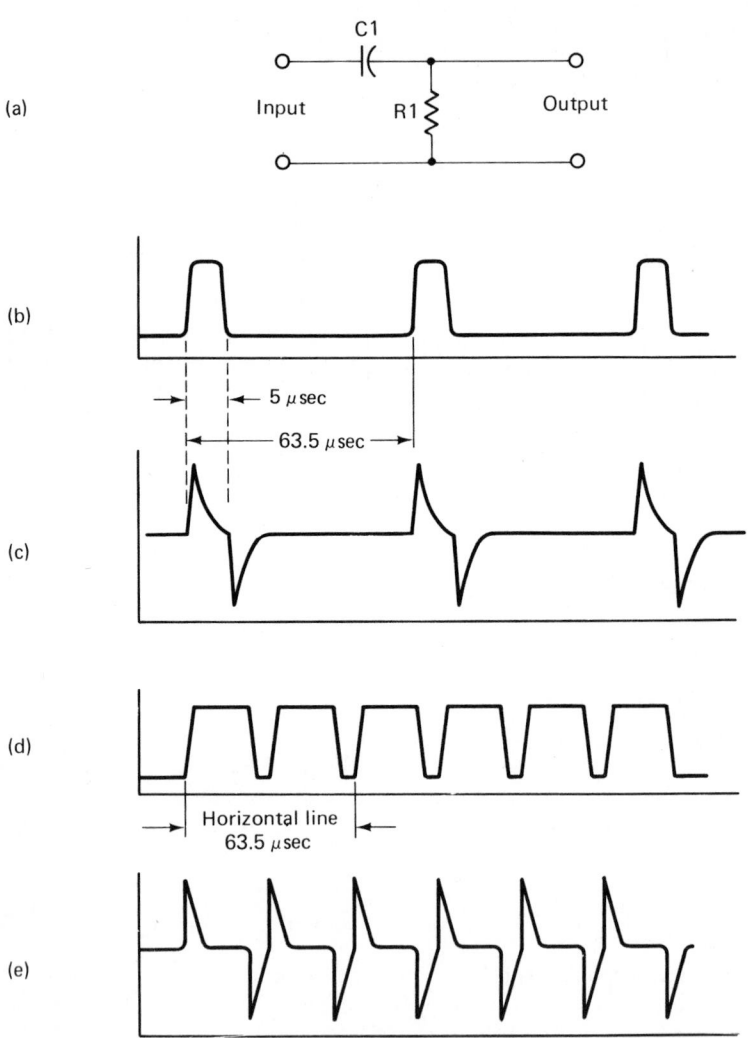

FIGURE 6-8   Horizontal sync-separator action. (a) Differentiating circuit; (b) incoming horizontal-sync pules; (c) differentiated horizontal-sync pulses; (d) incoming vertical-sync pulse; (e) differentiated vertical-sync pulse.

produces a small sawtooth-voltage output; an equalizing pulse produces a very small sawtooth output voltage. The comparatively wide vertical-sync pulse builds up a substantial charge on C1. The output waveform from the integrator circuit occurs at a 60-Hz rate and is utilized to lock the 60-Hz vertical-deflection oscillator. Equalizing pulses are included in the vertical-sync waveform to ensure that the integrator starts from practically zero volts at the beginning of the vertical sync-pulse interval. Due to interlaced scanning, there will be a half-line difference between successive vertical pulses and there would be a slightly different starting level in the integrator on successive pulses, unless the narrow equalizing pulses were provided. Thus, the equalizing pulses ensure good interlace (elimination of line pairing).

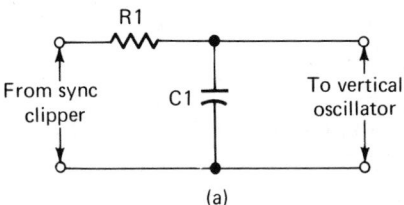

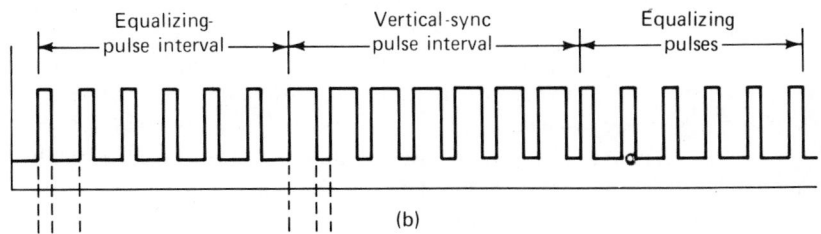

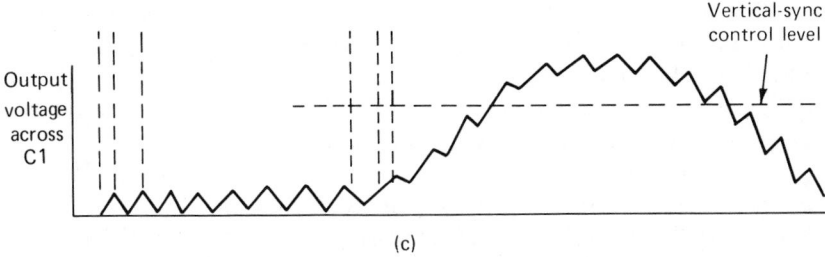

FIGURE 6-9   Vertical sync-separator action. (a) Integrating circuit; (b) incoming vertical-sync pulse; (c) integrated vertical-sync pulse.

A sync-circuit board configuration is depicted in Fig. 6-10. From the video detector, the complete color signal with negative-going sync pulses is coupled from terminal 3 via R474 and C461, and through the parallel combination of C462 and R475 to the base of Q458. This is an RC filtering arrangement that reduces the amplitude of random noise pulses without appreciable attenuation of sync pulses. Q458 operates as a clipper with signal-developed bias. That means that the transistor conducts on the sync tips and is cut off the remainder of the time. Output from the collector of Q458 is fed to horizontal- and vertical-oscillator circuit boards that contain the terminal subsections of the sync system in this example. (See Fig. 6-11.) Therefore, these terminal subsections are reserved for subsequent discussion.

Observe in Fig. 6-10 that sync clipper Q458 operates in combination with sync-clipper gate Q457. It has the function of gating out high-level noise pulses from the sync system. Note that the complete color signal is also fed from terminal 3 via diode D453 to the base of the noise-inverter transistor Q456. Diode D453 is reverse-biased by the noise-limiter control R472 through R470. This is a maintenance control, and it is set to a point where the diode conducts at a level that slightly exceeds the amplitude of the sync pulses. High-level noise pulses pass through D453 to the base of Q456, which inverts the pulse polarity from base to collector. Thus, positive-going noise pulses are coupled through C458 to the base of Q457. Bias is supplied to Q457 through R473, and the transistor is normally saturated. Therefore, there is an effective short-circuit in the emitter branch of Q458. However, when a positive-going noise pulse is applied to the base of Q457, the transistor comes out of saturation and turns off Q458. Thus, a "hole" is punched in the clipper output pulse train for the duration of the high-level noise pulse. The foregoing process is called the *noise-limiting function* of the sync system.

In the above example, the AGC function is also included on the sync and noise-limiter circuit board. Observe that the complete color signal from circuit-board terminal 3 in Fig. 6-10 is coupled via R453 to the base of the AGC inverter transistor Q451. Both the operating point and the gain of Q451 are determined by R452 and the AGC control (not shown), which together form the emitter resistance of the transistor. As the signal level varies, the inverter output also varies; and this output signal is applied from R454 and R455 to the base of the gated amplifier transistor Q452. This is a gated amplifier stage, and its conduction is determined by the state of the gate transistor Q453. It is instructive to consider this gated-AGC action.

Note that a flyback pulse from the horizontal-output circuit is coupled from circuit-board terminal 4 in Fig. 6-10 to the base of Q453. The base of Q453 is returned to ground through R460. Therefore, Q453 does not conduct unless the horizontal pulse is being applied to its base. Since Q453 is connected in series with the emitter of Q452, it follows that Q452 cannot conduct unless Q453 is conducting. The AGC control is set so that the bias on Q452 permits conduction only on the peaks of sync pulses during the time that Q453 is turned on. Thus, negative-going pulses are developed across collector-load resistor R456, and the amplitude of these pulses is proportional

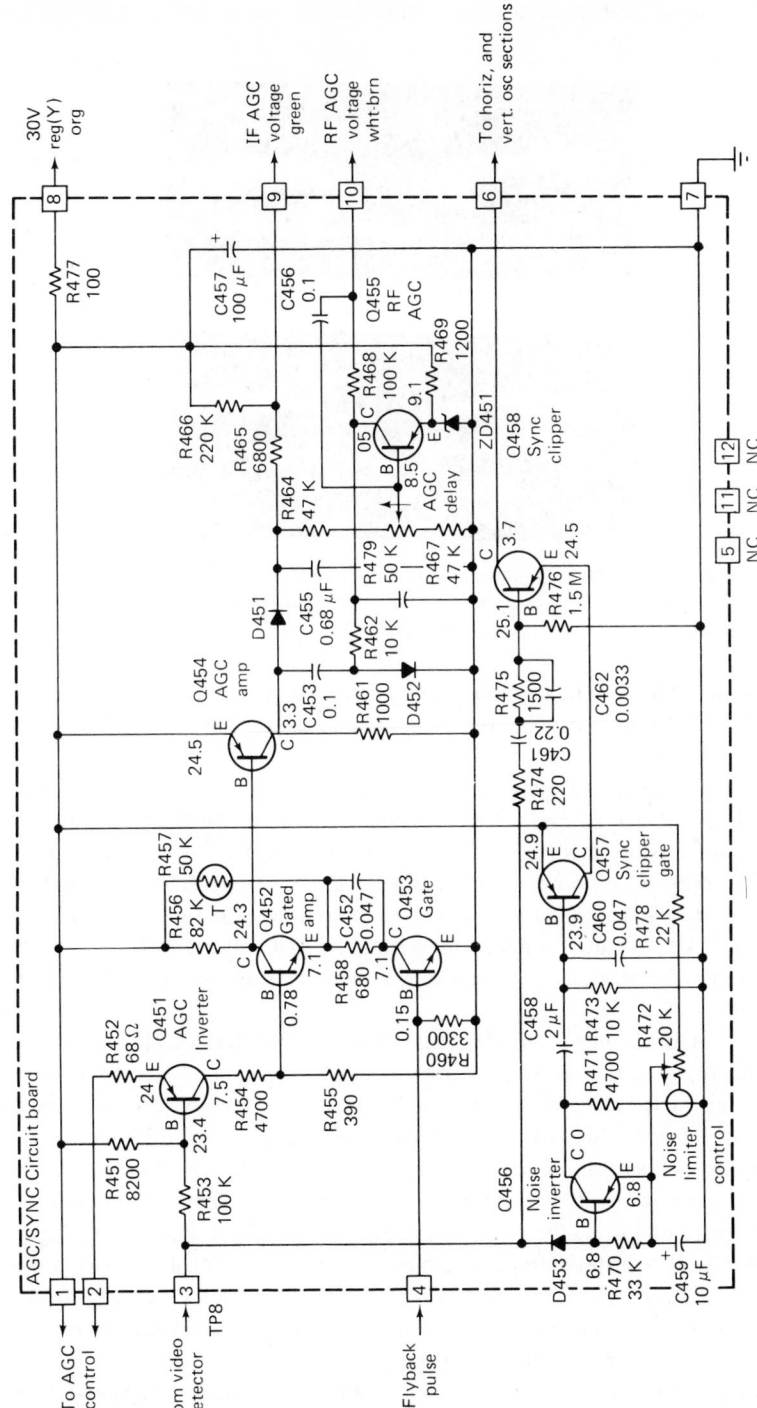

FIGURE 6-10 Configuration of a sync/AGC circuit board.

127

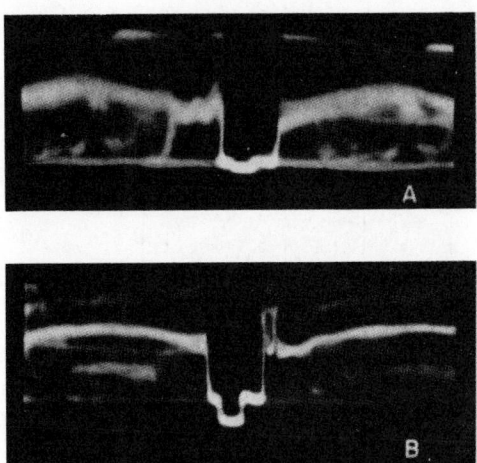

FIGURE 6-11   Clipping distortion of horizontal sync pulse. (a) Sync tip partially clipped; (b) sync tip completely clipped.

to the amplitude of the incoming sync pulses. These negative-going pulses are coupled to the base of the AGC amplifier transistor Q454, which steps up their amplitude.

The amplified pulses are then dropped across collector-load resistor R461 in Fig. 6-10. In case a weak station is tuned in, the IF AGC voltage is preset by resistors R466, R465, R464, R467, and the AGC delay control R470 in order to obtain maximum IF gain. On the other hand, in the case of a strong station being tuned in, the pulses at the collector of Q454 cause diode D451 to conduct, thereby increasing the AGC voltage and reducing the IF gain. C455 functions as a filter to smooth the pulse train into a steady DC voltage.

AGC action for the RF stage in the VHF tuner is employed to prevent overload of the mixer on strong incoming signals. Since an FET is employed in the RF stage in this example, positive bias is required for maximum gain, and negative bias for minimum gain. Observe in Fig. 6-10 that at the maximum gain setting, the emitter of RF AGC transistor Q455 is biased positive with respect to its base by the voltage drop across Zener diode ZD451. Under this condition, the transistor is saturated and its collector voltage is clamped to its emitter voltage, which is held constant by ZD451. Next, as the IF AGC voltage increases, the base voltage supplied by the voltage dividers R464, AGC delay control R470, and R467, also increases and brings the transistor out of saturation. At this point, the collector voltage becomes independent of the Zener voltage. The AGC pulses are coupled through C453 and cause current flow through D452, C454, and R462. This pulse current is rectified and produces a negative voltage at the junction of R462 and C454 with a value proportional to the amplitude of the AGC pulses. This voltage is filtered by R468 and C456 and then applied via terminal 10 to the RF tuner.

The sync-separator tube shown in Fig. 6-12 is biased by the signal peaks to a clipping level that passes the sync tips only. A scope check normally displays the composite video signal at test points A through D. Note that the signal amplitude is about 30% lower at D than at A. The 10k resistor prevents the separator input from loading the video amplifier excessively. Note also that the RC network between points B and D attenuates noise pulses to some extent. Accordingly, the sync pulse at D is displayed with a rounded top, compared with the display at A. In case the output signal at E is weak or zero, a scope signal-tracing test should be made from point A through to point D. The observed waveforms will usually serve to localize the faulty circuit. Normal signal amplitude at point F is about 15 volts p-p, for this configuration. Subnormal amplitude can be caused by a defective coupling capacitor.

Vertical-sync problems often are related to defects in the AGC system. As an illustration, if the AGC voltage is marginal, intermittent picture instability is a probable result. This instability may become pronounced only during commercials in the program. Note that if sync pulses are compressed in some overloaded circuit prior to the sync clipper(s), clipping action continues. Instead of slicing off sync pulses only, video-signal peaks are also passed. Consequently, both video signal and sync pulses are applied to the vertical-oscillator input, causing vertical-locking instability. It is not necessarily the value of AGC voltage that is directly related to vertical-locking instability. In other words, it is possible for the average AGC voltage to be correct and still develop vertical locking problems. This situation can occur if the AGC bypass capacitor becomes open-circuited. This is a comparatively common fault in circuitry that utilizes electrolytic capacitors.

A representative tube-type AGC circuit that includes a 1-μF electrolytic capacitor to bypass the AGC line is depicted in Fig. 6-13. Note that if this capacitor becomes open-circuited, the vertical-sync pulses increase the AGC voltage over the peak of each pulse, thereby varying the receiver gain at the vertical-sync rate. Another circuit action results in the attenuation of the pulses arriving at the vertical oscillator through the sync clipper. In turn, picture rolling or unstable and critical vertical locking occurs. By way of comparison, horizontal-sync response is not greatly affected because the paper bypass capacitors in the circuit provide relatively normal action at the horizontal frequency. A useful quick-check can be made by bridging the suspected electrolytic capacitor with a known good unit. If the vertical instability is corrected, the suspicion is confirmed.

Most solid-state receivers utilize electrolytic capacitors in the AGC circuitry. These capacitors are prime suspects when vertical instability is being

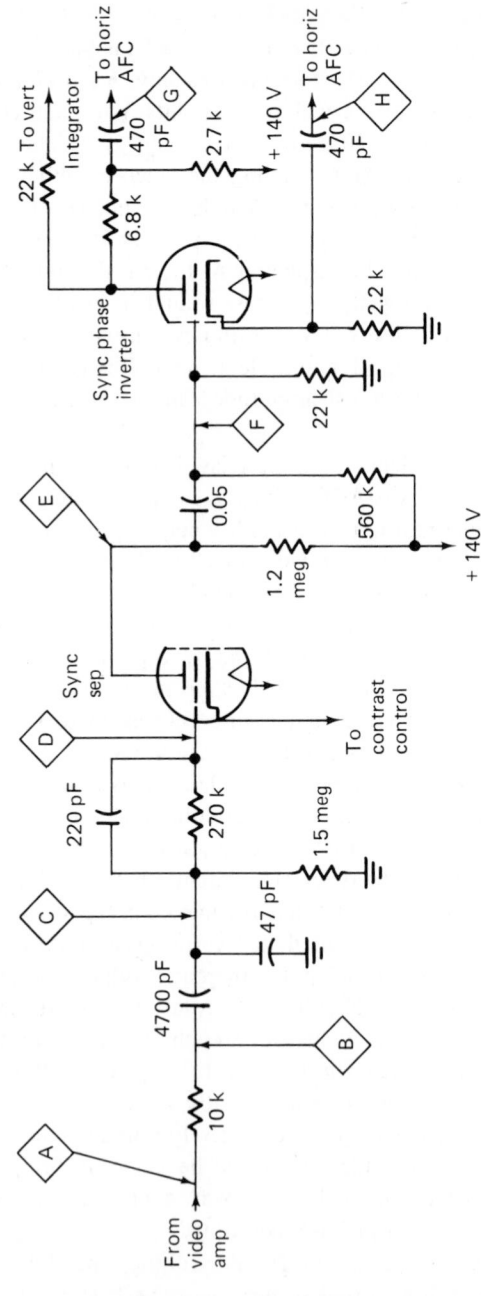

FIGURE 6-12 A tube-type sync separator and phase-inverter configuration, with test points indicated.

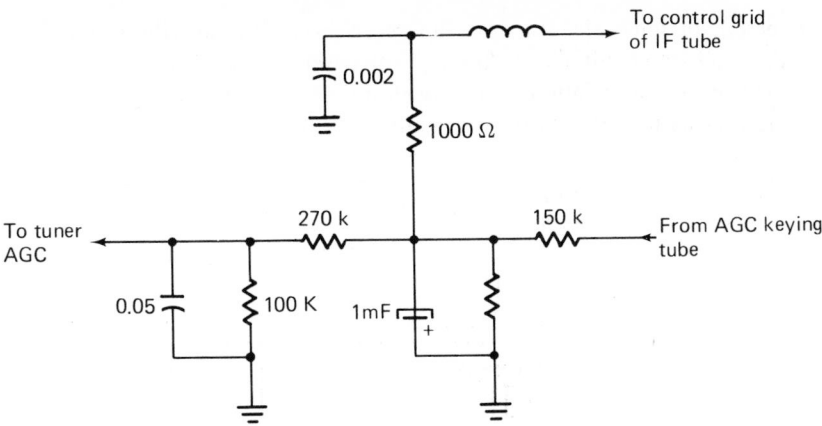

FIGURE 6-13   Typical tube-type AGC circuit that includes an electrolytic capacitor.

analyzed. A typical solid-state AGC circuit that includes an electrolytic capacitor is shown in Fig. 6-14.

As a practical troubleshooting note, open electrolytic capacitors in other circuits can also cause vertical locking instability. Thus, in some receivers an open electrolytic capacitor in a B+ line may permit a spurious "pip" to ride on the DC voltage and disturb circuit action in the sync section. Open electrolytic capacitors in video circuits, particularly screen bypass capacitors, can make "sync slippers" out of otherwise normal video stages by attenuating the sync pulses in the composite signal applied to the sync

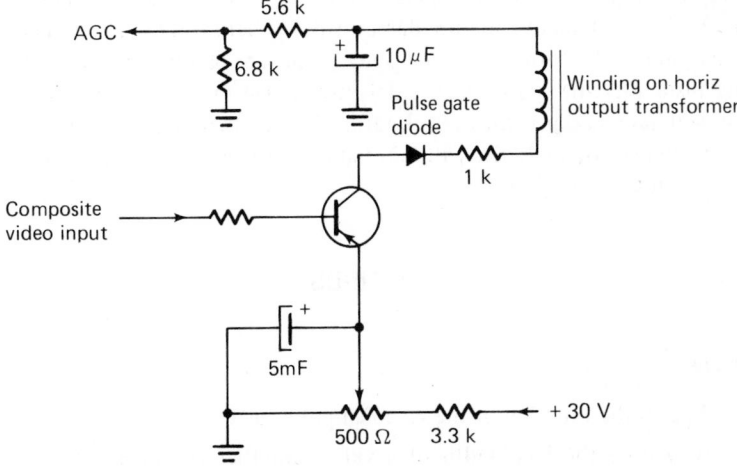

FIGURE 6-14   Representative solid-state AGC circuit that includes an electrolytic capacitor.

clipper(s). In some older-model color receivers, a 2-μF capacitor in the first video-amplifier circuit has a "history" of developing a poor power factor, causing serious attenuation of the vertical-sync pulse. Figure 6-15 shows a partial schematic of this particular circuit.

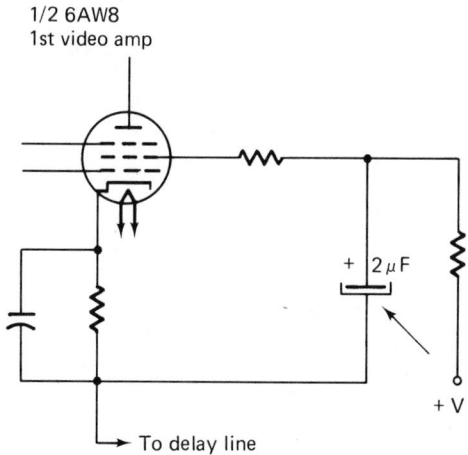

1/2 6AW8
1st video amp

+ 2 μF

+ V

To delay line

FIGURE 6-15   Arrow points to bypass capacitor that can cause vertical locking instability.

An unexpected source of vertical locking instability is a defective picture-detector diode. Occasionally, a diode develops an internal defect that causes clipping of high-amplitude signals. Inasmuch as the sync pulses have the highest amplitude of the signal components, they are attenuated or eliminated in the defective picture-detector diode before they are applied to the sync clipper. In turn, the video-signal peaks are passed, causing vertical locking instability. In many cases, a defective diode that causes this type of trouble will have a poor front-to-back ratio. On the other hand, the same malfunction can develop in diodes with normal front-to-back ratios. Accordingly, a substitution test is preferred.

EXERCISES

Questions

1. What is the function of the video amplifier in a color receiver?

2. How does the bandwidth of a video amplifier compare with the bandwidth of a Y amplifier?

3. Why is a color-subcarrier trap included in the Y-amplifier configuration?

4. Does the video amplifier process the same signal as the Y amplifier?

5. Explain the function of a *delay line.*

6. Why is DC coupling employed in video-amplifier and Y-amplifier circuitry?

7. Briefly discuss the function of a *brightness-control limiter.*

8. Describe the action of a retrace-blanking circuit.

9. How does a Darlington pair operate?

10. Explain the function of a *sync clipper.*

11. What is the advantage of signal-developed bias in sync-clipper operation?

12. Discuss the separation of horizontal sync pulses from vertical sync pulses.

13. Why are serrations provided in the vertical-sync pulse?

14. Explain the need for equalizing pulses.

15. How can the disturbing action of high-level noise pulses be minimized?

16. State a typical gain figure for a video-and-Y-amplifier arrangement.

17. Describe a simple procedure for measuring video-amplifier stage gain.

18. If malfunction occurs in a tube-type video amplifier, what component does the technician check first?

19. How does the DC-voltage distribution in a tube-type video amplifier compare with that of a solid-state amplifier?

20. Does a tube-type video-and-Y-amplifier arrangement provide more or less gain than its solid-state counterpart?

True-False

1. A typical bandwidth for a Y amplifier is 4.1 MHz.

2. Most video amplifiers are DC-coupled.

3. Video amplifiers are located between the IF amplifier and the video detector.

4. Y amplifiers are essentially video amplifiers with the addition of a delay line and a color-subcarrier trap.

5. A Darlington pair comprises two transistors series-connected in the CB mode.

6. Peaking coils are used more extensively in solid-state video amplifiers than in tube-type video amplifiers.

7. A Y amplifier includes a delay line, whereas a luminance amplifier contains a color-subcarrier trap.

8. The total gain of a video-and-Y-amplifier arrangement is typically 100 times.

9. Color picture tubes are extremely rugged and are not damaged by operation at excessive brightness.

10. Retrace-blanking circuits are usually provided in color receivers.

11. When malfunction occurs in a solid-state video amplifier, transistors are checked first.

12. Stage gain can be easily measured in a video amplifier or Y amplifier by comparing input/output waveform amplitudes with an oscilloscope.

13. Signal-developed bias has an advantage in sync-clipper operation in that the clipping level is made to "follow" the incoming signal amplitude.

14. A sync clipper that employs signal-developed bias has a greater dynamic range than a sync clipper that utilizes fixed bias.

15. High-level noise pulses can be minimized in a video amplifier by the use of series and shunt peaking coils.

16. An integrating circuit passes equalizing pulses, whereas a differentiating circuit passes serrations.

17. High-level noise pulses can be eliminated by cancellation, and the video-signal interruption generally passes unnoticed.

18. A Zener diode provides optimum picture-detector action, compared with a conventional germanium or silicon diode.

19. High-frequency video-signal boost action can be provided by suitable RC parallel circuitry.

20. Tubes are often connected as Darlington pairs in tube-type video amplifiers.

Multiple Choice

1. A video amplifier has full response at _____ .
   (a) 4.5 MHz
   (b) 3.58 kHz
   (c) 2 MHz
   (d) 45.75 MHz

2. A Y amplifier has full response at _____ .
   (a) 4.5 MHz
   (b) 3.58 MHz

(c) 2 MHz

(d) 45.75 MHz

3. Typical Y amplifiers have falling frequency responses past _____ .
   (a) 1 MHz
   (b) 2 MHz
   (c) 3 MHz
   (d) 455 kHz

4. Y amplifiers cannot use a 4.1-MHz bandwidth in practice, because of _____ .
   (a) inherent nonlinearities in the output system
   (b) phase shift in the video detector
   (c) retrace blanking action
   (d) DC restorer action

5. A delay line provides a Y-signal delay of approximately_____ .
   (a) 1 second
   (b) 1 millisecond
   (c) 1 microsecond
   (d) 1 radian

6. A color-subcarrier trap in a Y-amplifier circuit provides_____ attenuation of the color subcarrier.
   (a) complete
   (b) no
   (c) 10%, approximately
   (d) 70%, approximately

7. Darlington transistors consist of _____ .
   (a) two series-connected transistors in the CC mode
   (b) two series-connected transistors in the CB mode
   (c) three series-connected transistors in the CE mode
   (d) three series-connected transistors in the emitter-follower mode

8. The Y output transistor is usually driven in the_____ mode.
   (a) emitter-follower
   (b) common-emitter
   (c) common-base
   (d) common-collector

9. Negative feedback is utilized in video amplifiers for _____ .
   (a) elimination of high-level noise pulses
   (b) trapping the color-subcarrier signal
   (c) bias stabilization
   (d) deemphasis

10. Automatic brightness control is employed in color receivers to _____ .

(a) reinsert the DC component
(b) protect the color picture tube
(c) avoid eye strain
(d) cancel high-level noise pulses

11. Horizontal sync pulses have a width of _____ μsec.
    (a) 1.1
    (b) 3.1
    (c) 5.1
    (d) 63.5

12. Vertical sync pulses have a width of _____ μsec.
    (a) 0.08
    (b) 63.5
    (c) 190.5
    (d) 45.75

13. The basic functional device in a sync clipper is a _____ .
    (a) thermistor
    (b) diode or transistor
    (c) sync trap
    (d) DC restorer

14. A vertical integrator provides maximum output amplitude at the _____.
    (a) beginning of the vertical-sync pulse
    (b) half-way along the vertical-sync pulse
    (c) end of the vertical-sync pulse
    (d) none of the above

15. Serrated pulses are provided to _____ .
    (a) maintain horizontal locking during passage of the vertical-sync pulse
    (b) equalize sync phase shifts
    (c) minimize interference from noise pulses
    (d) blank retrace lines

16. A noise inverter functions by _____ .
    (a) trapping out noise pulses
    (b) blanking retrace lines
    (c) providing DC restoration
    (d) reversing the polarity of noise pulses

17. The noise-limiting function of a sync system _____ .
    (a) punches "holes" in the clipper-output pulse train
    (b) slices the tips off the blanking pedestals
    (c) provides bias stability
    (d) provides deemphasis for the high chroma frequencies

18. A gated amplifier stage _____.
    (a) operates continuously
    (b) operates only when enabled
    (d) operates only when disabled
    (d) none of the above

19. Retrace lines are blanked from the reproduced image by _____ .
    (a) AC restorer circuits
    (b) trap action
    (c) flyback pulses
    (d) automatic brightness control

20. When a transistor is in saturation,_____.
    (a) its collector voltage is very low and its base current is very high
    (b) its base voltage is very low and its collector voltage is very high
    (c) both its collector current and its base current are cut off
    (d) none of the above

Problems

1. If a video-and-Y-amplifier arrangement provides a signal-voltage gain of 100 times, how many dB of gain does it provide?

2. The rise time of a reproduced square wave is equal to: Rise Time = $1/(3F)$, where F is the high-frequency cutoff of the amplifier. If a video amplifier has a bandwidth of 4 MHz, what is its approximate rise time?

3. A video amplifier has a rise time of 0.1 μsec; what is the approximate bandwidth of the amplifier?

4. If the sound level is 20 dB down in a video amplifier, what is the comparative voltage level of the sound signal?

5. What is the frequency of the fifteenth harmonic in a 100-kHz square wave?

6. What is the period of a 100-kHz square wave?

7. Calculate the value of inductance required to resonate a peaking coil at 4 MHz with 30 pF of capacitance.

8. Calculate the reactance value of the inductance in problem 7.

9. Calculate the reactance value of the capacitance in problem 7.

10. What is the relative phase of the signal voltage and signal current at the half-power point on the resonance curve in problem 7?

Chapter 7

Bandpass Amplifier, Color Killer, And AGC Sections

7.1 FUNCTION AND OPERATION OF THE BANDPASS AMPLIFIER

As noted previously, the chroma bandpass amplifier has the function of passing the chroma signal and rejecting most of the Y signal. This is accomplished by means of a frequency-response characteristic of from 3.1 to 4.1 MHz, as shown in Fig. 7-1. Note that the bandpass amplifier in this example has a bandwidth of 1 MHz, or ±0.5 MHz. Practically all present-day color receivers employ this frequency characteristic. However, a greater chroma bandwidth (from 2.1 to 4.1 MHz) could have been utilized, if it were desired. That is, color-TV transmission provides chroma signals in a 2-MHz band, as depicted in Fig. 7-2. Old-model color receivers used the entire chroma signal, whereas modern color receivers reject part of the transmitted chroma signal. This design practice involves a trade-off, where the finer color detail is not reproduced for reasons of economy and reduction of the residual interference level.

Observe in Fig. 7-2 that a wide-band chroma signal and a narrow-band chroma signal are transmitted. These are called the *I signal* and the *Q signal*, respectively. The term "I" means "in-phase"; that is, in-phase with

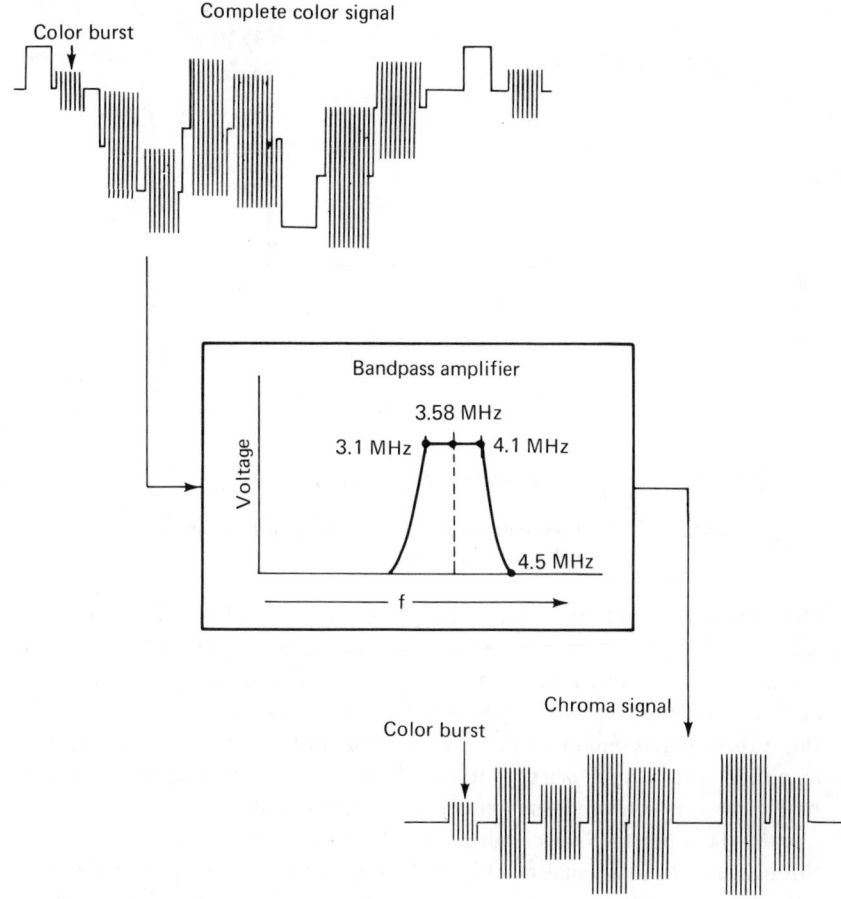

FIGURE 7-1   Separation of the chroma signal from the complete color signal by the bandpass amplifier.

the axis for maximum acuity of color vision. "Q" means "quadrature-phase"; that is, at right angles to the I axis. Note that the Q signal is a double-sideband signal, whereas the I signal is a vestigial-sideband signal from 2.1 to 3.1 MHz. Just as the R−Y and B−Y chroma signals contain the G−Y chroma signal, so do the I and Q chroma signals contain the R−Y, B−Y, and G−Y chroma signals. This fact is apparent from Fig. 7-3, which shows that any pair of chroma axes has components on any other pair of chroma axes. Observe also in Fig. 7-3 that the +I vector corresponds to orange hues, or flesh tones, to which the eye is most sensitive.

To summarize briefly, color-TV transmitters provide wide-band chroma information (the I signal) so that flesh tones can be reproduced with

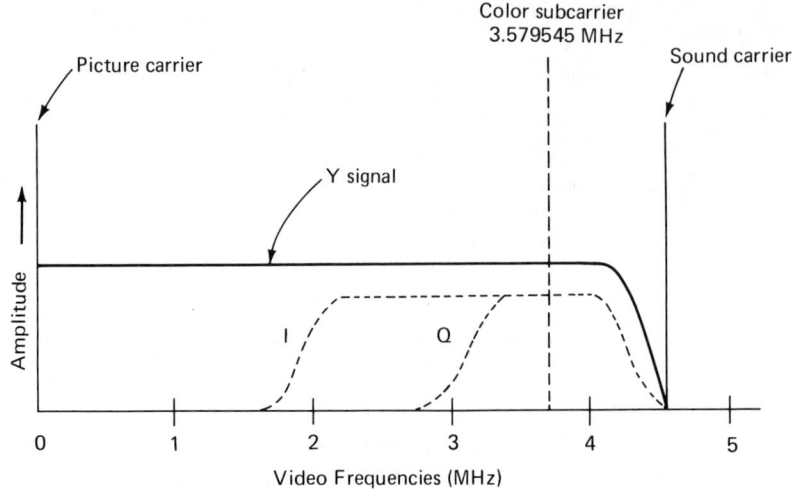

FIGURE 7-2 Color-television transmission provides chroma signals in a 2-MHz band.

relatively fine detail. However, color-TV receivers employ only the double-sideband chroma information and operate on the R−Y and B−Y axes. From a practical viewpoint, a color receiver processes the incoming chroma signal as if it were an R−Y/B−Y double-sideband transmission. The first aspect of this processing is found in the 1-MHz bandwidth of the chroma bandpass amplifier. The second aspect is found in the chroma decoding or demodulation process, which will be subsequently explained in detail.

In case the IF amplifier should have a flat frequency response through the chroma-signal region, the bandpass amplifier will also have a flat frequency-response curve from 3.1 to 4.1 MHz, as depicted in Fig. 7-1. However, few color receivers have a flat IF response, and in turn, the bandpass amplifier is aligned to provide frequency compensation, as Fig. 7-4 demonstrates. Note that the slope of the bandpass response curve is opposite to that of the IF response curve. In turn, the overall IF-and-bandpass response is uniform from 3.1 to 4.1 MHz. When a uniform system response is obtained by the net response of non-uniform subsystem responses, a specialized sweep-alignment technique is employed, called *video sweep modulation* (VSM). Details of VSM arrangements and operating procedure will be explained in the discussion of color-TV test equipment (Chapter 15).

7.2 BANDPASS-AMPLIFIER CIRCUITRY

Figure 7-5 shows the configuration of a typical chroma circuit board. Note that the complete color signal is fed from the video detector to terminal 11 on the circuit board. Coupling capacitor C353, coil L351, resistor R359, and

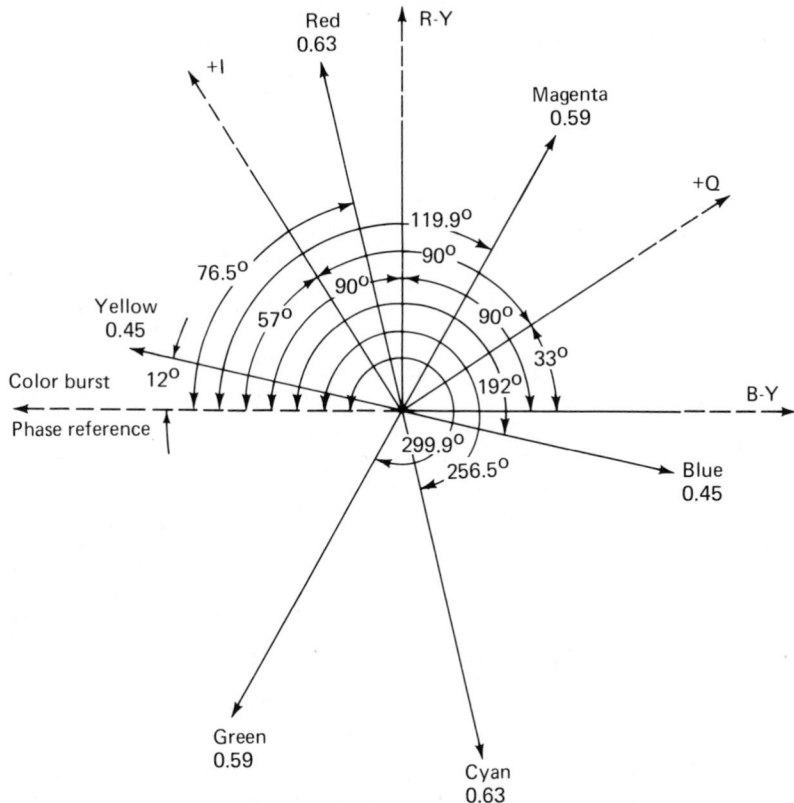

FIGURE 7-3   Relation of the I and Q axes to the R−Y and B−Y axes.

capacitor C354 form a bandpass filter with a center frequency of 3.58 MHz. Thus, most of the Y signal is eliminated, and the chroma signal is passed. Just as AGC is employed in the RF and IF sections to obtain a wide dynamic range, so is automatic chroma control (ACC) utilized in the chroma section. That is, the gain of Q354 is controlled by the collector voltage of the ACC amplifier Q352. Observe that a voltage divider made up of resistors R356, R355, R354, and ACC amplifier transistor Q352 sets the base bias of Q354 for maximum gain when no chroma signal is present.

The ACC detector, which will be discussed later, develops a negative voltage when a chroma signal is present. This negative voltage is coupled to the base of Q351 through resistor R351 and terminal 5 of the circuit board in Fig. 7-5. In turn, when the amplitude of the chroma signal increases, Q351 and Q352 conduct less, which increases the voltage at the base of Q354. This circuit action reduces the amplification of the chroma signal due to the forward automatic gain characteristics of Q354. The amplified chroma signal

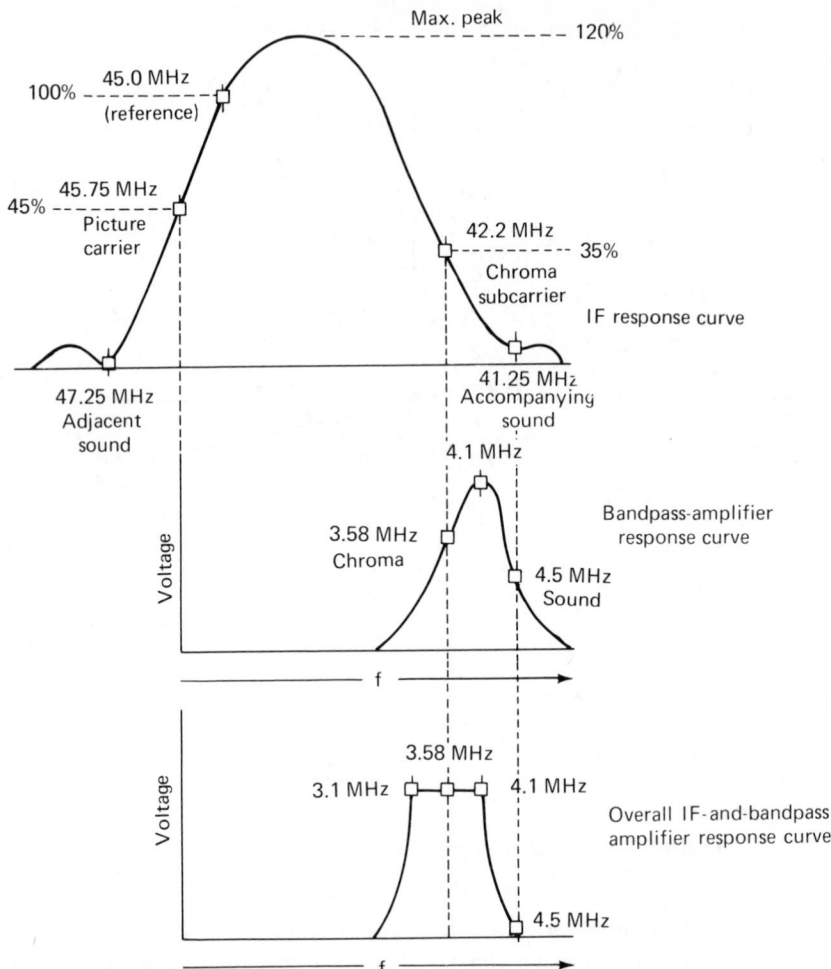

FIGURE 7-4   Frequency compensation in bandpass-amplifier response.

is then coupled through C355 to the base of Q355. Base bias for this stage is provided by the voltage-divider resistors R363 and R364. In turn, the signal at the collector of Q355 is connected through terminal 8 on the circuit board to the color-sync section. At the same time this chroma signal is coupled through C359 to the color control (not shown) via terminal 9 on the circuit board. From the arm of the color control, the chroma signal is fed via terminal 7 on the circuit board through C360 to the base of Q356.

Transistor Q356 is biased by the voltage divider consisting of resistors R369 and R370 in Fig. 7-5. Blanking pulses are fed via terminal 3 on the

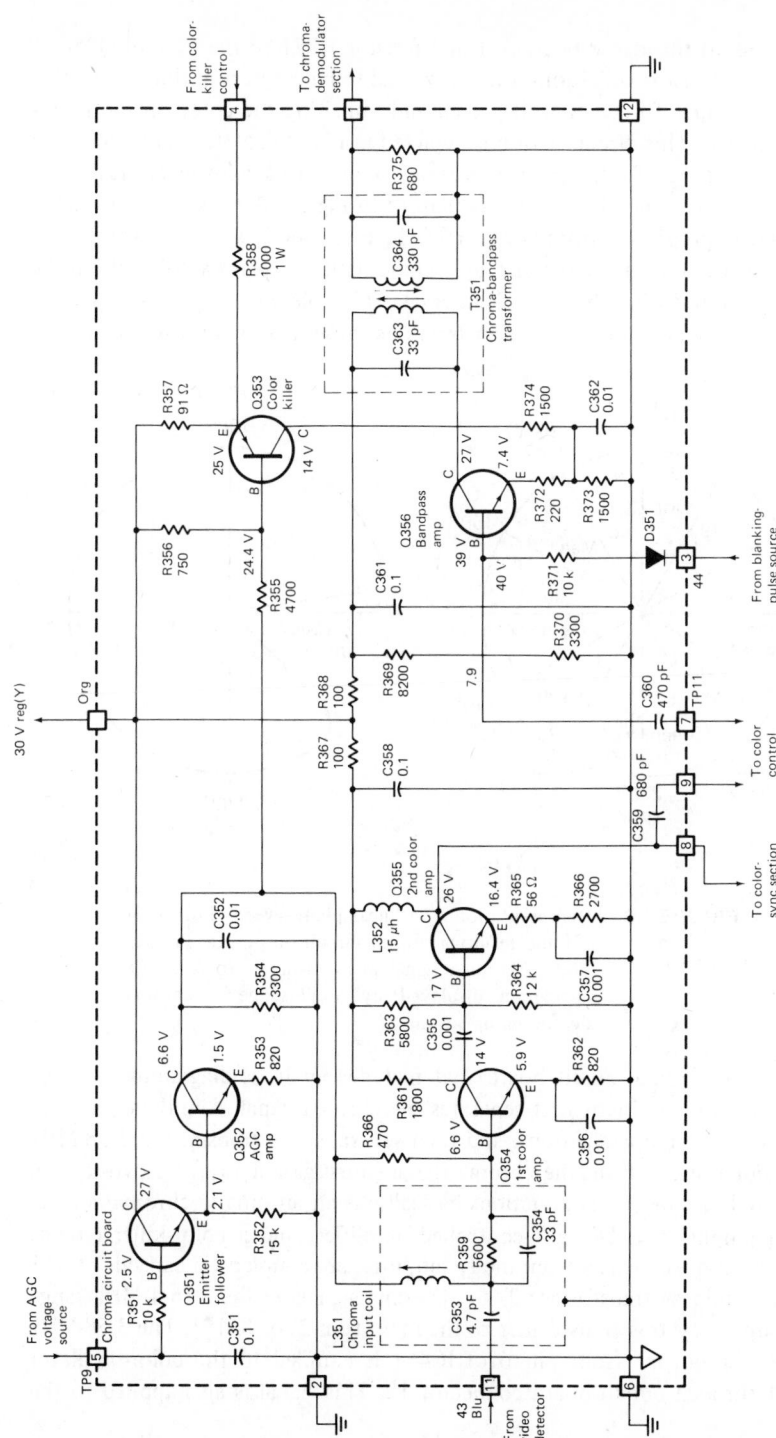

FIGURE 7-5 Configuration of a chroma circuit board.

143

circuit board through diode D351 and resistor R371 to the base of Q356 to turn the transistor off during the horizontal flyback interval, thus preventing the color burst from being displayed and impairing the colors in the reproduced image. This circuit action is depicted in Fig. 7-6. Note that if a chroma signal consisting of a linear phase sweep is applied to a color receiver, there is no output from the chroma demodulators during the flyback interval. As the test signal progresses from 0° to 360° in Fig. 7-6(a), an oscilloscope connected at the output of the chroma demodulators displays a "pie cut" in the signal pattern during the flyback interval. This "pie cut" is produced by the blanking pulse. Chroma sweep signals will be subsequently discussed in detail.

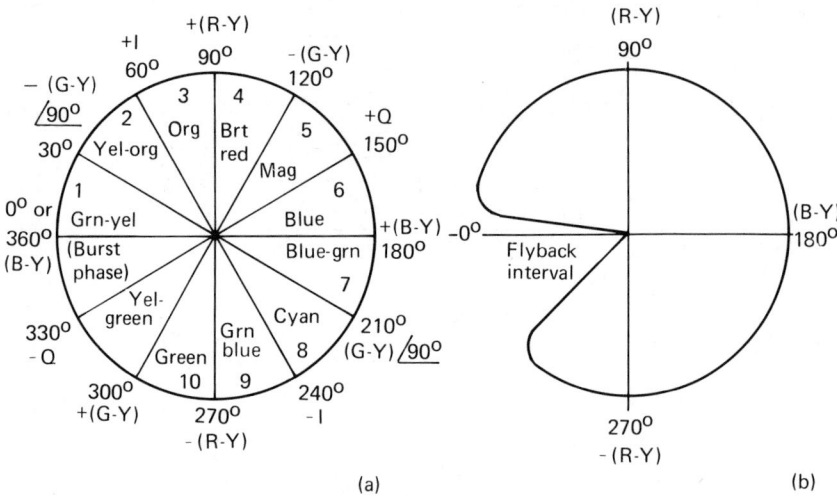

(a)

(b)

FIGURE 7-6   Representation of a linear phase-sweep chroma signal and resulting chroma-circuit output. (a) Progression of chroma signal phase from 0° to 360°; (b) cyclogram displayed by oscilloscope at chroma-demodulator outputs.

A chroma circuit board that includes an IC is diagrammed in Fig. 7-7. The chroma circuit first separates the chroma signal from the composite video signal (bandpass action), and then separates the transmitted 3.58-MHz color burst signal from the chroma signal (burst-gate action). A two-section integrated circuit IC451 functions basically as an automatic chroma control (ACC) amplifier, a DC gain-controlled amplifier, and a color-killer circuit. Composite video signals from the IF amplifier are coupled via capacitor C451 to the bandpass transformer T451. Chroma signals in the 3-to-4 MHz range are coupled by this transformer to the input (pin 2) of IC451. The 3.58-MHz color-burst output from pin 6 of IC451 is coupled to the color-oscillator circuit through circuit-board connector 12. This signal is also applied to the

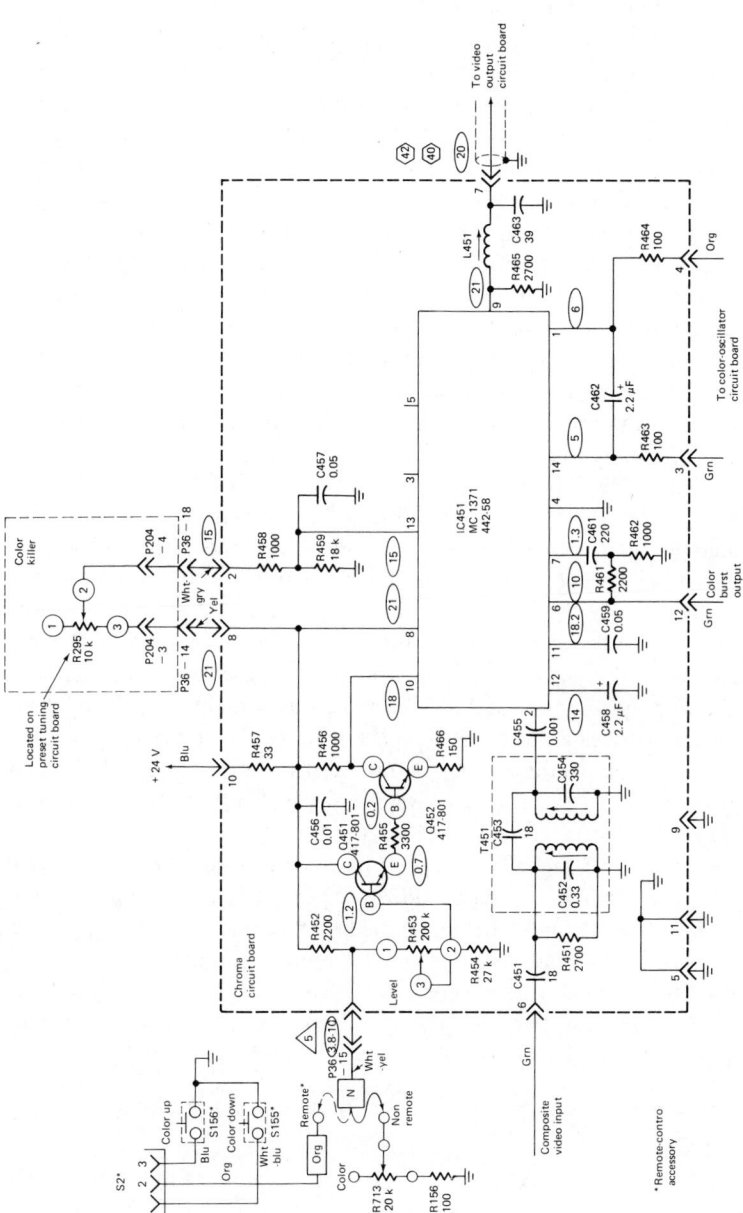

FIGURE 7-7 A chroma circuit-board configuration that includes an IC. *(courtesy of Heath Co.)*

voltage divider of R461 and R462, and connected via C461 to the input (pin 7) of the second amplifier contained in IC451.

Amplification in the second amplifier of IC451 is determined by the DC voltage at pin 10 of IC451. This DC voltage is controlled by the color control R713 and is amplified by DC amplifier Q451 and Q452 before it is applied to pin 10. Control R453 sets the range of the DC control voltage. The color-killer control R295 is connected to +24 volts through resistor R457. Resistors R458 and R459 form a voltage divider for the voltage present at the wiper (lug 2) of the control. The color-killer control thus sets the voltage at pin 13 of IC451. This is a threshold voltage that is compared internally to the ACC voltage to provide automatic cutoff of the second amplifier when there is no incoming color signal. The ACC voltage applied to pins 1 and 14 of IC451 from the color-oscillator circuit is proportional to the amplitude of the 3.58-MHz color-burst signal applied to the oscillator. This voltage automatically controls the gain of the chroma amplifier to maintain uniform chroma strength for each received channel.

Next, the chroma-output signal from pin 9 of the second amplifier in Fig. 7-7 is coupled through coil L451 and the circuit board connector 7 to the video-output circuit board. Since the hue of a picture is controlled by the phase relationship between the chroma signal and the color-oscillator reference signals (from the color-oscillator circuit board), coil L451 is initially adjusted to place the phase of the chroma signal within the range of the color-oscillator reference signals. The phase of the color-oscillator reference signals is established by tint control R721. Thus, because the tint control adjusts the relationship between the chroma signal and the color-oscillator reference signals, it controls the hue of the picture. Resistor R465 is a fixed load at the output of IC451 to prevent excessive amplitude changes in the chroma signal as L451 is adjusted.

Figure 7-8 depicts a representative tube-type bandpass-amplifier configuration. This arrangement utilizes a pentode bandpass-amplifier tube and a triode-follower tube. In case of malfunction, check or replace the tubes first. If normal operation is not restored, circuit tests must be made. (Be sure that the color-killer control is not set so high that the bandpass amplifier is biased off.) Many technicians prefer to make signal-injection tests with a television analyzer, as shown in Fig. 7-9. The procedure is as follows:

1. A VHF test signal is injected at the antenna-input terminals of the tuner.

2. Turn up the color control on the analyzer to its maximum position.

3. The Gnd lead (black test lead) must be connected from the Gnd jack on the analyzer to the receiver chassis.

4. Connect the red test lead from the color jack on the analyzer to the input of the chroma demodulators.

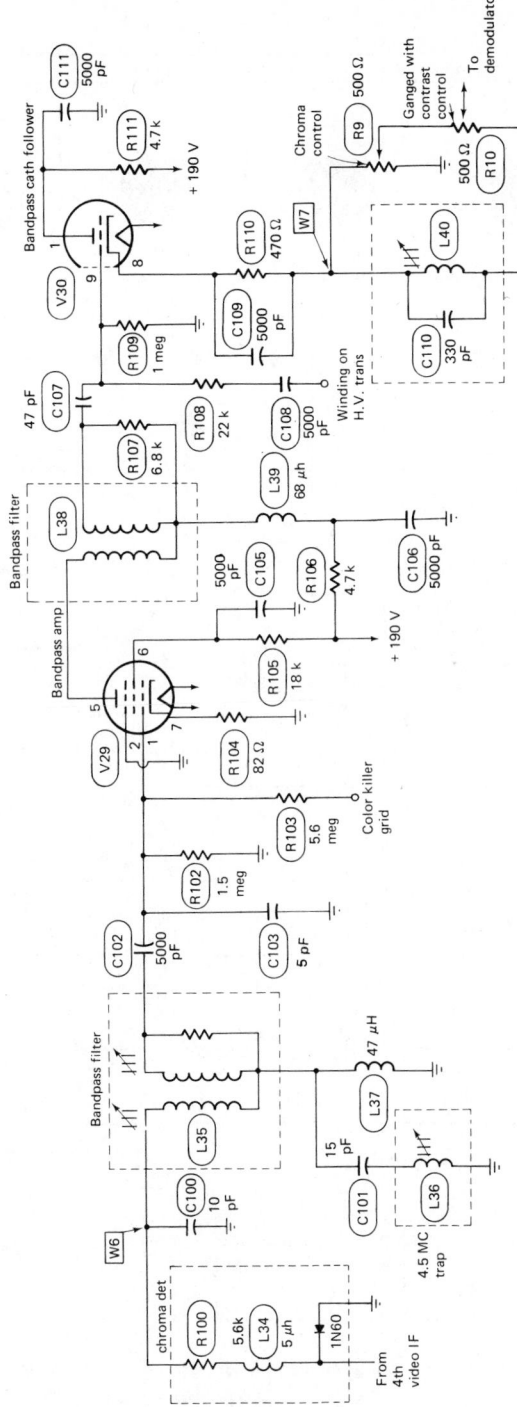

FIGURE 7-8 Representative tube-type bandpass-amplifier configuration.

147

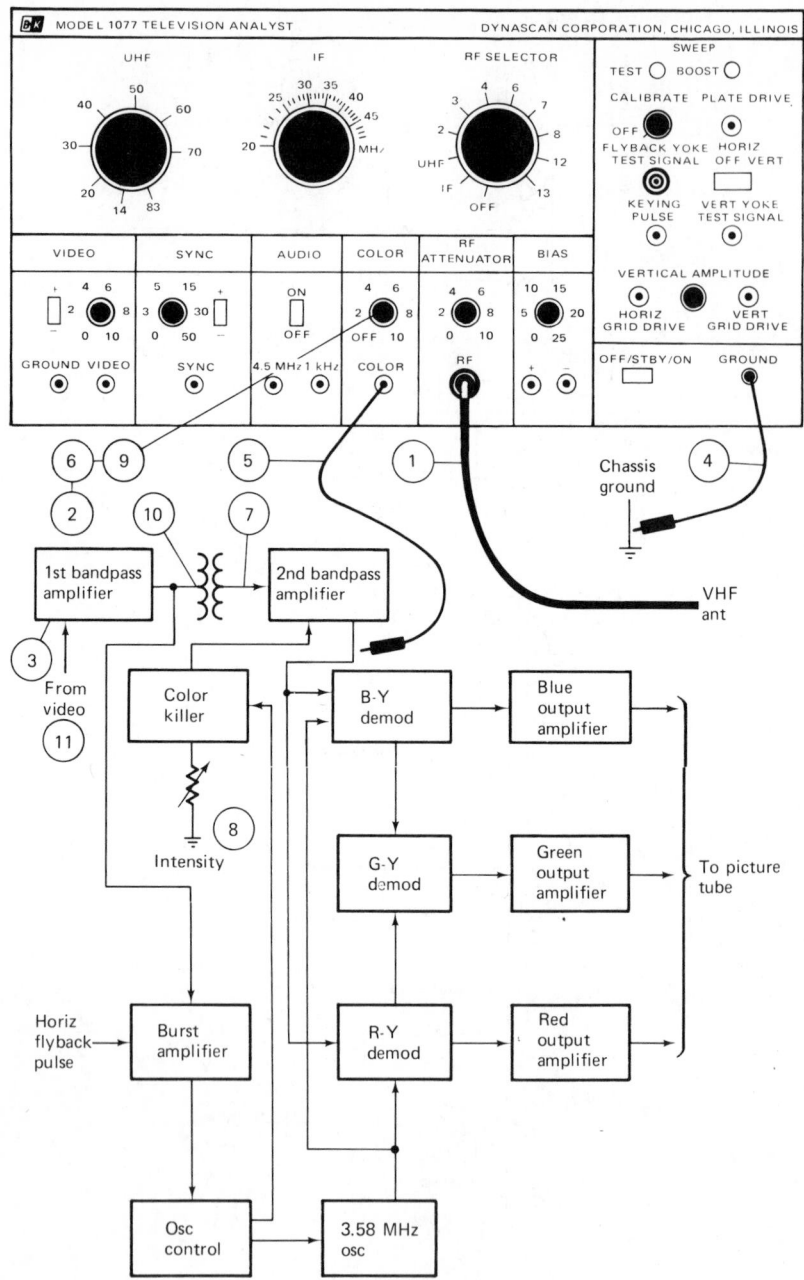

FIGURE 7-9 Bandpass-amplifier tests with a television analyzer. *(courtesy of B&K Mfg. Co., Division of Dynascan Corp.)*

5. A rainbow color-bar pattern displayed on the picture-tube screen is an indication that the chroma channel is workable past the band-pass amplifier.

6. Move the red test lead of the analyzer to the input of the second bandpass-amplifier tube. (The color-killer control must be turned to its "off" or minimum position.) If a rainbow color-bar pattern is displayed on the picture-tube screen, it is an indication that the second bandpass-amplifier stage is workable.

7. Reduce the setting of the color control on the analyzer to avoid overloading. Move the red test lead of the analyzer to the input of the first bandpass-amplifier tube. If no rainbow color-bar pattern is displayed, it is indicated that the trouble is in the first bandpass-amplifier stage.

7.3   COLOR KILLER AND AUTOMATIC CHROMA CONTROL

The color killer has the function of disabling the bandpass amplifier when the color receiver is tuned to a black-and-white transmission. This switching action is desirable because it stops any random noise pulses that might other-wise proceed through the chroma section. A manual control is provided to set the threshold of color-killer operation. This is a maintenance control that is adjusted during receiver set-up procedures. If the color-killer control is set to its minimum position, the bandpass amplifier will not be switched off when the receiver is tuned to a black-and-white transmission. If the noise level is comparatively high, colored snow (confetti) will be displayed on the pic-ture-tube screen. If the color-killer control is set to its maximum position, the bandpass amplifier will not switch on when the receiver is tuned to a color-TV transmission. Accordingly, the color program will be displayed as a black-and-white image. Therefore, it is essential to adjust the color-killer threshold correctly.

Figure 7-10 shows the plan of a color-killer and automatic chroma-control system. Note that the color killer switches the output bandpass-am-plifier stage on or off, whereas the ACC section controls the gain of the input bandpass-amplifier stage. Observe that the ACC section also energizes the color-killer section. Control action in the system is determined by the presence or absence of a color burst and also by the amplitude of the color burst. The burst amplifier picks out the color burst from the complete color signal, as will be subsequently explained in greater detail. Output from the burst amplifier is fed to the ACC section, which in turn drives the color killer and the input bandpass-amplifier stage. Consider how this system operates both in the presence of and in the absence of a burst signal.

If the color receiver is tuned to a black-and-white transmission, there will be no output from the burst amplifier in Fig. 7-10. Accordingly, there

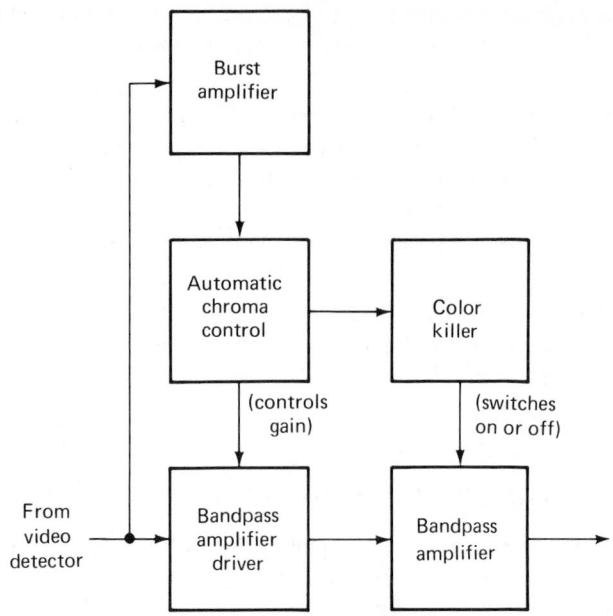

FIGURE 7-10 Plan of the color-killer and automatic chroma-control system.

will also be no output from the ACC section. Under this condition, the band-pass-amplifier output stage is cut off by the color killer. That is, the chroma section is disabled. If the receiver is tuned to a color-TV transmission, there will be an output voltage from the burst amplifier. This output voltage is stepped up through the ACC section and actuates the color killer, which, in turn, switches the bandpass-amplifier output stage into operation. By this means, the chroma section is enabled. At the same time, the output voltage from the ACC section is also applied as a control potential to the bandpass-amplifier input stage. As the color burst varies in peak-to-peak voltage, the ACC control voltage varies accordingly. Therefore, the gain of the bandpass-amplifier section is maintained in practically a constant state.

Referring to Fig. 7-5, we note that the color killer controls the emitter circuit of Q356. Base bias for the killer transistor Q353 is provided by the same voltage-divider network that sets the bias on Q354. An adjustable bias is applied to the emitter of Q353 by the voltage divider R357, R358, and the color-killer control (not shown). The emitter voltage of Q353 can then be adjusted either positively or negatively with respect to its base voltage, which is applied by the ACC amplifier. Since Q353 is a PNP-type transistor, collector current will flow when the emitter is slightly positive with respect to the base. Therefore, when no chroma signal is present, the color-killer control is adjusted so that sufficient collector current flows through R374 and R373 to make the base of Q356 positive with respect to its fixed base bias. Since the bandpass amplifier employs an NPN-type transistor, no collector current flows. However, when a chroma signal is present, the ACC voltage at the junction of R356 and R355 goes positive. In turn, the color-killer transistor is cut off, and the bandpass amplifier goes into operation.

7.4   TRANSIENT RESPONSE OF THE BANDPASS AMPLIFIER

An amplifier has a steady-state response and a transient response. Its steady-state response is defined as the output that results from application of a sinusoidal (CW) continuous-wave signal to the input of the amplifier. A steady-state response is given in the form of a frequency-response curve, as is depicted in Fig. 7-1. The transient response of an amplifier is defined as the time that is required for the output to rise from zero to maximum value when a step function is applied to the input of the amplifier. A step function is exemplified by the leading edge of a square wave. In the case of a bandpass amplifier, a step function is represented by the leading edge of a color-burst signal or of a color-bar signal.

   With reference to Fig. 7-11, two bandpass-amplifier responses are illustrated for a 3.58-MHz signal consisting of a color burst and a color-bar signal. In Fig. 7-11(a) the bandpass amplifier has a normal bandwidth. In Fig. 7-11(b), the bandpass amplifier has been misaligned and has subnormal bandwidth. Observe that the output waveform of Fig. 7-11(a) rises with comparative rapidity and that the corners of the waveform are not unduly rounded. However, in Fig. 7-11(b) the output waveform rises slowly, and the corners are excessively rounded. Thus, the transient response of the bandpass amplifier is related to its frequency response. Conversely, the frequency response of the amplifier is related to its transient response.

   Since the chroma signal in Fig. 7-11 has a frequency of 3.58 MHz,

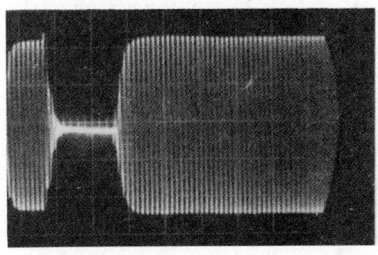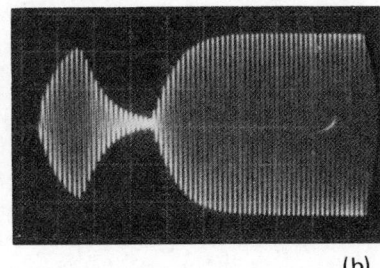

(a)   (b)

FIGURE 7-11   Effect of subnormal bandwidth on chroma-signal reproduction. (a) Normal bandwidth; baseline is reasonably straight; (b) subnormal bandwidth; baseline is seriously curved. *(courtesy of Howard W. Sams & Co., Inc.)*

it might seem puzzling that appreciable bandwidth is required for satisfactory transient response. However, the chroma signal has a single frequency (3.58 MHz) only after the waveform has risen to maximum. That is, during the rise interval of the waveform, a wide band of frequencies is present. This is the result of amplitude modulation of the 3.58-MHz signal by a pulse or square waveform. Amplitude modulation produces sideband frequencies, and, if a sufficient number of sideband frequencies are not accepted by the bandpass amplifier, its transient response will be poor. Figure 7-12 depicts how a square wave can be synthesized from a fundamental sine wave and its odd harmonics.

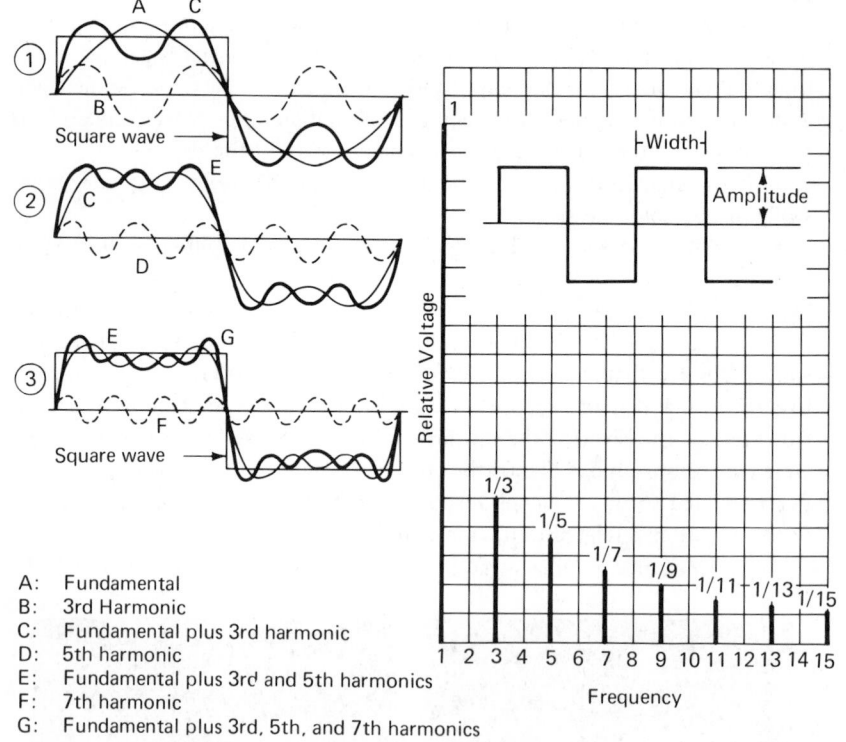

A: Fundamental
B: 3rd Harmonic
C: Fundamental plus 3rd harmonic
D: 5th harmonic
E: Fundamental plus 3rd and 5th harmonics
F: 7th harmonic
G: Fundamental plus 3rd, 5th, and 7th harmonics

FIGURE 7-12   Synthesis of a square wave from a fundamental sine wave and its odd harmonics.

Since the transient response of a bandpass amplifier improves as its bandwidth is increased, it might seem that it would be advantageous to utilize a very considerable bandwidth, such as 4 MHz. However, unless a bandpass amplifier has a bandwidth that is limited to 1 MHz, or at most to 2 MHz, objectionable interference from the Y signal becomes apparent. When excessive amounts of Y signal gain entry into the chroma section, the frequency-interleaving process becomes inadequate to cancel the interference satisfactorily. Therefore, the frequency limits of the bandpass amplifier are designed to provide optimum transient response, while maintaining the interference level sufficiently low so that the viewer is unaware of residual crawl and spurious patterns.

When a carrier is amplitude-modulated by a sine wave, a pair of sidebands are generated. For example, if a 1-MHz carrier is modulated by a 1-kHz sine wave, the modulated waveform has frequencies of 1 MHz, 1.001 MHz, and 0.999 MHz.

Next, consider the amplitude modulation of a 1-MHz carrier by a 1-kHz square wave. Referring to Fig. 7-12, note that the modulated waveform will have frequencies of 1 MHz, 1.001 MHz, 0.999 MHz, 1.003 MHz, 0.997 MHz, and so on. In practice, it is not essential to pass an infinite number of sideband frequencies. It is, however, necessary to pass a sufficient number of sideband frequencies so that the output waveform is a reasonable facsimile of the modulation envelope. Therefore, the alignment of a bandpass amplifier entails the compromise that was noted previously.

EXERCISES

Questions

1. What is the basic function of the chroma bandpass amplifier?
2. State the frequency limits of a typical bandpass amplifier.
3. Can an I/Q signal be processed as an (R−Y)/(B−Y) signal?
4. Explain the relationship of the I and Q axes to the R−Y and B−Y axes.
5. How is the slope of the IF frequency-response curve compensated by the frequency characteristic of the bandpass amplifier?
6. Describe the operation of an *ACC circuit*.
7. Discuss the operation of a *color-killer circuit*.
8. How is the color killer enabled and disabled?
9. Is the I signal a vestigial-sideband signal? The Q signal?
10. What bandwidth would be required in the bandpass amplifier to process the entire I signal?
11. Explain the trade-off that is involved in limiting the bandpass-amplifier pass band to 3.1-4.1 MHz.
12. Can the G−Y signal be recovered from I and Q signals?
13. What are the chief differences between a tube-type bandpass amplifier and a solid-state bandpass amplifier?

14. Is the I signal centered on the color-subcarrier frequency? the Q signal?

15. Are adjustable tuned circuits provided in a bandpass amplifier?

16. Why is a blanking pulse applied to the bandpass-amplifier transistor?

17. Do typical bandpass amplifiers contain more than one stage?

18. How could the technician quickly determine whether signal stoppage is being caused by color-killer trouble or by bandpass-amplifier trouble?

19. Is the color-killer threshold adjustable?

20. Explain the distinction between the ACC section and the AGC section.

True-False

1. A bandpass amplifier permits the Y signal to flow through to the chroma demodulators.

2. Most bandpass amplifiers have a frequency response from 2.1 to 4.1 MHz.

3. "I" means "in phase" and "Q" means "quadrature phase."

4. The I signal is a vestigial-sideband signal, whereas the Q signal is a double-sideband signal.

5. Any pair of chroma axes has components on any other pair of chroma axes.

6. R—Y and B—Y signals are centered on the color-subcarrier frequency.

7. The Q signal is centered on the color-subcarrier frequency, but the I signal is not.

8. A G—Y signal is centered on the color-subcarrier frequency.

9. Bandpass-amplifier frequency-response curves are generally tilted to compensate for the slope in the IF response curve.

10. Burst amplifier and bandpass amplifier are equivalent terms.

11. An I axis has a phase angle of $57°$ with respect to the $-(B-Y)$ axis.

12. A Q axis has a phase angle of $33°$ with respect to the B—Y axis.

13. I and Q axes are in phase quadrature to each other.

14. An ACC section has the same basic action as an AGC section.

15. ACC bias voltage is applied to the bandpass amplifier and to the IF amplifier.

16. Blanking pulses are applied to the bandpass amplifier to prevent the color burst from proceeding into the chroma demodulators.

17. If the color burst were displayed on the picture-tube screen, the colors in the image would be impaired.

18. The color-killer threshold is not adjustable.

19. Color-killer action is enabled or disabled by the presence or absence of the color burst.

20. ACC bias-voltage variation "follows" the amplitude of the color burst.

Multiple Choice

1. A chroma bandpass amplifier _____ .
   (a) reinserts the color subcarrier
   (b) processes the burst signal
   (c) steps up the Y signal
   (d) steps up the chroma signal

2. Early color-TV receivers used the _____ chroma system.
   (a) (R–Y)/(B–Y)
   (b) I/Q
   (c) AM/FM
   (d) phase-modulated

3. Present-day color receivers employ a bandpass-amplifier frequency response from _____ .
   (a) 550 to 1500 kHz
   (b) 88 to 108 MHz
   (c) 20 Hz to 20 kHz
   (d) 3.1 to 4.1 MHz

4. The color-burst signal is_____ by the bandpass amplifier.
   (a) stepped up
   (b) ACC controlled
   (c) rejected
   (d) combined with the Y signal

5. An I signal is a _____ type of signal.
   (a) narrow-band FM
   (b) double-sideband
   (c) single-sideband
   (d) vestigial-sideband

6. A Q signal is a _____ type of signal.
   (a) narrow-band FM
   (b) double-sideband

    (c) single-sideband
    (d) vestigial sideband

7. An R–Y signal is a _____type of signal.
    (a) narrow-band FM
    (b) double-sideband
    (c) single-sideband
    (d) vestigial sideband

8. A chroma bandpass amplifier is a _____ configuration.
    (a) gated
    (b) nonlinear
    (c) untuned
    (d) zero-gain

9. Bandpass-amplifier frequency-response curves are generally contoured to provide _____ .
    (a) uniform overall response
    (b) peak response at 3.58 MHz
    (c) peak response at 4.5 MHz
    (d) peak response at 455 kHz

10. An ACC section functions by _____ of the bandpass amplifier.
    (a) varying the bandwidth
    (b) varying the gain
    (c) varying the phase
    (d) none of the above

11. A color-killer section has the function of _____ .
    (a) reconstituting the chroma signal
    (b) trapping out the chroma signal
    (c) disabling the bandpass amplifier during black-and-white reception
    (d) disabling the Y amplifier during black-and-white reception

12. Output from the burst amplifier is applied to the _____ .
    (a) bandpass amplifier
    (b) ACC section
    (c) AGC section
    (d) video detector

13. When the color-killer transistor is cut off, its collector_____ .
    (a) operates as an emitter
    (b) voltage is very low
    (c) voltage rises to the supply-voltage value
    (d) current is maximum

14. A color-burst signal has _____ .
    (a) no sidebands
    (b) significant sidebands
    (c) frequency modulation
    (d) phase modulation

15. A chroma signal has _____ .
    (a) no sidebands
    (b) significant sidebands
    (c) zero rise time
    (d) no transient component

Problems

1. A bandpass-amplifier coupling capacitor has a value of 4.7 pF. What is its reactance at 3.58 MHz? At 0.358 MHz?

2. If a 3.58-MHz resonant coil is shunted by 40 pF of capacitance, what is the inductance value of the coil?

3. What is the reactance of the coil in problem 2 at 3.58 MHz?

4. What is the reactance of the capacitor in problem 2 at 3.58 MHz?

5. If a 3.58-MHz resonant coil is shunted by 330 pF of capacitance, what is the inductance value of the coil?

6. The Q value of the coil in problem 5 is 12; what is the value of the effective resistance in series with the coil?

7. A bandpass amplifier has a bandwidth of ±0.5 MHz. Calculate the percentage of bandwidth to the center frequency (3.58 MHz) of the amplifier.

8. If a bandpass-amplifier choke has an inductance of 15 μH, what is its inductive reactance at 3.58 MHz?

9. An emitter partial-bypass capacitor in a bandpass amplifier has a value of 0.001 μF. Calculate the reactance of the capacitor at 3.58 MHz.

10. A bandpass-amplifier transistor has a voltage gain of 20 times and a current gain of 5 times. What is its power gain in dB?

Chapter 8

Burst Amplifier, Oscillator, And Color Synchronization

8.1   COLOR SYNCHRONIZATION REQUIREMENTS

Before the chroma signal can be decoded, the 3.58-MHz color-subcarrier frequency, which was suppressed at the transmitter, must be inserted at the receiver. Reconstitution of the chroma signal requires that the inserted subcarrier be locked in both frequency and phase to the suppressed subcarrier. For this purpose, a sample of the suppressed subcarrier is provided on the back porch of the horizontal sync pulse. This color burst serves as a phase and frequency reference for the color synchronizing system, which keeps the 3.58-MHz subcarrier oscillator locked precisely in step with the suppressed subcarrier. It follows that the first function of the color-sync system entails separation of the color burst from the complete color signal. This is the function of the *burst amplifier*.

8.2   BURST-AMPLIFIER OPERATION AND CHARACTERISTICS

The color burst has a frequency of 3.58 MHz and contains 8 or 9 cycles, as Fig. 8-1 shows, and the burst-amplifier section operates at a center frequency of 3.58 MHz. Figure 8-2 shows the plan of a burst-amplifier arrangement.

Signal input is obtained either from the video amplifier or the chroma amplifier. If the video-amplifier output signal is utilized, the burst-takeoff circuit is energized by the complete color signal. If the chroma-amplifier output signal is utilized, the burst-takeoff circuit is energized by the chroma signal. As depicted in Fig. 8-3, the burst-takeoff circuit has a center frequency of 3.58 MHz, and a typical bandwidth of approximately 0.5 MHz. Although the bandwidth value is not critical, it is undesirable to employ excessive bandwidth because the amount of noise that is passed by a signal channel is proportional to the channel bandwidth. That is, a better signal-to-noise ratio is obtained by restricting the bandwidth. On the other hand, it is undesirable to employ very narrow bandwidth because with it the color burst becomes attenuated and its waveshape becomes distorted.

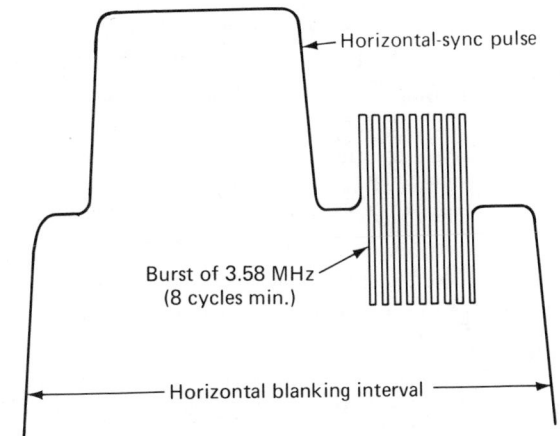

FIGURE 8-1   Burst waveform comprises approximately 8 cycles of the 3.58-MHz color subcarrier.

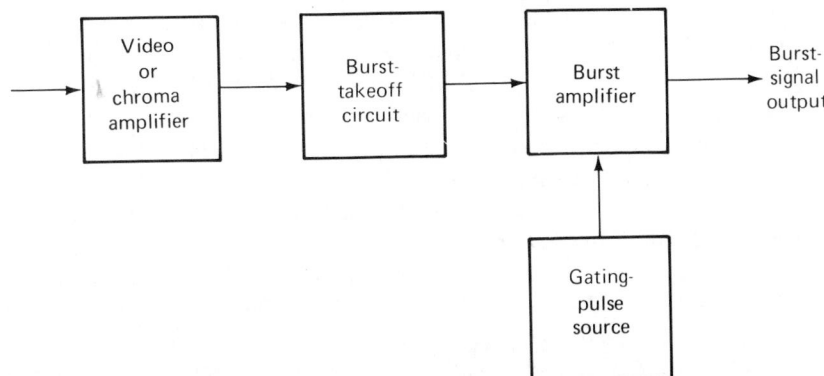

FIGURE 8-2   Plan of the burst-amplifier section.

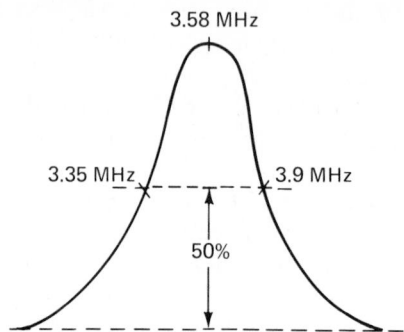

FIGURE 8-3   Typical frequency response of a burst amplifier.

Observe in Fig. 8-2 that the burst amplifier is *gated*; this is an essential feature of the arrangement. This means that the burst amplifier must remain cut off until the arrival of the color burst, at which time the burst amplifier must be gated on for the duration of the color burst. At the end of the color-burst interval, the burst amplifier must go promptly into cutoff. Unless the burst amplifier is properly gated, a "clean" burst signal will not be separated from the incoming chroma signal, or complete color signal. The output from the burst amplifier will be contaminated with spurious signal voltages. In turn, color sync action will be impaired or lost. Therefore, it is essential that the gating action proceed correctly whenever the receiver is tuned to a color-TV transmission.

Figure 8-4 depicts the action of the gating pulse with respect to the color burst. Note that the gating pulse has a slightly greater width than the color burst, and that it is so precisely timed that all of the 3.58-MHz signal is picked out of the incoming waveform. It is evident that if the gating pulse becomes sufficiently mistimed, no burst signal will be passed by the burst amplifier. Again, if the gating pulse is partly mistimed, only part of the burst signal will be passed by the burst amplifier. It is not desirable for any portion of the burst signal to be eliminated in the gating process. The reason for this is that, as noted in Chap. 7, this type of signal contains sideband frequencies. When part of the burst is passed, more or less of the sideband frequencies are rejected, and new sideband frequencies are introduced. Therefore, if the timing of the gating pulse drifts slightly, the frequency content of the partial burst signal varies; and, as a result, the picture displayed on the screen is plagued with incorrect and drifting colors.

The timing of the burst-gating pulse often depends upon the setting of the horizontal-hold control. That is, if the hold control is set near one end of its range, the picture tends to be pulled off-center on the screen. In turn, the horizontal-flyback pulse tends to lead or lag the horizontal-sync pulse. Since the burst-gating pulse is derived from the horizontal-flyback pulse, the

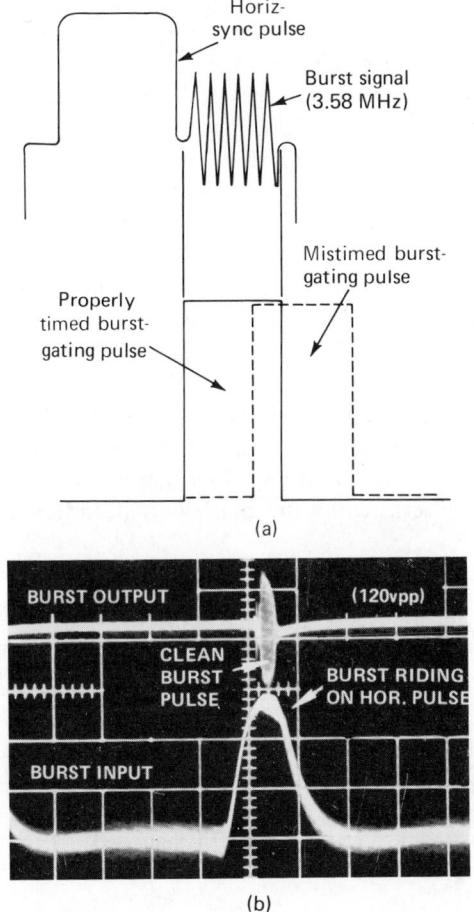

FIGURE 8-4   (a) Gating out the color burst from the horizontal-sync pulse; (b) test of color-burst timing displayed on a dual-trace oscilloscope. *(courtesy of Sencore)*

result is the mistiming of the gating pulse. Therefore, good color sync lock requires that the horizontal-hold control be adjusted to some point approximating the midpoint of its range.

8.3   SUBCARRIER-OSCILLATOR OPERATION AND CONTROL

There are two basic types of subcarrier-oscillator configurations. The simplest arrangement consists of a 3.58-MHz quartz crystal which is shock-excited into oscillation by the color burst. A more elaborate arrangement employs a *crystal-controlled, free-running* 3.58-MHz oscillator, which is locked to the frequency and phase of the color burst by an automatic phase control (APC) subsection. A shock-excited oscillator does not require APC because the ringing crystal automatically keeps in phase with the burst signal. Both arrangements have their advantages and disadvantages. The shock-excited oscillator is economical, but it does not maintain tight synchronization when the incoming signal is comparatively weak. Therefore, the free-running oscillator with APC is used in the majority of color receivers.

Figure 8-5 shows the plan of a *ringing-crystal subcarrier-oscillator section*. From the gated burst amplifier, the color-burst voltage is applied to a 3.58-MHz quartz crystal, which is shock-excited into oscillation. Since the ringing waveform tends to decay in amplitude between successive bursts, a limiter is employed to clip the peaks at a uniform amplitude. This regenerated 3.58-MHz subcarrier is then fed to a phase-shifter network, which develops a two-phase output. These outputs have the R–Y and B–Y phases in this example. Utilization of these R–Y and B–Y subcarrier phases is reserved for subsequent discussion. Note that when a color burst is present, the color-killer is actuated by the limiter output. In turn, the chroma circuits are enabled.

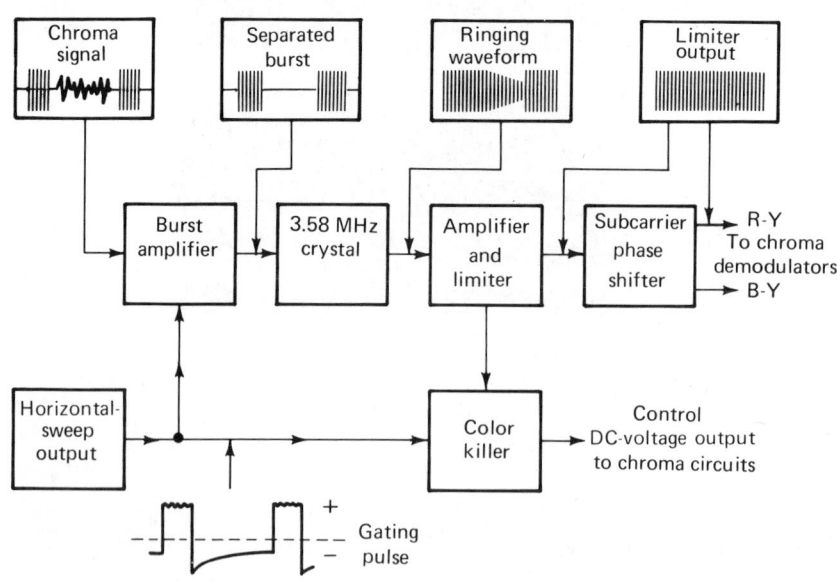

FIGURE 8-5   Plan of a ringing-crystal, subcarrier-oscillator section.

Output waveforms for a ringing crystal are illustrated in Fig. 8-6. When a color burst arrives, the combination of the burst voltage and the ringing voltage causes the waveform to rise to a peak amplitude. Then, as the burst signal terminates, the ringing waveform begins to decay in amplitude.

However, since a quartz crystal has a very high Q value, the decay rate is slow. To obtain a uniform amplitude, this ringing waveform must be passed through a limiter stage to clip the 3.58-MHz peaks at a predetermined level. In Fig. 8-6(a) the scope sweep rate is set to 7875 Hz. Two burst intervals are displayed in turn, but the individual 3.58-MHz cycles are not visible. However, in Fig. 8-6(b) the scope sweep rate is approximately 10 times greater, and individual 3.58-MHz cycles are clearly visible.

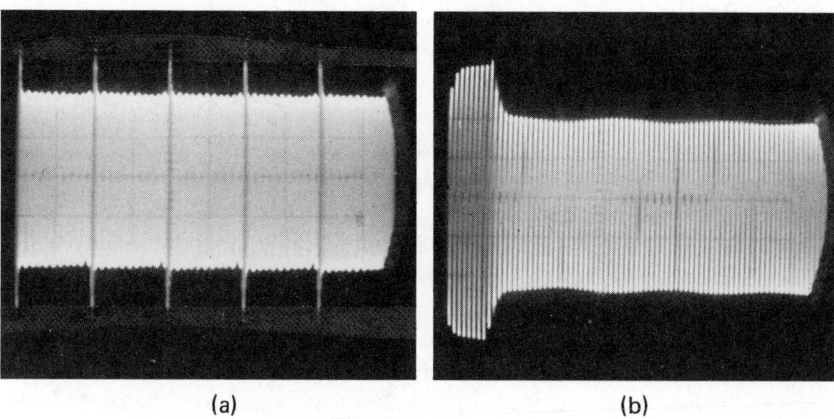

(a)   (b)

FIGURE 8-6   Output waveform from a 3.58-MHz ringing crystal.
(a) Displayed on 3,150-Hz sweep; (b) displayed on
high-speed sweep.

Next consider the operation of an *APC-controlled, free-running* sub-carrier oscillator. Figure 8-7 depicts the plan of this method for subcarrier regeneration. Note that the output from the burst amplifier is fed to the APC section, which is basically a discriminator. This means that the APC circuit compares the burst phase with the phase of the 3.58-MHz oscillator output. If the oscillator output tends to lead or lag the burst phase, the discriminator responds by developing a positive or negative DC control voltage. This control voltage is applied to a varactor diode, which responds with a change in effective capacitance. The varactor diode is connected in parallel with the 3.58-MHz quartz crystal. Since the resonant frequency of a quartz crystal can be changed over a limited range by shunt-capacitance variation, the end result is the pulling of the crystal oscillator back into phase with the color burst.

8.4   SUBCARRIER-OSCILLATOR CIRCUIT BOARD CONFIGURA-TION

Figure 8-8 shows the configuration for a typical subcarrier-oscillator circuit board. The 3.58-MHz subcarrier oscillator, IC401, operates as a basic Colpitts circuit. This integrated circuit functions as an amplifier and has the internal

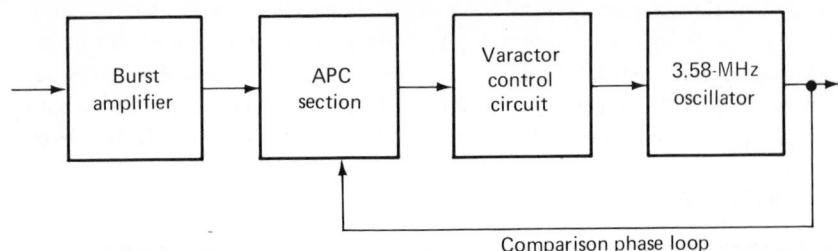

Comparison phase loop

FIGURE 8-7   Plan of APC-controlled free-running subcarrier oscillator.

circuitry depicted in Fig. 8-9. Here, the feedback energy required to sustain oscillation of the quartz crystal is coupled from output terminal 7 of IC401 to input terminal 3 by the capacitance voltage divider C415 and C413. Note that the frequency of oscillation is determined by the resonant frequency of the quartz crystal CR401 as modified by the capacitance of varactor diode D406. In turn, the automatic phase control voltage and the voltage established by the divider network comprising resistors R416 and R417 determine the frequency and phase of the oscillator output. Note that resistors R418 and R419 serve to isolate the oscillator-input AC signal from the DC control voltages.

The output from the oscillator circuit in Fig. 8-8 is then coupled through C418 to the base of Q402, which operates as an amplifier. Base bias for this stage is established by the voltage divider consisting of R426 and R427. In turn, the amplified reference signal at the collector of Q402 drops across the primary of T402 and is coupled by the secondary through terminal 11 on the circuit board to the chroma-demodulator section. Observe that a portion of the signal at the collector of Q402 is fed back through the voltage divider comprising C421 and C409 through R414 to the junction of the APC diodes D402 and D403. This signal at the junction of the capacitance-divider network is shifted 90° in phase by the network comprising C409 and C420 with coils L402 and L403. This signal then feeds through R415 to the junction of ACC diodes D404 and D405.

Observe that the chroma signal from the chroma (bandpass) amplifier is coupled via C401 to the base of Q401 in Fig. 8-8. Transistor Q401 operates as a gated burst amplifier with a positive-going gating pulse applied through R401. The gated-out pulse voltage drops across the primary of T401. Note that during conduction, R403 sets the correct operating point for the emitter of Q401, and C402 provides bypass action. Resistor R402 introduces emitter degeneration sufficient for operating stability and also fixes the stage gain. Cutoff bias for Q401 between successive bursts is supplied by the discharging of C402 through R403; that is, the transistor is reverse biased by signal-developed bias voltage. Diode D401 is included to prevent this reverse

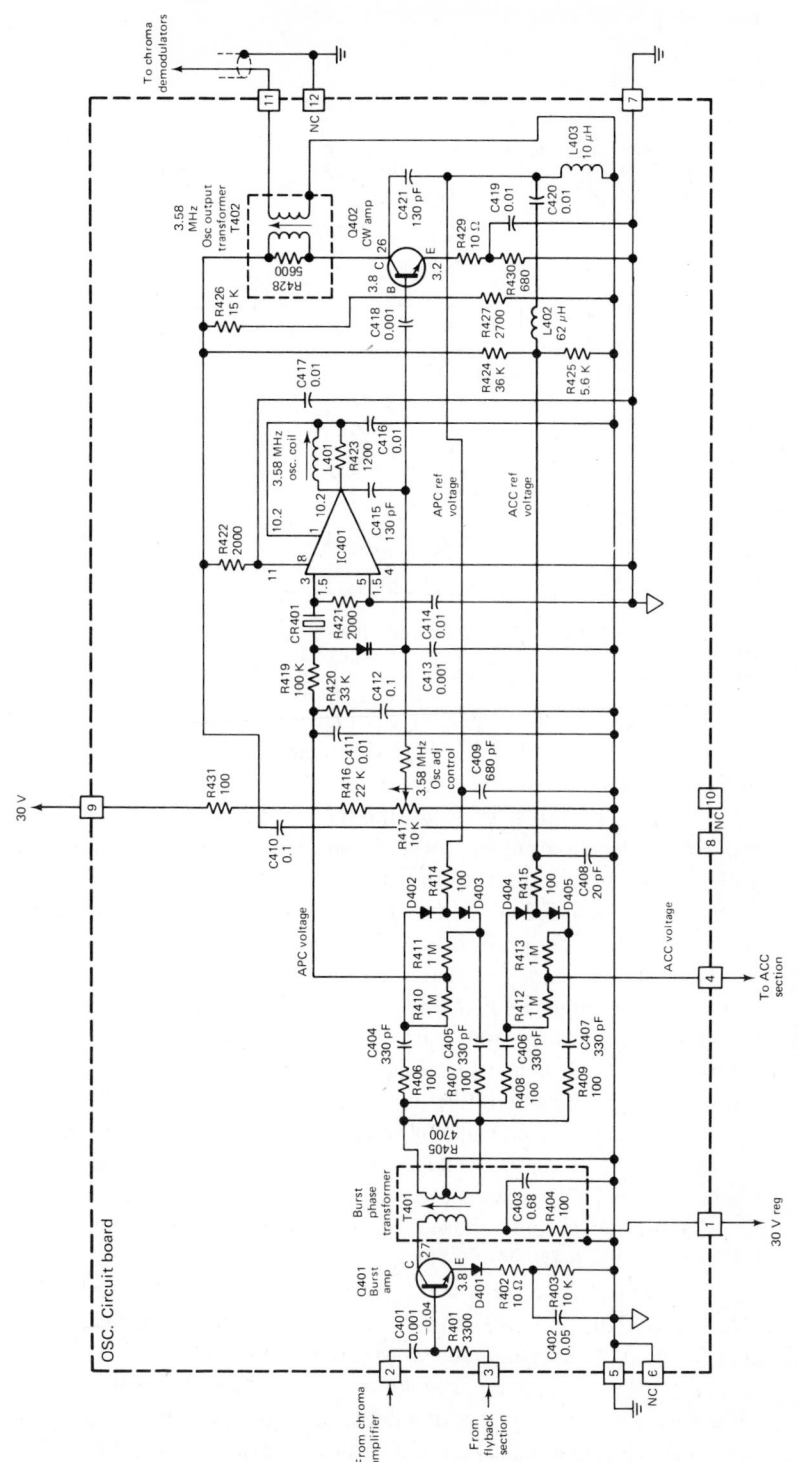

FIGURE 8-8 Configuration of a typical subcarrier-oscillator circuit board.

165

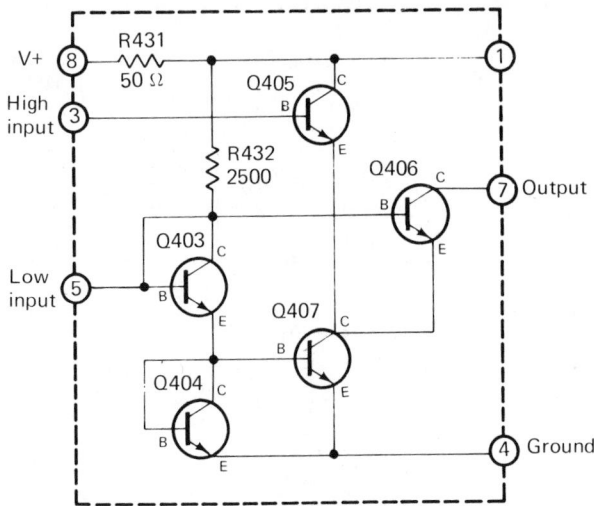

FIGURE 8-9   IC internal circuitry employed in Figure 8-8.

bias from exceeding the reverse emitter-base breakdown voltage of the transistor. Correct bandwidth for T401 is obtained by connection of the 4700-ohm resistor R405 across its secondary terminals.

A center tap is provided for the secondary of T401 in Fig. 8-8 so that a two-phase output is provided with equal 3.58-MHz voltages that are 180° out of phase with each other. One phase is fed through R406 and C404 to the anode of D402; the other phase is coupled through R407 and C405 to the cathode of D403. In turn, when the burst signal and the reference signal from the subcarrier oscillator are compared in the diodes, an output voltage will be produced if there is any difference in frequency or phase between the two signals. Any output voltage will appear at the junction of R410 and R411. If a control voltage is developed by the phase detector, its polarity will be either positive or negative, depending upon the sign of the phase or frequency error, while its amplitude will depend upon the magnitude of the error. This control voltage will be applied to the anode of varactor diode D406, which produces a corresponding frequency or phase correction in the subcarrier output voltage. The APC voltage will be filtered into steady DC by C411 and R420 with C412.

Next observe that the burst signal is also supplied through R408 and R409 with C406 and C407 to the ACC diodes D404 and D405. The operation of the ACC detector is the same as that of the APC detector, except that the 3.58-MHz reference input is shifted 90° in phase. This detector will produce a negative DC voltage at the junction of R412 and R413 with an amplitude that is proportional to the amplitude of the burst signal. A fixed DC bias potential, which is provided by R424 and R425, is applied also through D405 to this junction. The fixed DC voltage and the prevailing ACC voltage are coupled from terminal 4 on the circuit board to the chroma circuit board, as was explained previously.

It is helpful to study the features of the representative tube-type color-sync configuration shown in Fig. 8-10. A pentode burst-amplifier tube is utilized, followed by a shock-excited quartz-crystal section, and a pentode limiter output. In case of trouble symptoms, the tubes are checked or replaced first. However, if the malfunction persists, it is advisable to check waveforms through the network with an oscilloscope and low-capacitance probe. As an illustration, it might be observed that there is no color-burst input to the 3.58-MHz crystal. In such a case, it is indicated that the trouble will be found in the burst-amplifier section. Again, if there is no 3.58-MHz input to the limiter tube, it is indicated that there is a fault in the subcarrier-amplifier section. Or, in case there is no 3.58-MHz output from the limiter, it is indicated that there is a defect in the limiter section.

Following preliminary localization by means of waveform checks, the defective component can often be pinpointed by DC-voltage and resistance measurements. Receiver service data often includes resistance charts such as that shown in Fig. 8-11. These charts tabulate resistance values that are specified from various tube terminals to chassis ground. A resistance chart is of considerable assistance in evaluating in-circuit resistance measurements. From a statistical standpoint, electrolytic capacitors such as C20 in Fig. 8-10 are ready suspects. Otherwise, paper capacitors are probable troublemakers. In older sets, do not overlook the possibility of poor contacts between tube pins and socket spring. It is good practice to make voltage and resistance measurements from the tube pins—not the socket terminals.

8.5   SYMPTOMS OF COLOR-SYNC LOSS

Screen symptoms of color-sync loss are illustrated in Fig. 8-12. It follows from previous discussion that in case the color-subcarrier oscillator breaks color sync, the black-and-white sync action is not affected. Thus, the loss of color sync only produces the screen pattern illustrated in Fig. 8-12(a). In this example, a keyed-rainbow signal is being used to drive the color receiver. On the other hand, if the horizontal oscillator breaks sync, this loss of black-and-white sync action causes loss of color sync also. In other words, the burst-gating pulse becomes mistimed when the horizontal oscillator operates off-frequency, and concomitantly there is erratic output or no output from the burst amplifier. The screen symptom appears as shown in Fig. 8-12(b). It will now be instructive to consider what is to be learned from an out-of-sync screen pattern.

When horizontal-sync lock is lost, the picture breaks up into a group of diagonal lines. The slope of these diagonal lines indicates whether the horizontal oscillator is running too fast or too slow. That is, if the diagonal lines

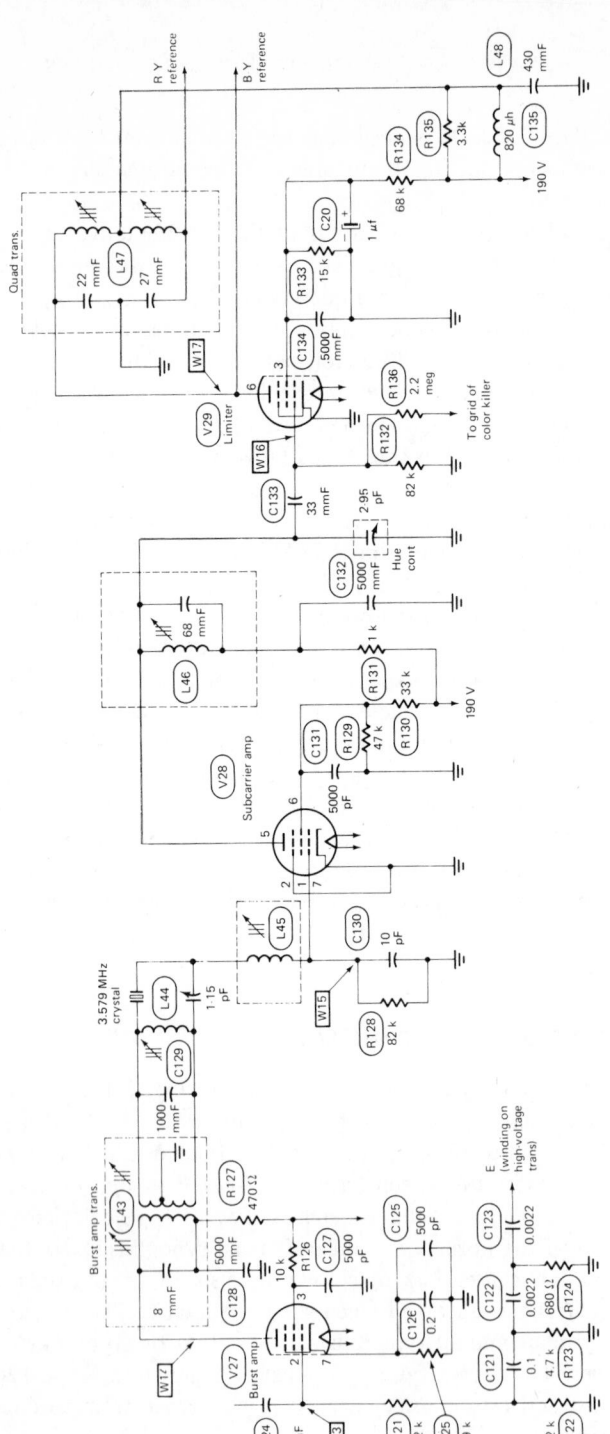

FIGURE 8-10 Representative tube-type color-sync configuration.

Resistance measurements

Item	Pin 1	Pin 2	Pin 3	Pin 4	Pin 5	Pin 6	Pin 7	Pin 8	Pin 9	Pin 10	Pin 11	Pin 12	Pin 13	Top cap
V41	FIL	4 meg†	NC	294 Ω*	NC	2.5 meg	2.5 meg	2200 Ω	750 Ω	800 K	0 Ω	FIL		
V42	10 K*	150 K	6800 Ω*	FIL	FIL	45 K*	27 Ω	750 Ω	1.8 meg					
V91	18 K*	18 K*	18 K*	FIL	FIL	1 meg	130 Ω	1 meg	1 meg					
V92	850 Ω	270 K	850 Ω	FIL	FIL	FIL	4500 Ω*	6800 Ω*	NC					
V200	FIL	0 Ω	4200 Ω	0 Ω	650 K	NC	NC	NC	650 K	0 Ω	4200 Ω	FIL		5 Ω†
V201	FIL	NC	NC	27 Ω*	NC	NC	800 K	NC	NC	27 Ω*	NC	FIL		
V202	Pins 1 thru 9 have infinite resistance													590 Ω†
V203	FIL	7500 Ω*	350 K	500 K†	500 K†	7500 Ω*	350 K	NC	45 meg	NC	7500 Ω*	350 K	500 K†	Pin 14 FIL

FIGURE 8-11   A typical resistance chart. *(courtesy of Howard W. Sams & Co., Inc.)*

slope downhill, the oscillator is operating at a frequency higher than 15,750 Hz. If, on the other hand, the diagonal lines slope uphill, the oscillator is operating at a frequency lower than 15,750 Hz. It is left as an exercise for the student to calculate the operating frequency of the oscillator from the number of diagonal lines that are displayed in the out-of-sync screen pattern. For example, two downhill diagonal lines are displayed in Fig. 8-12(b); in turn, we can state whether the oscillator is operating too fast or too slow, and its actual operating frequency.

> **FIGURE 8-12**   Screen symptoms of color-sync loss. (a) Loss of color sync only; (b) loss of both black-and-white and color sync. (See art on inside back cover.)

The same general analysis may be applied to the pattern shown in Fig. 8-13, to determine whether the color-subcarrier oscillator is operating above or below 3.58 MHz. In this situation, it is the sequence of the rainbow colors that indicates whether the oscillator is running too fast or too slow. As we look down one of the bars, we observe that the color sequence is green-blue-red. This means that the color-subcarrier oscillator is operating above 3.58 MHz. If we should observe a green-red-blue color sequence as we look down one of the bars, this would indicate that the color-subcarrier oscillator is operating below 3.58 MHz.

EXERCISES

Questions

1. What is the basic function of the *color sync system*?
2. How is color sync related to the frequency and phase of the subcarrier-oscillator output?
3. Explain the action of the *burst amplifier*.
4. State a typical bandwidth for a burst amplifier.
5. Why is the burst amplifier gated?
6. If the burst-gating pulse is partially mistimed, what trouble symptom results?
7. Describe the relation of burst-gate timing to the horizontal-hold control setting.
8. Name two basic types of *subcarrier-oscillator configurations*.
9. Why is a ringing crystal followed by a limiter?
10. Briefly discuss the chief subsections in an APC network.

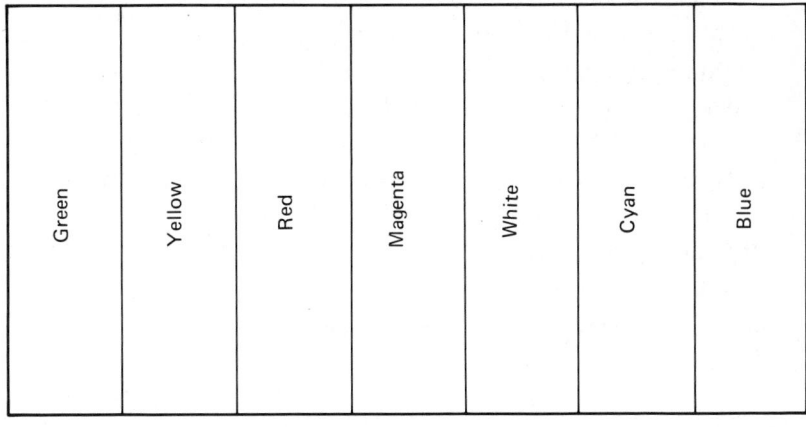

(a)

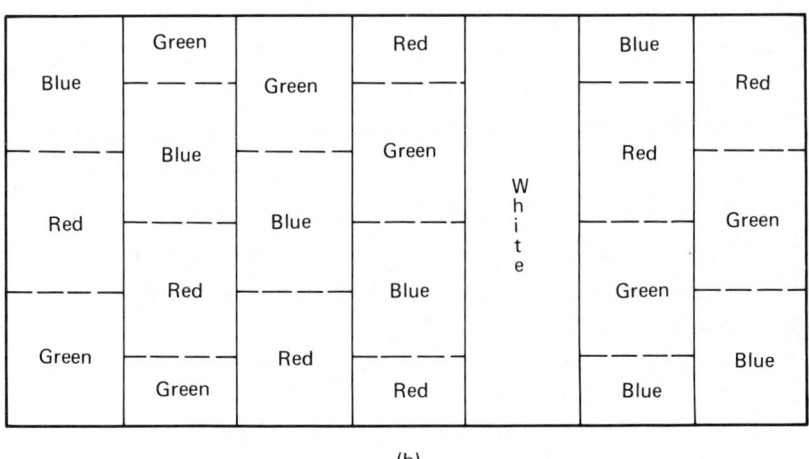

(b)

FIGURE 8-13   NTSC color-bar patterns. (a) Pattern in color sync;
(b) pattern out of color sync.

11. How does the sequence of rainbow colors in an out-of-sync color picture indicate the subcarrier-oscillator frequency?

12. Explain how a two-phase 3.58-MHz output is developed from the subcarrier oscillator.

13. What are the chief differences between tube-type and solid-state color-sync systems?

14. State two possible sources for the burst-amplifier input signal.

15. Would an oscilloscope or a signal generator be more useful in troubleshooting the color sync system?

16. How many cycles does a color burst contain?

17. Why is it undesirable to employ a very narrow burst-amplifier bandwidth?

18. If the burst amplifier has excessive bandwidth, what trouble symptom results?

19. Compare the operation of an ACC detector with that of an APC detector.

20. How can the subcarrier-oscillator frequency be measured?

True-False

1. Under some trouble conditions, color sync lock is maintained although horizontal sync lock is lost.

2. A color-sync system keeps the frequency and phase of the color burst the same as that of the subcarrier oscillator.

3. The burst amplifier is energized from the Y amplifier.

4. A color burst contains from 8 to 11 cycles of 3.58-MHz signal.

5. Burst amplifiers have a bandwidth of approximately 0.5 MHz.

6. Chroma amplifiers have less bandwidth than burst amplifiers.

7. Timing of the burst-gating pulse depends upon a normal setting of the vertical-hold control.

8. Both ringing-crystal subcarrier oscillators and APC-controlled free-running subcarrier oscillators are in extensive use.

9. Tube-type color-sync systems do not employ varactor diodes.

10. An APC circuit is essentially a phase detector.

11. The oscillating frequency of a quartz crystal depends to some extent upon the value of capacitance in shunt to the crystal.

12. Ceramic crystals are often used in subcarrier-oscillator circuits.

13. Some subcarrier-oscillator circuits employ integrated circuits instead of transistors.

14. Subcarrier-oscillator output transformers are tunable.

15. An electron tube can serve the same function as a varactor diode.

16. A subcarrier-oscillator tank coil is fixed-tuned.

17. Bias stability is provided by an unbypassed emitter resistor.

18. APC and ACC discriminators typically employ 3.58-MHz reference inputs that have a $90°$ phase difference.

19. AFC and APC are equivalent terms.
20. An ACC network is part of the AGC system.

Multiple Choice

1. Output from the video amplifier is called the _____ signal.
   (a) complete color
   (b) chroma
   (c) monochrome
   (d) luminance

2. A burst amplifier has _____ bandwidth, compared with a video amplifier.
   (a) more
   (b) less
   (c) the same
   (d) none of the above

3. The color-burst waveform has _____ cycles.
   (a) 3.58
   (b) 4.5
   (c) 15,750
   (d) 8 to 11

4. A burst-takeoff circuit is located between the _____ and _____ amplifiers.
   (a) video; chroma
   (b) video; burst
   (c) IF; video
   (d) video; Y

5. Typical frequency limits for the burst amplifier are _____ MHz.
   (a) 3.1 to 4.1
   (b) 2.1 to 4.1
   (c) 3.58 to 4.5
   (d) 3.35 to 3.90

6. Burst amplifiers are always _____.
   (a) AGC-controlled
   (b) ACC-controlled
   (c) APC-controlled
   (d) gated

7. A burst-gating pulse is _____than the color burst.
   (a) narrower
   (b) wider
   (c) higher in frequency
   (d) none of the above

8. Burst-gating pulses are derived from_____ .
   (a) vertical-flyback pulses
   (b) shock-excited quartz oscillators
   (c) the APC circuit
   (d) horizontal-flyback pulses

9. Timing of the burst-gating pulse depends upon the setting of the _____ control.
   (a) color-killer
   (b) contrast
   (c) horizontal-hold
   (d) vertical-hold

10. A color burst contains_____ .
    (a) significant sideband frequencies
    (b) no sideband frequencies
    (c) an encoded chroma signal
    (d) a decoded chroma signal

11. Automatic-phase-control discriminators develop _____ output voltage.
    (a) positive
    (b) negative
    (c) zero
    (d) either positive or negative

12. Automatic chroma-control discriminators develop _____ output voltage.
    (a) positive
    (b) negative
    (c) zero
    (d) either positive or negative

13. Typical ACC and APC circuits have 3.58-MHz reference inputs that differ_____ in phase.
    (a) $0°$
    (b) $45°$
    (c) $90°$
    (d) $180°$

14. Most subcarrier oscillators employ_____ .
    (a) LC tuned circuits
    (b) quartz crystals
    (c) Rochelle-salt crystals
    (d) ceramic crystals

15. APC and ACC discriminator circuits contain _____ .
    (a) power-type transistors
    (b) silicon controlled rectifiers

(c) thermistors

(d) small-signal diodes

16. Tube-type color-sync systems operate at _____ , compared to their solid-state counterparts.
    (a) low DC voltages
    (b) high DC voltages
    (c) the same DC voltages
    (d) none of the above

17. A varactor diode is always _____ .
    (a) reverse-biased
    (b) forward-biased
    (c) zero-biased
    (d) signal-developed bias-controlled

18. A bias-level control is _____provided with a varactor in a subcarrier-oscillator configuration.
    (a) always
    (b) never
    (c) occasionally
    (d) none of the above

19. The tank coil in a subcarrier-oscillator circuit resonates at the _____ frequency of the quartz crystal.
    (a) second harmonic
    (b) third harmonic
    (c) subharmonic
    (d) fundamental

20. Subcarrier-oscillator output transformers are _____ :
    (a) non-resonant
    (b) varactor-tuned
    (c) slug-tuned
    (d) untuned

Problems

1. A typical quartz crystal has a Q value of 8000. In a ringing circuit, the Q value is equal to $n\pi$, where n is the number of cycles over which the ringing waveform decays to 37% of its initial amplitude. If the ringing frequency is 3.58 MHz, how many cycles will pass before the waveform amplitude decays to 37%?

2. What is the period of the ringing interval calculated in problem 1?

3. An out-of-sync color bar has the downward sequence green-blue-red, and one rainbow is displayed. What is the subcarrier-oscillator frequency?

4. Another out-of-sync color bar has the downward sequence green-red-blue, and one rainbow is displayed. What is the subcarrier-oscillator frequency?

5. If an out-of-sync color bar has the downward sequence green-blue-red, and two rainbows are displayed, what is the subcarrier-oscillator frequency?

6. A 3.58-MHz series-resonant circuit has a Q value of 80. If the circuit is energized by a 0.5-volt sine-wave signal with a frequency of 3.58 MHz, what is the voltage drop across the coil?

7. What is the voltage drop across the capacitor in problem 6?

8. A parallel-resonant 3.58-MHz circuit has a 50-μH coil shunted by a 40-pF capacitor. What is the reactance of the coil?

9. What is the reactance of the capacitor in problem 8?

10. If the inductance and capacitance values in problem 8 are employed in a series-resonant circuit, what is the resonant frequency?

Chapter 9

Chroma Demodulation And Matrixing

9.1   CHROMA DEMODULATION REQUIREMENTS

Chroma demodulation is one of the basic processes involved in the decoding of the complete color signal. A typical chroma-demodulator section reconstitutes the chroma signal by inserting the 3.58-MHz subcarrier and processes the reconstituted signal by developing its R—Y and B—Y components with a pair of synchronous detectors. A *synchronous detector*, also called a *product detector*, is a combined phase and amplitude detector. An R—Y signal or a B—Y signal has a certain phase with respect to burst and also varies in amplitude. As noted previously, chroma phase corresponds to hue, and chroma amplitude corresponds to saturation. A synchronous detector may also develop the G—Y component of the reconstituted chroma signal. Still other demodulation axes are employed in various receiver designs, as will be explained later.

Figure 9-1 depicts the plan of an R—Y/B—Y demodulator section. Note that the chroma signal from the bandpass amplifier is applied to both of the demodulators. Also, the output from the subcarrier oscillator is applied to the B—Y demodulator in the B—Y phase and to the R—Y demodulator in

177

the R—Y phase. The chroma signal is reconstituted and its R—Y and B—Y components are developed. This quadrature synchronous-detector action is shown in Fig. 9-2. Note that each demodulator conducts only for a brief interval at the peak of the injected 3.58-MHz voltage. This is due to the fact that synchronous detection entails a sampling of the chroma signal in a stipulated phase, such as the R—Y phase. Also, the amplitude of successive samples will vary in accordance with the varying amplitudes of the given chroma-signal component. (See Fig. 9-3.)

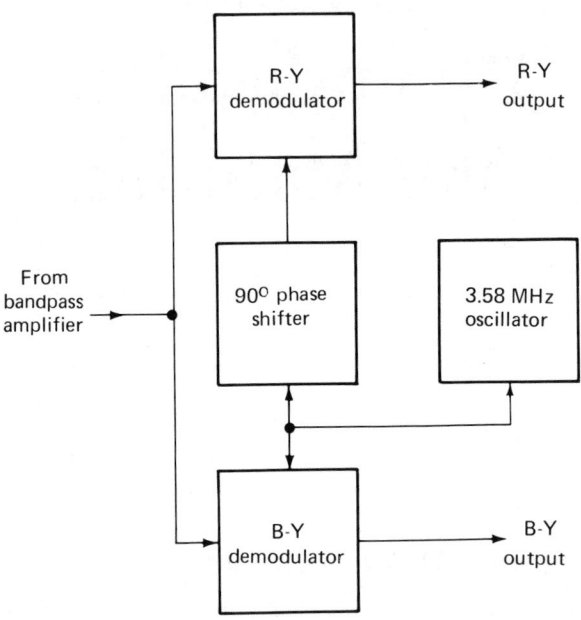

FIGURE 9-1   Plan of an R—Y/B—Y demodulator section.

As the R—Y chroma-signal component is going through its peak value, the B—Y component is going through zero. Therefore, the R—Y demodulator output consists of samples of the R—Y component only. As the B—Y chroma-signal component is going through its peak value, the R—Y component is going through zero. Accordingly, the B—Y demodulator output consists of samples of the B—Y component only. Observe that —(R—Y) and —(B—Y) signal components are sampled in a similar manner. Thus the R—Y and B—Y components are separated from the incoming chroma signal in this part of the decoding process. As noted previously, a G—Y component must also be developed or separated from the incoming chroma signal. In many color receivers, this G—Y component is demodulated directly and thereby separated from the chroma signal in much the same manner as are the R—Y and B—Y components.

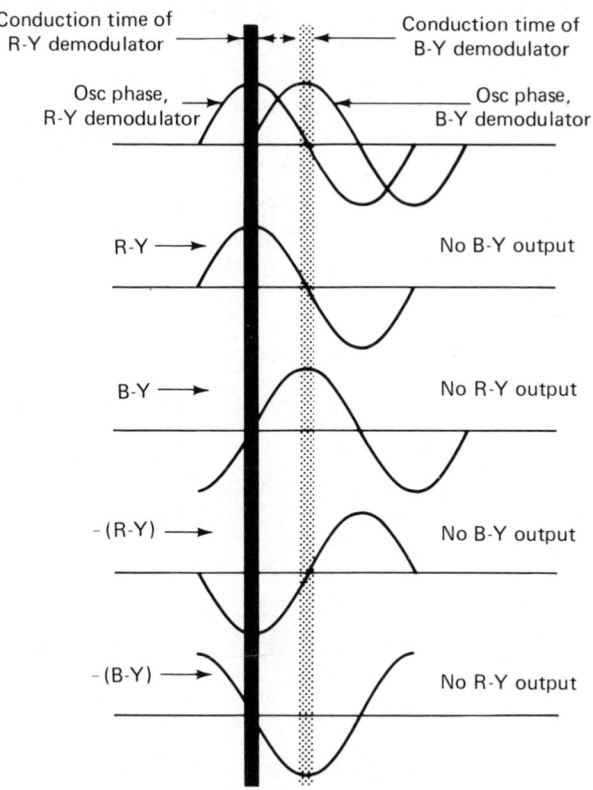

FIGURE 9-2   Quadrature synchronous-detector action.

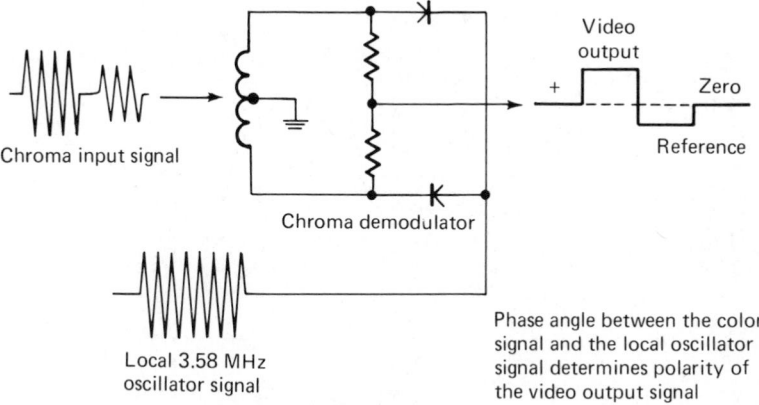

FIGURE 9-3   Basic synchronous-detector action.

Figure 9-4 depicts the plan of a G—Y demodulator. Observe that this is not an instance of quadrature demodulation, and the analysis of demodulator action is somewhat different from that of the foregoing example. The subcarrier injection phase is such that neither the —(R—Y) nor the —(B—Y) signal is going through zero at the time of sampling. Accordingly, both signals are sampled simultaneously. The G—Y demodulator output consists of samples of the G—Y signal because a G—Y vector has components on the R—Y and B—Y axes, as is evident from the phase diagram depicted in Fig. 9-4(a). When the subcarrier is injected into a synchronous detector with the G—Y phase, the required amounts of —(R—Y) and —(B—Y) signals are combined to produce a G—Y signal output.

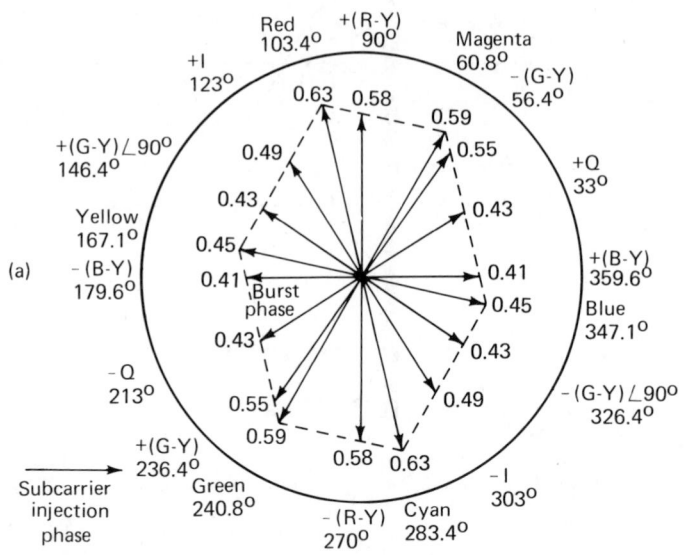

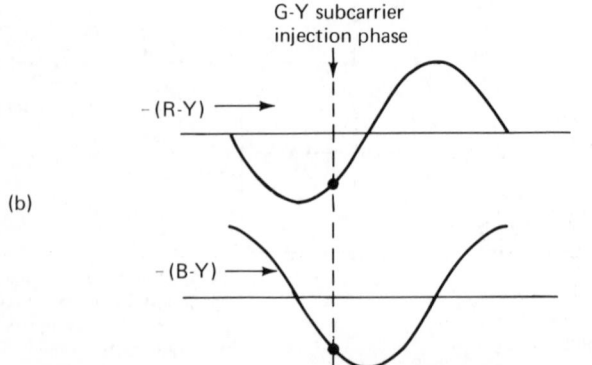

FIGURE 9-4   G—Y synchronous-detector action. (a) Phase of injected subcarrier; (b) simultaneous R—Y and B—Y sampling process.

9.2   CHROMA MATRIXING REQUIREMENTS

Instead of decoding the G–Y signal in the demodulation process, the outputs from the R–Y and B–Y demodulators can be combined in proportions and polarity suitable to the formation of the G–Y signal as depicted in Fig. 9-5. This method of chroma-signal recovery is called a *matrixing process*. In this example, the G–Y matrix inverts the polarity of the R–Y signal and of the B–Y signal. Then, these signals are combined in the proportion of 0.51 (R–Y) and 0.19 (B–Y). The end result is the same as if the –(R–Y) and –(B–Y) signals were sampled in a G–Y synchronous detector.

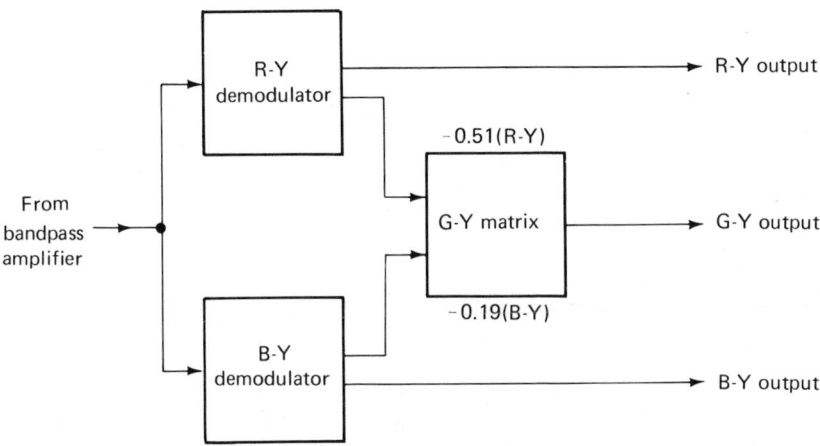

FIGURE 9-5   Arrangement of a G–Y matrix.

Other chroma matrix arrangements are also employed. For example, R–Y and G–Y may be demodulated, and B–Y may be matrixed. Some older designs demodulated I and Q signals, from which R, G, and B signals were matrixed. Although this arrangement has become obsolete, in a widely used variation R, G, and B chroma signals are simultaneously demodulated and matrixed with the Y signal. This process will be discussed in detail subsequently. Still another widely used method demodulates X and Z signals, from which R–Y, B–Y and G–Y signals are matrixed. The terms X and Z represent somewhat arbitrary chroma phases that are defined by the individual manufacturer. Most X and Z demodulator configurations employ demodulation axes that are 105° apart, as illustrated in Fig. 9-6. The outputs from the X and Z demodulators are matrixed to form R–Y, B–Y, and G–Y signals.

Note that R is the matrix resistor in Fig. 9-6(b). In the absence of R, the R–Y matrix would provide an X output signal, the B–Y matrix would provide a Z output signal, and there would be no output from the G–Y matrix. However, with R connected into the ground-return circuits of the R–Y and B–Y matrices, both the X and Z signals are dropped across R.

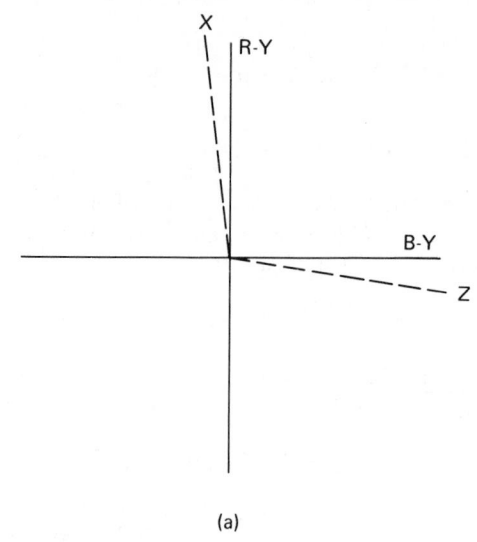

(a)

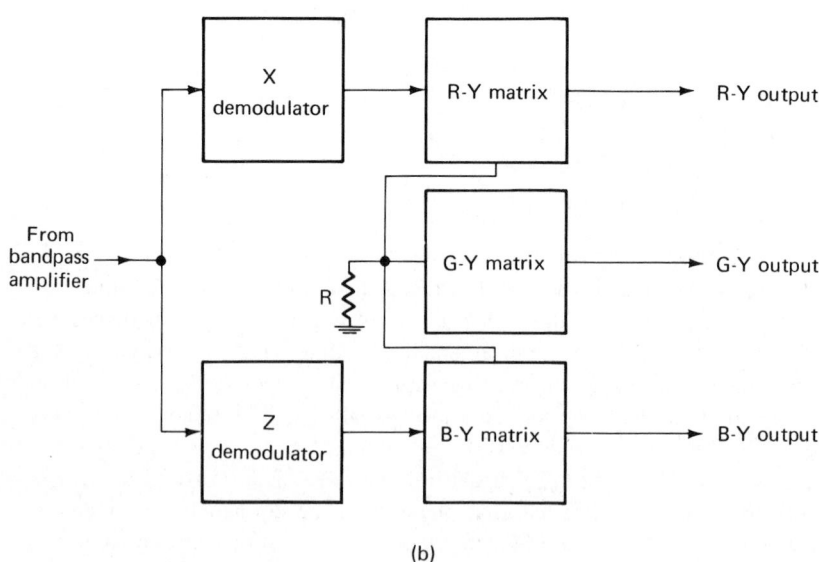

(b)

FIGURE 9-6

The two signals are combined in suitable proportions to form R–Y and B–Y output signals. This is also the proportion that results in formation of a –(G–Y) signal across R. This –(G–Y) signal is inverted by the G–Y matrix section, and a G–Y output signal is provided. The chief advantage of X and Z demodulation systems is manufacturing economy.

Regardless of the demodulation phases that are employed in various color receivers, all chroma demodulators can be classified as either low-level or high-level types. A *low-level demodulator* is followed by active devices to step up the signal amplitude before it is applied to the color picture tube. A *high-level demodulator* energizes the picture tube directly. Modern design practice generally favors low-level or medium-level demodulation.

There are also two basic modes of color picture tube operation. The most widely-used method employs the picture tube as an RGB matrix as well as a display device. However, there is a marked trend toward the employment of external RGB matrixing which utilizes the picture tube only as a display device. It is instructive to observe the matrixing principles that are involved in these two approaches.

Figure 9-7 depicts the utilization of a color picture tube as an RGB matrix. Observe that the Y signal is applied to all three cathodes, and that R–Y, B–Y, and G–Y signal voltages are applied to the individual grids. In case a red hue is to be displayed at full saturation and brightness, only the red gun is operative. Thus, the combined outputs from the Y amplifier and the B–Y demodulator cut off the blue gun. Similarly, the combined outputs from the Y amplifier and the G–Y matrix cut off the green gun. Thus, the color picture tube provides the final red, green, and blue signal voltages by functioning as an RGB matrix.

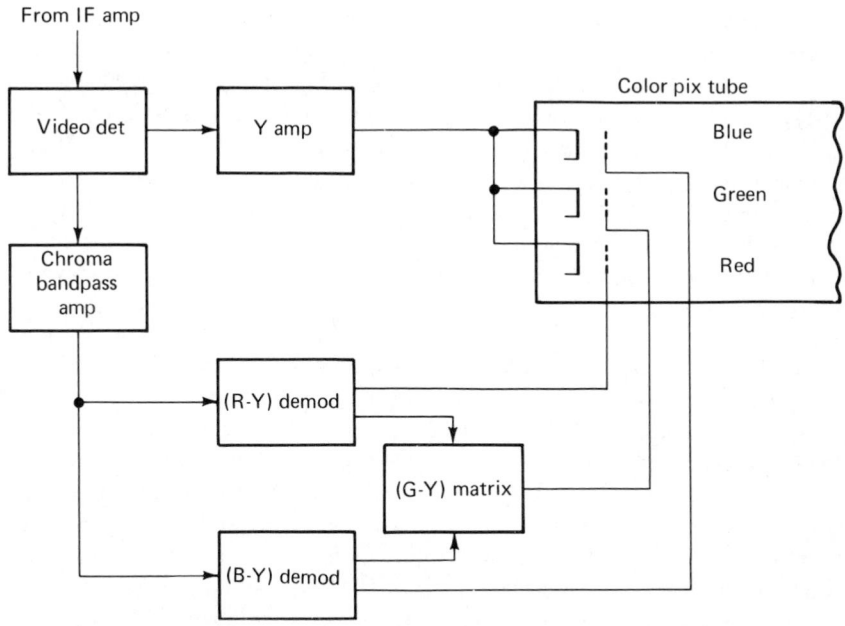

FIGURE 9-7   Color picture tube operation as an RGB matrix.

Next consider the operation of an external RGB matrix arrangement in which simultaneous demodulation and matrixing are employed, as shown in Fig. 9-8. In this arrangement, the demodulator outputs are applied to the individual cathodes of the color picture tube, and the grids are not driven. Note that the Y signal is combined or matrixed with the chroma signal prior to demodulation. Thus, each demodulator processes both the chroma signal and the Y signal. In turn, the demodulator phases are on the primary color axes instead of on the color-difference axes. These demodulator phases are depicted in Fig. 9-8(a). Note that the diodes in the green-demodulator circuit are polarized in a sequence opposite to that of the diodes in the red-demodulator and blue-demodulator circuits. This is a design factor that is employed to automatically cancel out spurious demodulation products called "blips." As would be anticipated, the subcarrier injected into the green-demodulator circuit necessarily has the magenta, or −G phase.

Observe in Fig. 9-8 that the phase of the subcarrier voltage from the 3.58-MHz oscillator can be manually shifted by adjustment of the hue control. This is an operating control that is adjusted by the viewer to provide optimum reproduction of flesh tones. It is necessary to readjust the setting of the hue control from time to time because of various residual chroma phase errors that are present in the color-TV system. For example, the receiving antenna often has a slightly different phase characteristic from one channel to another, or the color-TV transmission system may introduce a small phase shift in the chroma signal with respect to the burst signal. Anomalies in propagation that result in close-in color ghosts can also introduce phase errors. Receiver components may drift slightly in characteristics as the receiver warms up or may acquire a permanent shift as components age. Therefore, manual control of the reproduced flesh tone hue is required.

As Fig. 9-8 shows, the color intensity control determines the amount of chroma signal that is applied to the color demodulators. This is also an operating control that is adjusted by the viewer to provide normal color saturation in the image. The brightness of the color image is determined by the setting of the contrast and brightness controls, as in a black-and-white receiver.

A tube-type chroma-demodulator configuration is shown in Fig. 9-9. The DC voltages are much higher than in corresponding solid-state configurations. However, the circuit actions are essentially the same. In case of circuit malfunction, tubes are checked or replaced at the outset. Then, if the trouble symptom is not eliminated, it is good practice to check the output waveforms from the X and Z demodulators. This test can be helpful in preliminary trouble localization. If a keyed-rainbow generator input signal is employed, the normal output waveforms from the X and Z chroma demodulators appear as those depicted in Fig. 9-10. Note that an output waveform might have correct waveshape, with subnormal amplitude. Therefore, the

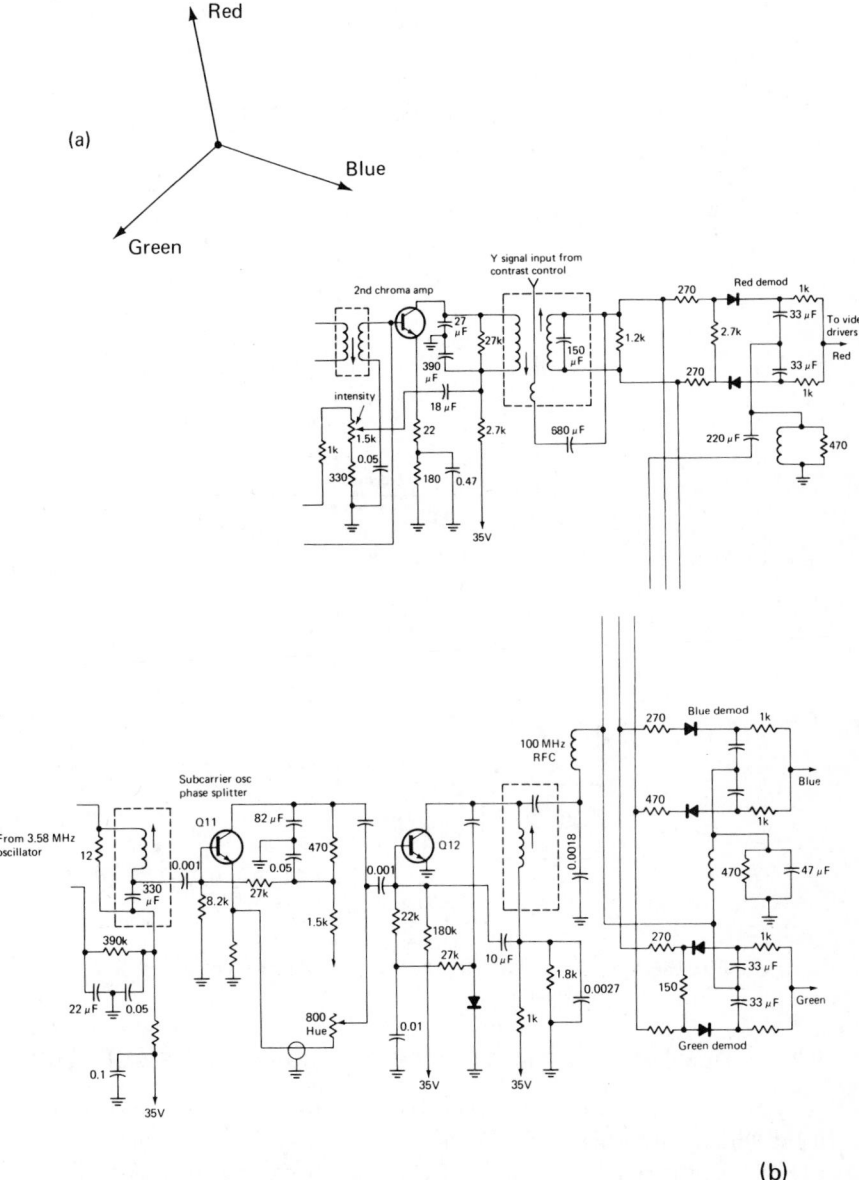

FIGURE 9-8   Simultaneous RGB demodulation and matrixing
arrangement. (a) Demodulator phases; (b) typical
configuration.

peak-to-peak voltage of the displayed waveform should be checked against
the specified value in the receiver service data.

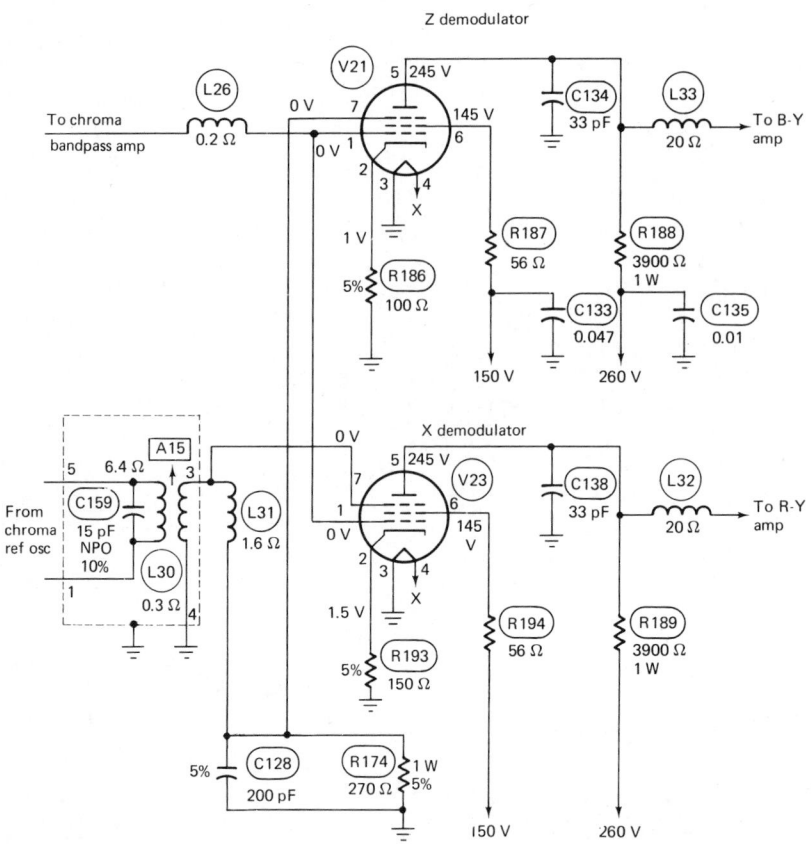

FIGURE 9-9   A widely used tube-type chroma-demodulator configuration.

9.3   DEMODULATOR-MATRIX, CIRCUIT-BOARD CONFIGURATION

A typical *demodulator-matrix, circuit-board* configuration is shown in Fig. 9-11. Observe that the 3.58-MHz subcarrier voltage from terminal 2 on the circuit board is coupled to the phase-shift network comprising L301, R301, C301, and the tint (hue) control (not shown). The phase of the injected subcarrier at the junction of L301 and C301 can be shifted approximately 70° by the viewer. The subcarrier is coupled via C302 to terminal 4 of the demodulator integrated circuit IC301. Next, the phase of the subcarrier is shifted 103° by L302, R302, and C304. This subcarrier phase is coupled by C303 to terminal 5 of IC301. Note that the chroma signal is available at pin 1 on the circuit board and is coupled via C305 to terminal 3 on IC301.

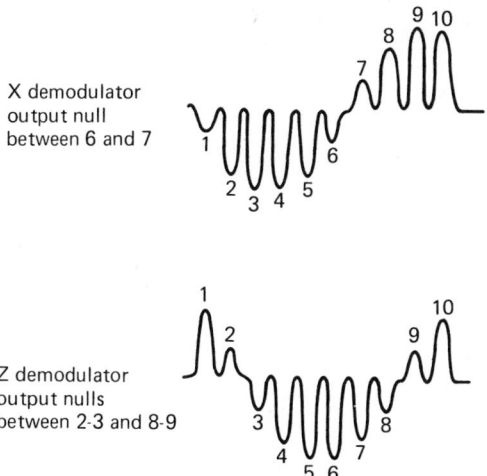

X demodulator
output null
between 6 and 7

Z demodulator
output nulls
between 2-3 and 8-9

FIGURE 9-10   Normal output waveforms from the X and Z
chroma demodulators.

As Fig. 9-12 shows, this integrated circuit contains two doubly-balanced synchronous detectors. That is, the IC configuration is balanced with respect to both of the input signals—the subcarrier signal and the chroma signal. The subcarrier cancels itself out in the demodulators and does not feed through into the R—Y, B—Y, and G—Y output circuits. This eliminates the need for including low-pass filters in the output leads. Note in Fig. 9-12 that the chroma signal at terminal 3 of the IC is connected to the parallel inputs of Q316 and Q318. Opposite-phase chroma signals from the differential amplifier, Q315 and Q316, are fed to transistor switches, Q307, Q308, Q309, and Q310. Opposite-phase chroma signals from the differential amplifier, Q317 and Q318, are fed to transistor switches Q311, Q312, Q313, and Q314. These switches are controlled by two signals: the subcarrier signal at pin 4, which is also applied to the bases of Q311 and Q314; and the second subcarrier signal at pin 5, which is also applied to the bases of Q307 and Q310. Note that the chroma signal currents from each differential amplifier will flow into one of the two outputs, depending on the instantaneous state of the individual switches.

Observe in Fig. 9-12 that the outputs from each of the two synchronous detectors are fed to a resistive chroma matrix consisting of resistors R318 through R326, where the three color-difference signals are formed. A regulated 6-volt supply for the switching transistors is provided by Zener diode ZD302 and transistor Q306. This supply voltage flows through a resistive divider and is applied to the base of Q305 to establish a 3-volt supply for the differential amplifiers. Transistors Q320 and Q321, which are controlled by the diode-connected transistor Q319, are regulated current sources for the differential amplifiers. Finally, the R—Y, B—Y, and G—Y output voltages

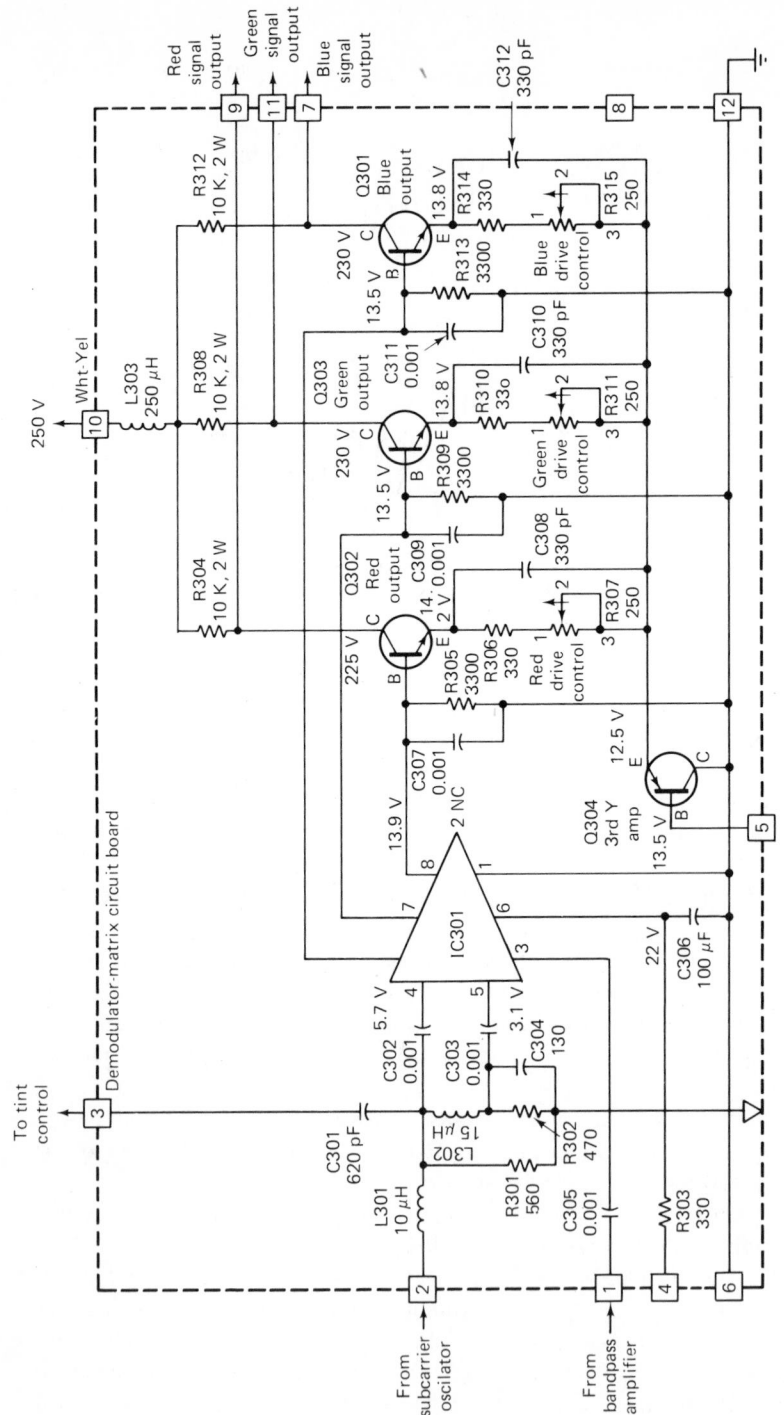

FIGURE 9-11 Configuration of a demodulator-matrix circuit board.

188

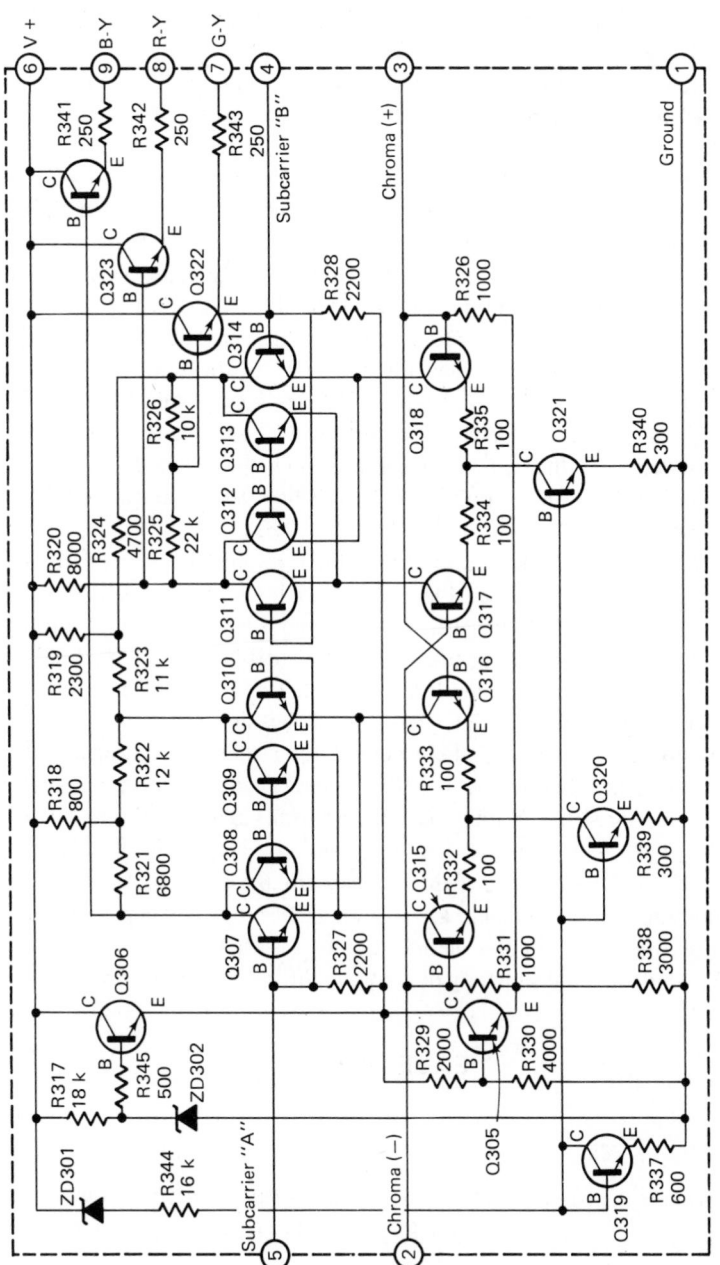

FIGURE 9-12 IC pair of doubly-balanced synchronous detectors.

189

from the chroma matrix are fed through the emitter-follower resistors Q322, Q323, and Q324. Each emitter lead contains a 250-ohm resistor, which protects the IC against damage from accidental short-circuits.

Referring to Fig. 9-11, note that DC bias voltage and the color signal for the red output amplifier, Q302, are developed across R305, which is bypassed by C307. Similarly, DC bias and the color signal for the green output amplifier, Q303, are developed across R309, which is bypassed by C309. Also, DC bias and the color signal for the blue output amplifier, Q301, are developed across R313, which is bypassed by C311. Carefully note that the Y signal is injected into the emitter circuits of the color output amplifiers from the third Y amplifier, Q304. Thus, the color output amplifiers operate as an RGB matrix. The red, green, and blue signal outputs are applied respectively to the red, green, and blue cathodes of the color picture tube. Observe that post-demodulator RGB matrixing is employed in Fig. 9-11, whereas pre-demodulator matrixing is utilized in Fig. 9-8. It is due to this distinction between the two decoding processes that demodulation takes place along the R—Y, B—Y, and G—Y axes in Fig. 9-11, whereas demodulation proceeds along the R, G, and B axes in Fig. 9-8.

Observe the chief features of the tube-type chroma-demodulator and matrix configuration shown in Fig. 9-13. Duo-diode chroma demodulators apply R—Y and B—Y output signals to triode amplifiers, followed by a triode G—Y matrix tube. The matrix tube combines suitable portions of the R—Y and B—Y signals to reconstitute the G—Y signal. This arrangement is found in many of the older tube-type color receivers. Tubes are suspected first in the event of malfunction, but if the troubleshooter must "dig deeper," it is advisable to make oscilloscope waveform checks with an incoming keyed-rainbow color-bar signal. Normal output waveforms appear as those shown in Fig. 9-14. Note that both the waveshape and the amplitude of an output waveform should be checked in preliminary trouble analysis. Normal waveform amplitudes are specified in the receiver service data and depend to some extent upon the type of color picture tube that is used. Figure 9-15 provides a helpful tabulation and summary of color-circuit troubleshooting procedures.

EXERCISES

Questions

1. What is the basic function of the *chroma-demodulation section*?

2. How do chroma demodulators separate R—Y, B—Y, and G—Y signals?

3. Explain how the G—Y signal can be matrixed from R—Y and B—Y signals.

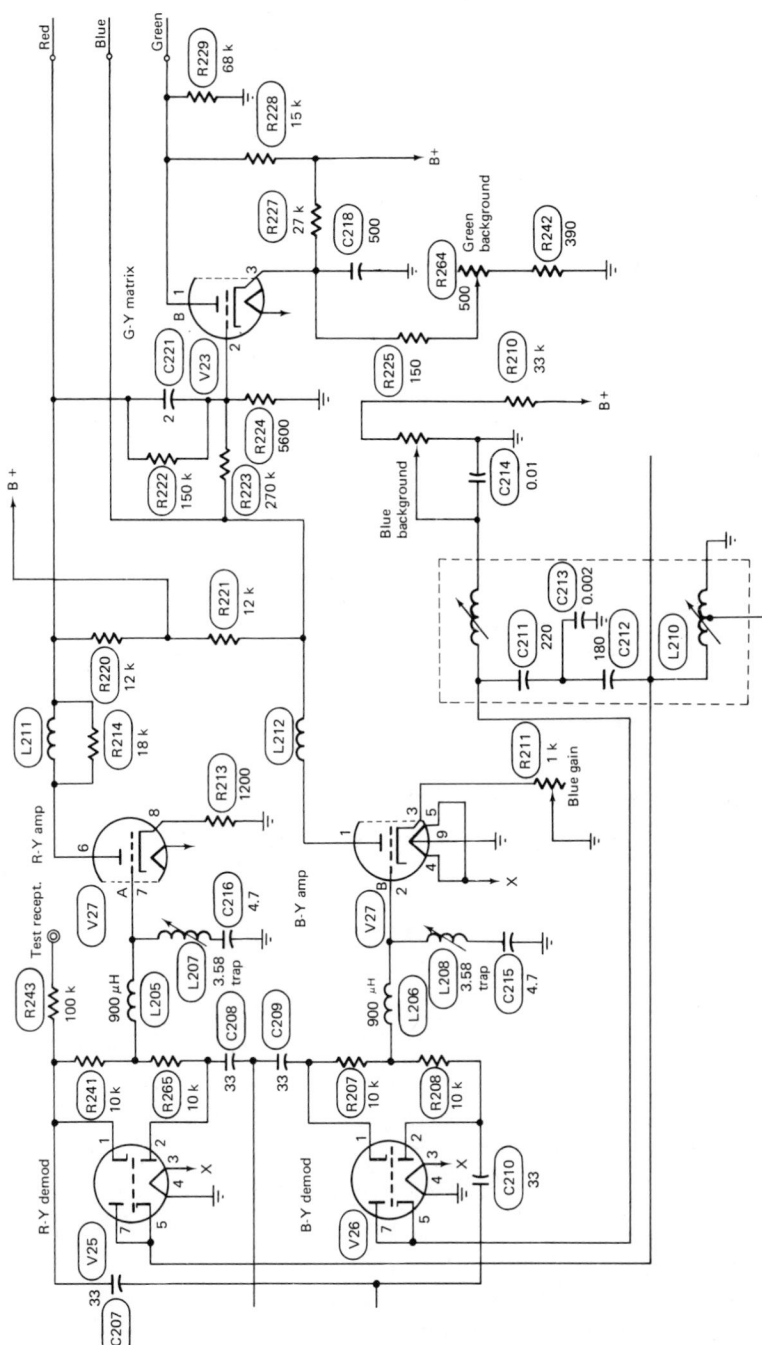

FIGURE 9-13 A tube-type chroma demodulator and matrix configuration.

191

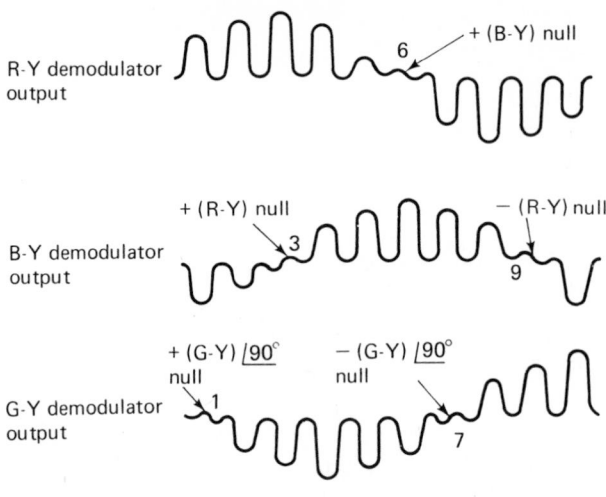

R-Y demodulator output

B-Y demodulator output

G-Y demodulator output

FIGURE 9-14   Normal output waveforms from the R—Y, B—Y and G—Y chroma demodulators.

4. Can R, G, and B signals be matrixed from I and Q signals?

5. Why is X/Z demodulation utilized extensively?

6. Briefly describe how a color picture tube is employed as an RGB matrix.

7. Discuss the essentials of simultaneous RGB demodulation and matrixing action.

8. When malfunction occurs in a tube-type chroma-demodulation and matrix system, what components or devices does the technician check first?

9. Are integrated circuits utilized as chroma-demodulator devices?

10. Is it feasible to demodulate R—Y and G—Y signals, and to matrix the B—Y signal?

11. Why are low-pass filters included in chroma-demodulator output circuits?

12. Explain the chief distinction between demodulation of the G—Y signal and matrixing of the G—Y signal.

13. Are the G—Y and the G demodulation axes identical?

14. Is there any technical difference between a synchronous detector and a product detector?

15. Briefly describe the relation of chroma phase to hue, and of chroma amplitude to saturation.

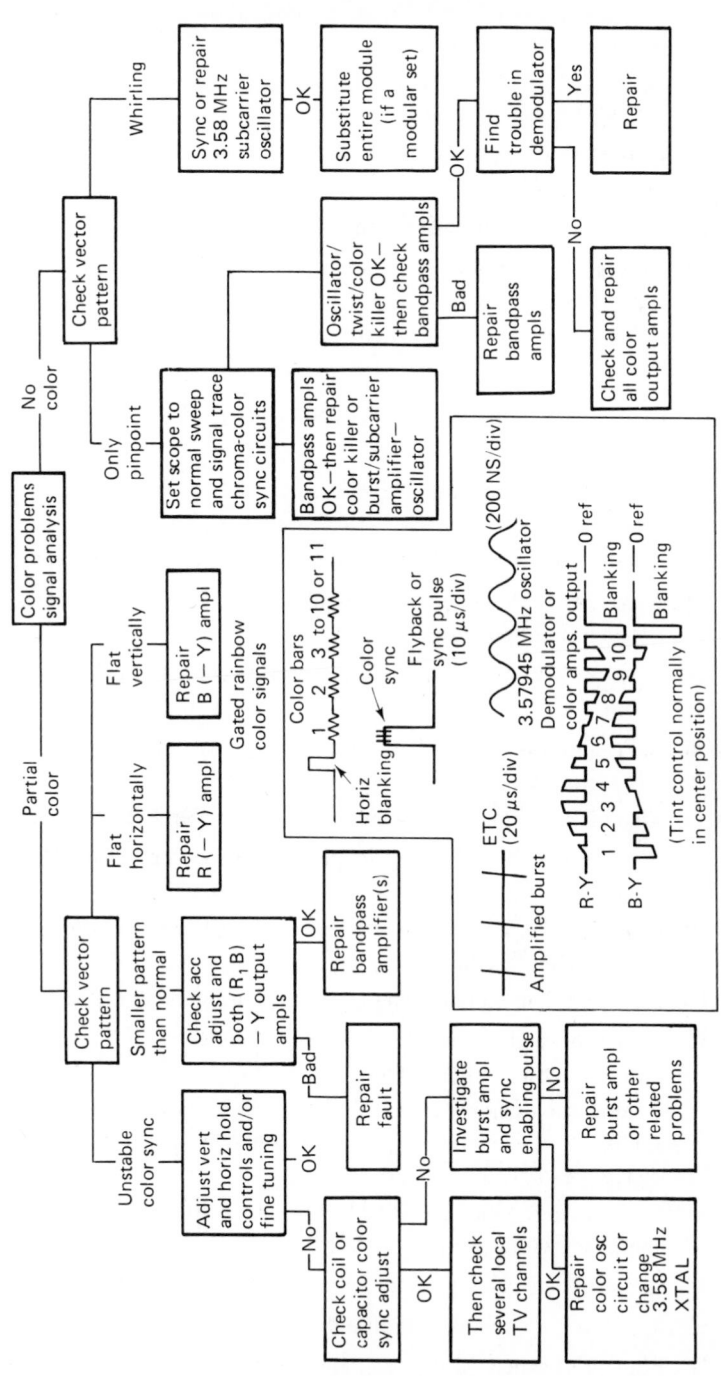

FIGURE 9-15 Color-circuit troubleshooting procedures. *(courtesy of Radio-Electronics)*

193

16. Does the output from a chroma demodulator always have the same polarity?

17. Does the output from a chroma matrix always have the same polarity?

18. Does the output from an R demodulator-matrix always have the same polarity?

19. Do X and Z signals correspond to certain hues?

20. Distinguish between low-level chroma demodulation and high-level chroma demodulation.

True-False

1. Chroma demodulation involves both phase and amplitude detection.

2. Reconstitution of the chroma signal takes place in a chroma demodulator.

3. Synchronous detection and product detection are equivalent terms.

4. When the G−Y signal is demodulated, R−Y and B−Y signals are sampled simultaneously.

5. A G−Y demodulator employs the same configuration as a G−Y matrix.

6. R−Y and B−Y demodulators utilize typical subcarrier reference phases separated by $62°$.

7. X and Z demodulators use quadrature subcarrier reference phases.

8. R−Y and R demodulators employ the same subcarrier reference phase.

9. Low-level demodulators are followed by active devices to step up the signal amplitude.

10. High-level demodulators energize the picture tube directly.

11. Tube-type and solid-state chroma demodulators have completely different basic configurations.

12. To operate a color picture tube as an RGB matrix, the Y signal is applied to the picture-tube cathodes and R−Y, B−Y, and G−Y signals are applied to the grids.

13. Many modern color receivers have external RGB matrices that are combined with color demodulators.

14. A color demodulator is another name for a chroma demodulator.

15. Spurious color-demodulation products are called *blips.*

16. Blips are cancelled by suitable polarizations of the chroma-demodulator diodes.

17. LCR phase-splitter circuits are employed to develop $-Y$ signals from Y signals.

18. Tube-type chroma demodulator-matrix systems include one or more integrated circuits.

19. Traps are utilized following chroma demodulators to eliminate feedthrough subcarrier signals.

20. Post-demodulator RGB matrixing and pre-demodulator RGB matrixing are equivalent terms.

Multiple Choice

1. Chroma demodulation is a basic process in _____ .
   (a) decoding the chroma signal
   (b) regenerating the color subcarrier
   (c) deemphasizing the color burst
   (d) delaying the complete color signal

2. A chroma signal is reconstituted by _____ .
   (a) trapping out the Y signal
   (b) insertion of the color subcarrier
   (c) elimination of the color burst
   (d) inverting the Y signal

3. If the color subcarrier is applied to an $R-Y$ demodulator in the $B-Y$ phase, the output from the $R-Y$ demodulator is a/an _____ .
   (a) $R-Y$ signal
   (b) DC level
   (c) $B-Y$ signal
   (d) matrixed $R-Y/B-Y$ signal

4. Polarity of the output signal from a chroma demodulator depends upon the _____ of the input signal.
   (a) amplitude
   (b) DC component
   (c) power factor
   (d) phase

5. Amplitude of the output from a chroma demodulator depends upon the _____ of the input signal.
   (a) phase
   (b) power factor
   (c) DC component
   (d) amplitude

6. If a $G-Y$ demodulator develops a $B-Y$ output, the malfunction is a result of _____ .
   (a) a phase error in the subcarrier reference voltage
   (b) an amplitude error in the chroma signal

(c) incomplete trapping of the Y signal

(d) incomplete trapping of the sound signal

7. An R—Y demodulator is often followed by a _____ .

(a) B—Y demodulator

(b) G—Y demodulator

(c) G—Y matrix

(d) bandpass amplifier

8. R—Y, B—Y, and G—Y signals are often matrixed from_____ signals.

(a) X and Z

(b) X and Y

(c) Z and Y

(d) Y and —Y

9. Low-level chroma demodulators are followed by _____ .

(a) high-level demodulators

(b) delay lines

(c) LCR phase splitters

(d) active devices

10. High-level demodulators are followed by _____ .

(a) a color picture tube

(b) synchronous demodulators

(c) product detectors

(d) LCR phase splitters

11. The magenta reference phase is the same as the _____ phase.

(a) R—Y

(b) —G

(c) I—Y

(d) Q—Y

12. A chroma-demodulator IC typically contains _____ transistors.

(a) 3

(b) 13

(c) 20

(d) 30

13. Discrete-component chroma-demodulator arrangements generally employ _____ .

(a) semiconductor diodes

(b) thermistors

(c) electron multipliers

(d) phase-lag networks

14. A diode-type chroma demodulator has a basic configuration similar to a _____ .
    (a) ratio detector
    (b) video detector
    (c) radiation detector
    (d) none of the above

15. The color picture tube is driven by _____ .
    (a) unadjusted chroma values
    (b) readjusted chroma values
    (c) AM and FM signals
    (d) AM and PM signals

Problems

1. A chroma-demodulator output circuit has a typical bandwidth of 0.5 MHz. What are the frequency limits of the output circuit?

2. A chroma-demodulator input circuit has a typical bandwidth of 1 MHz. What are the frequency limits of the input circuit?

3. A chroma-matrix circuit has typical input and output bandwidths of 0.5 MHz. What are the frequency limits of the input and output circuits?

4. By what factors must the outputs from the R–Y and B–Y demodulators be multiplied to drive the color picture tube with unadjusted chroma values?

5. When two amplifiers, for example, are connected in series, the total rise time is equal to the square root of the sum of the squares of the individual rise times. If the bandpass amplifier has a rise time of 0.33 μs, and a chroma demodulator has a rise time of 0.67 μs, what is their total rise time?

6. Subcarrier suppression results in doubling the envelope frequency. When the incoming chroma signal has an envelope frequency of 0.25 MHz, calculate the signal frequency at the chroma-demodulator output.

7. Subcarrier suppression in a 100% modulated chroma signal has the effect of dividing the waveform amplitude in half. What is the corresponding reduction in chroma-signal power?

8. If an oscilloscope has a sweep speed of 1 microsecond per cm, how many cycles of subcarrier voltage will be displayed per cm?

9. If an oscilloscope has a sweep speed of 1 millisecond per cm, how many cycles of subcarrier voltage will be displayed per cm?

10. How many cycles will be displayed in problem 9 if the oscilloscope is switched for X10 horizontal expansion?

Chapter 10

Static And Dynamic Convergence

10.1   CONVERGENCE REQUIREMENTS

As noted previously, a shadow-mask type of color picture tube has the basic construction depicted in Fig. 10-1. Each electron beam must be precisely directed so that it strikes a corresponding phosphor dot, as demonstrated in Fig. 10-2. A color picture tube that operates in this manner is said to be *converged*. This relationship must be maintained at all points on the viewing screen. After the picture tube has been converged for correct color reproduction at the center of the screen, additional convergence procedures are required to obtain correct color reproduction at the top and bottom, and at the sides of the screen. The position of the *deflection yoke* (Fig. 10-3) is quite critical, and the yoke must be positioned to obtain satisfactory color purity. This means that when only one electron gun is operating, a pure color results. For example, when only the red gun operates, a pure red color field is displayed. Figure 10-4 illustrates impure and pure red fields.

As the electron beams are deflected toward the outer edges of the screen, the distance from the electron guns to the screen becomes greater than it is while the central region is being scanned. This varying distance must

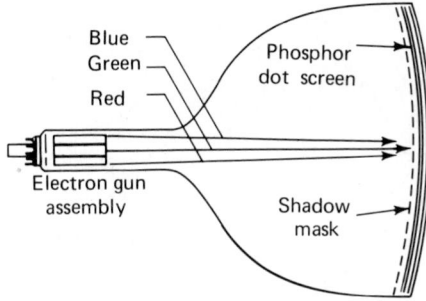

FIGURE 10-1   Basic construction of a shadow-mask type of color picture tube.

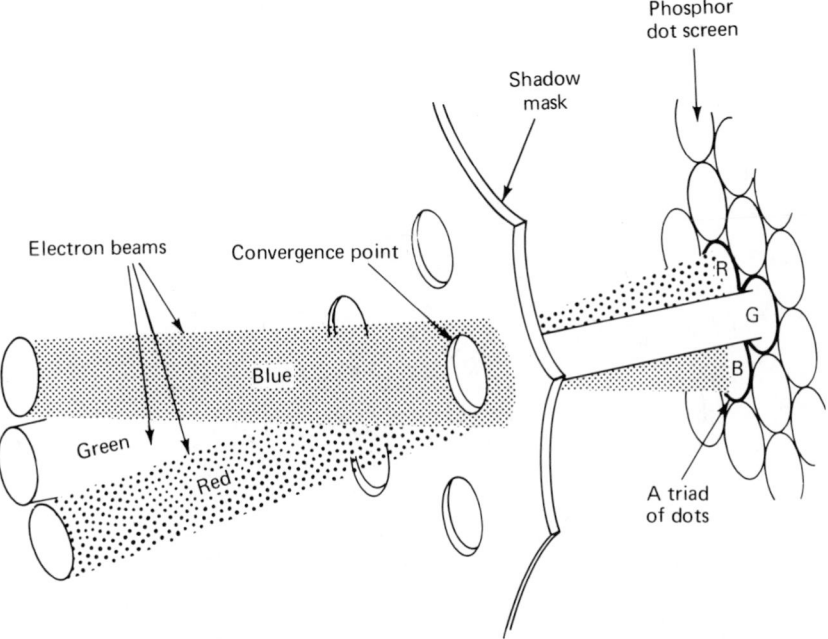

FIGURE 10-2   Each electron beam must move in a precise direction.

be progressively compensated for as an electron beam moves away from center screen. Center-screen convergence is determined by the adjustment of permanent magnets mounted over the electron guns. These magnets are called *static-convergence controls*. Edge-screen convergence is controlled by horizontal and vertical electromagnets mounted over the electron guns. These are called *dynamic-convergence controls*. Figure 10-5 illustrates pole-piece assemblies.

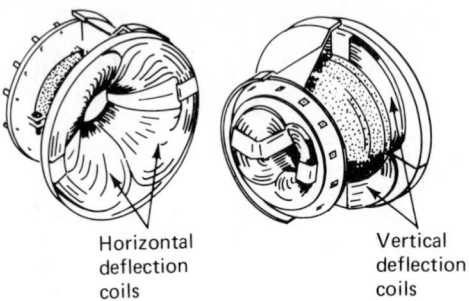

Horizontal
deflection
coils

Vertical
deflection
coils

FIGURE 10-3   Arrangement of the deflection coils in the yoke.

FIGURE 10-4   Effect of purity adjustment. (a) Impure red field;
(b) pure red field. (See art inside back cover.)

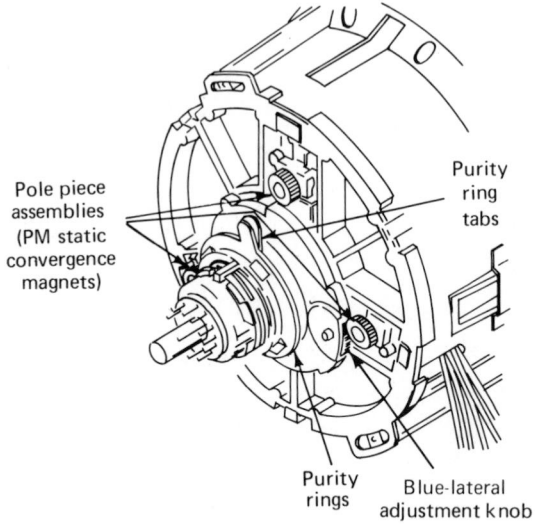

Pole piece
assemblies
(PM static
convergence
magnets)

Purity
ring
tabs

Purity
rings

Blue-lateral
adjustment knob

FIGURE 10-5   Pole-piece locations in the convergence assembly.

Current waveforms in step with the horizontal and vertical sweep voltages are passed through the horizontal and vertical dynamic-convergence coils. These electromagnetic fields correct the beam directions before they are deflected by the yoke. Dynamic-convergence waveforms necessarily have a critical amplitude, shape, and phase. Figure 10-6 shows conditions of horizontal misconvergence and vertical misconvergence. Observe that before corrective dynamic-convergence fields are applied, the amount of misconvergence increases progressively from center screen toward the edges of the picture tube. Basically, the corrective current waveform that is required has the shape of a parabola.

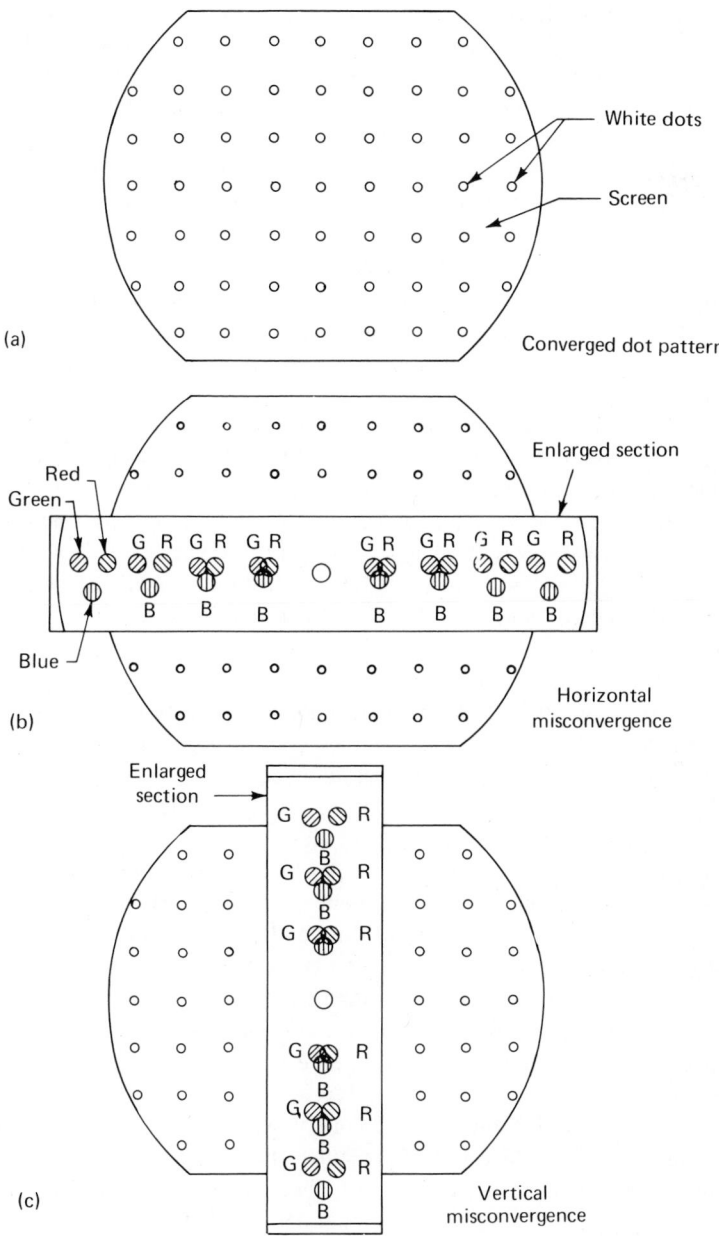

FIGURE 10-6   Examples of horizontal and vertical misconvergence.

Vertical parabolic current waveforms are obtained by processing the vertical sawtooth deflection voltage. Partial integration of this sawtooth waveform produces an approximation of a parabolic waveform. A sectional schematic diagram is shown in Fig. 10-7. Currents flowing through C652, R651, differential resistor R652, the red and green vertical-convergence coils L804 and L805, diode D653, and to the B+ supply converge at the upper half of the raster. As taught in black and white television courses, a *raster* is a display of horizontal scanning lines on the picture tube screen. Note that capacitor C652 and resistors R654 and R655 operate in a waveshaping circuit that produces the required convergence waveform. Current flowing through resistor R654, differential resistor R653, the convergence coils, amplitude resistor R658, shaping network diodes D651 and D652 and their associated capacitor and resistor C651 and R656 converge at the lower half of the raster. This second circuit operates during the time that diode D653 is not conducting.

Observe that the amplitude controls R654 and R658 in Fig. 10-7 control the amount of correction in convergence of the vertical lines, whereas the differential controls R652 and R653 govern the amount of correction in vertical convergence of the horizontal lines at the top and bottom of the raster. Figure 10-8 shows the basic convergence circuit for vertical convergence of the blue horizontal lines. Controls R659 and R660 affect convergence at the bottom and top halves of the raster. This circuit is energized from the same source utilized by the red and green convergence circuits.

Next consider the plan of the horizontal convergence circuitry. Figure 10-9 shows a basic schematic diagram of this section. Flyback pulses are fed to coil L651, controls R667 and R668, and capacitors C659 and C660 to form sawtooth waveforms, which are applied to the red and green convergence coils. In turn, integrating action produces parabolic current waveforms through the coils. Note that clamping diodes D655 and D656, and the associated resistors rectify a portion of the parabolic current waveform, thereby adding a DC component to the convergence waveform. This DC component is required to ensure that the current through the convergence coils is zero as the scanning beam passes through the center of the screen. Observe that coil L654 and control R667 provide a difference in current flow through the red and green convergence coils. This differential current flow corrects unsymmetrical errors in horizontal-line convergence on the left and right sides of the screen.

Figure 10-11 shows a basic configuration for the blue horizontal-convergence circuits, which operate in much the same manner as the red and green horizontal-convergence circuits described above. However, an additional wave-shaping network consisting of coil L653, capacitor C657, and resistor R663 is included. This waveshaping circuit adds a small amount of second-harmonic sine-wave current to the blue convergence-coil current. Therefore, the blue horizontal-convergence waveform is optimized. Details of receiver set-up procedure will be discussed subsequently.

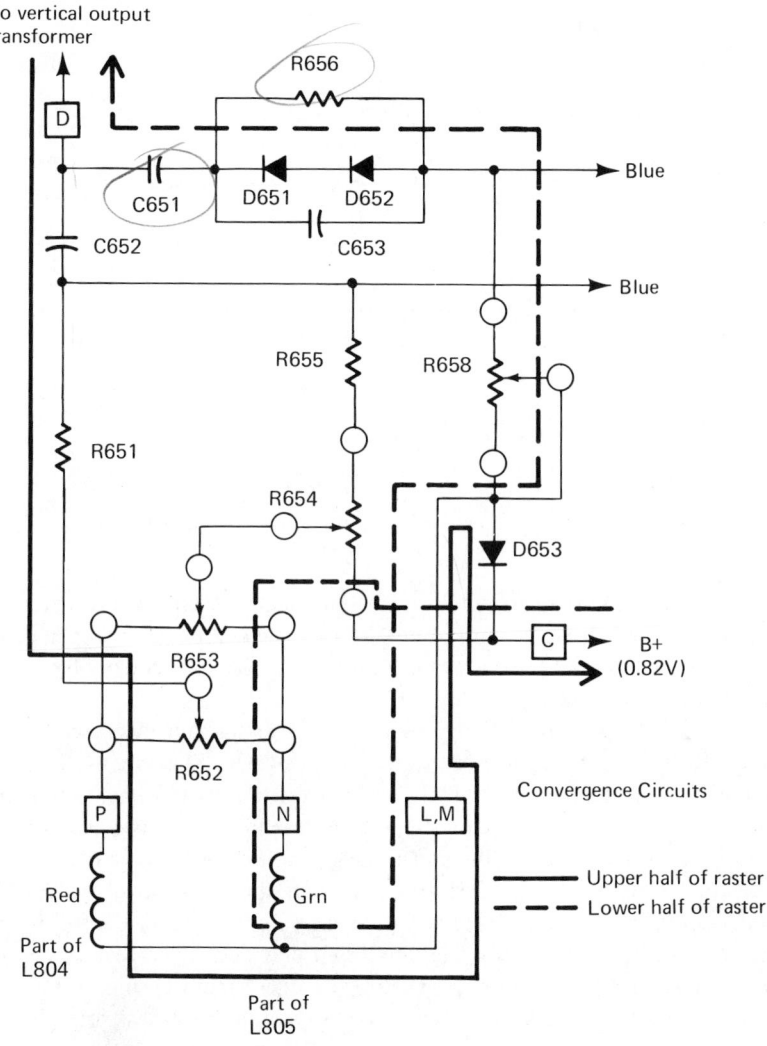

FIGURE 10-7   Vertical parabolic waveforms for convergence are produced by integrating a sawtooth waveform.

A complete convergence circuit-board configuration is shown in Fig. 10-10. This assembly employs the partial circuit diagrams, with the same component identifications, as those assemblies that were previously discussed. Note that thirteen maintenance controls are provided. Nine of these are resistive controls; four are inductive controls. As anticipated, there is considerable interaction among the convergence controls because the fields from the three pole pieces affect the directions of all three electron beams to a

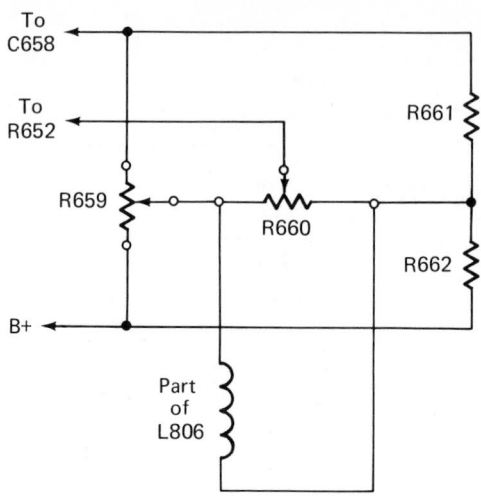

FIGURE 10-8   Basic convergence circuit for vertical convergence
of blue horizontal lines.

greater or lesser extent. In practice, this interaction results in a relatively involved convergence set-up procedure; an overview will be given in Chap. 16.
Skill and speed in convergence of a color picture tube are acquired by a combination of study and experience.

In the example of Fig. 10-10, six other maintenance controls are
mounted on the convergence panel assembly. However, these are not included
in the convergence circuitry. Thus, the dot control operates in conjunction
with the horizontal sweep circuit to generate white-dot pulses that are utilized in convergence procedures. The height control is part of the vertical-
deflection system; the AGC control is part of the gain-control system; the
tone control is part of the sound system; and the color-killer control is part
of the chroma system, as has been discussed previously. The peaking control
provides an adjustable load resistance for the video amplifier so that the high-
frequency response can be trimmed as required during alignment procedures.

10.2   TROUBLESHOOTING INCORRECT COLOR DISPLAY

Incorrect color displays can be grouped into two categories. The first is comprised of displays with no normal color in the picture, although there are
colors on the screen with blob or bar forms. The second category is comprised of displays with color in the correct locations, but with incorrect hues.
A subgroup consists of color-tinted screens. Note that the second group of
incorrect color displays usually originates from a defective color picture tube,
although an incorrect setting of a screen control can also be responsible. For

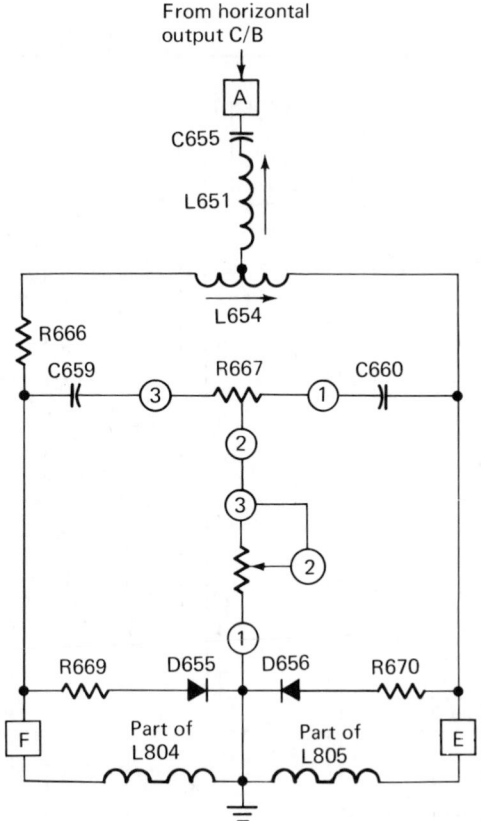

FIGURE 10-9   Plan of the horizontal-convergence circuitry.

example, in a faulty color picture tube, the screen could display a bright green hue for five minutes after the receiver is turned on and then proceed to display normal screen colors thereafter. In this situation, the red and blue guns in the color picture were slow to develop normal emission, whereas the green gun was operating normally. A similar trouble symptom is sometimes caused by a heater-cathode short-circuit in one of the guns; however, this fault is likely to produce a permanent trouble symptom. It can often be corrected by installation of an isolation-type "brightener."

Sometimes the screen-circuit control for one of the guns drifts out of normal range. In a typical case history, the green raster could not be extinguished at any setting of the green screen control. In this situation, it may be possible to compensate for the incorrect green range by setting the red and blue screen voltages somewhat higher. However, even though essentially normal color reproduction is obtained by means of this expedient, it is good

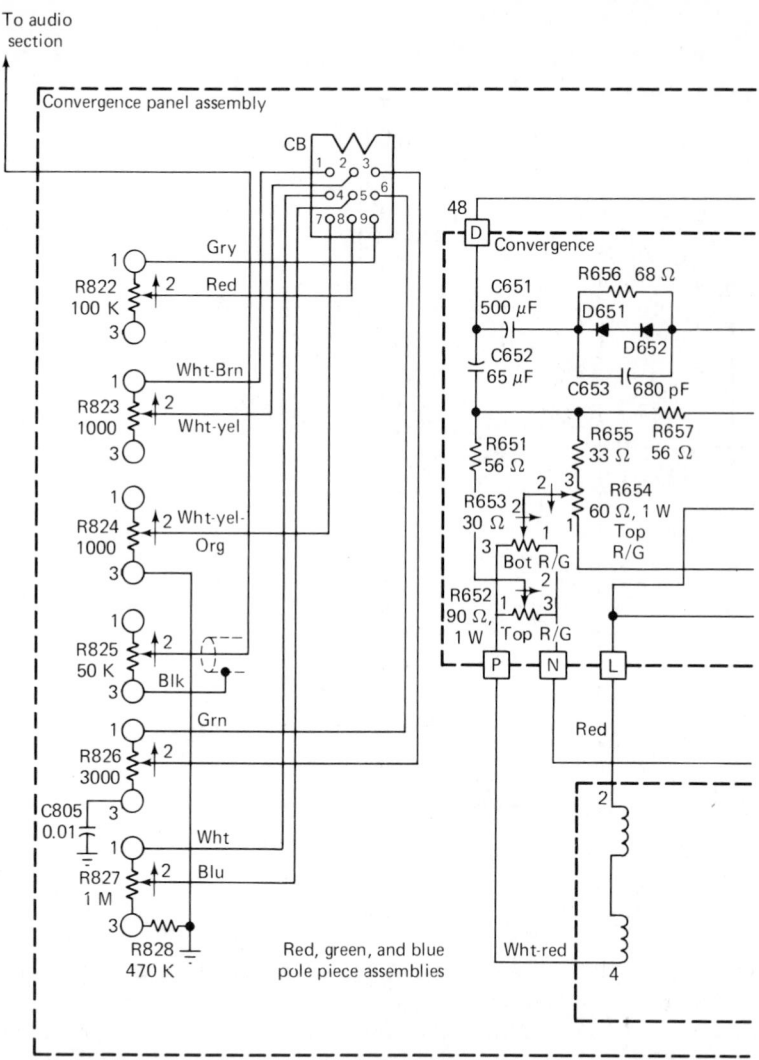

FIGURE 10-10   Configuration for a convergence circuit board.

practice to find out why the screen range is incorrect and to replace the defective component. We will find that most of the incorrect color-reproduction problems originate in 3.58-MHz oscillator phasing, distortion in the chroma demodulators, and miscellaneous defects in components associated with the chroma demodulators.

Intermittent color reproduction can be caused by faulty operation in the horizontal-oscillator/AFC output section. Trouble in this circuitry can change the amplitude and shape of the keying pulses utilized in the chroma section. In addition to waveshape and amplitude, the phasing of the keying

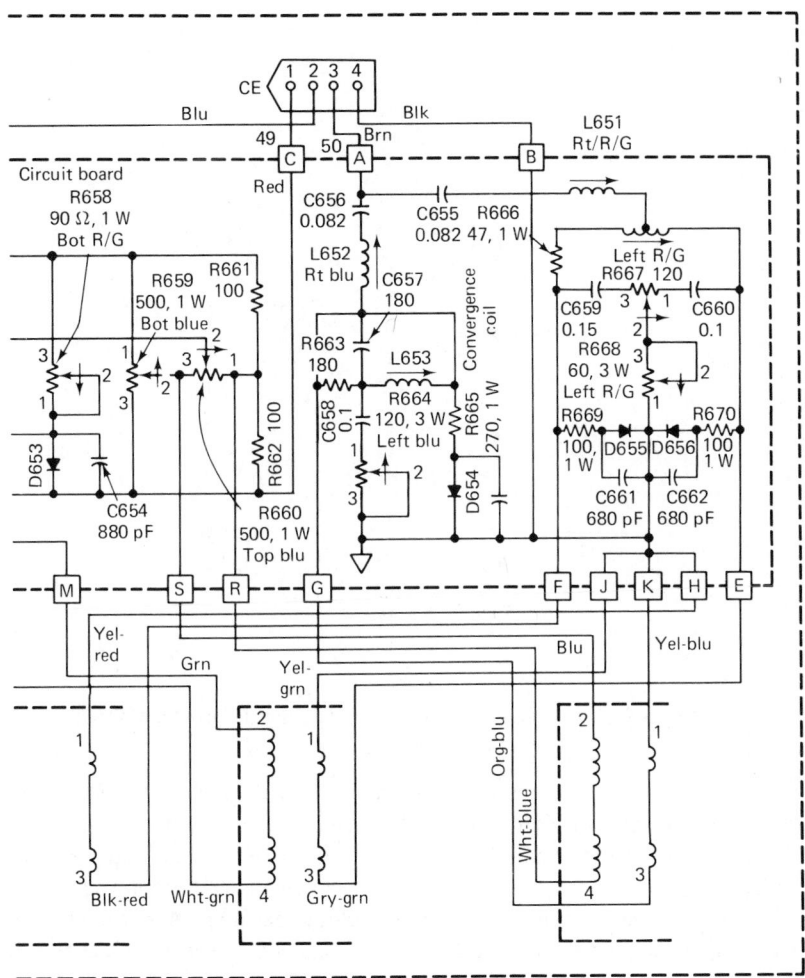

FIGURE 10-10   (Continued.)

pulse is a very important requirement. Although the picture display appears to be reasonably stable horizontally, the keying pulse can be sufficiently mis-phased to disturb the burst-amplifier operation. In turn, a weak or intermit-tent burst output results, with the trouble symptom of "colors jumping in and out." It is advisable, therefore, to check the setting of the horizontal-hold control first. Incorrect setting of the fine-tuning control can cause a similar trouble symptom. If the technician must "dig deeper," the AFT sec-tion should be checked. Incorrect AFT alignment or a leaky transistor can cause drift in the local-oscillator frequency.

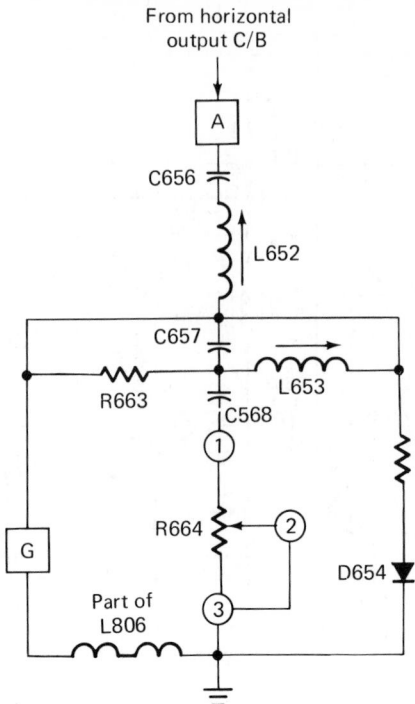

FIGURE 10-11   Basic circuitry for the blue horizontal-convergence section.

Sometimes the color display is correct, with a tendency for the color to "jump in and out" very suddenly. In such a case, there is likely to be a fault in the bandpass-amplifier section. Make DC-voltage and resistance measurements, replace suspected open-circuited capacitors and check the alignment of the bandpass-tuned circuits. Some baffling color trouble symptoms can be caused by a defective tube or semiconductor diode in the color AFC/ACC section. The diodes must be matched precisely to operate correctly. Many of the older-model color receivers utilize a 6JU8 quadruple diode tube. In case this tube develops interelectrode leakage, puzzling color trouble symptoms are produced. Of course, the primary rule in servicing tube-type receivers is to "replace tubes first."

Defects in the Y amplifier or video amplifier can simulate, in some cases, chroma trouble symptoms. This situation is more likely to be encountered in hybrid receivers that utilize both tubes and transistors. For example, if a transistor develops a short-circuit between its input and output elements, the signal will feed through without phase reversal. Negative-going signal voltages become positive-going voltages, and vice versa. The result is a negative-picture display, in which white areas are black and black areas are white. When this malfunction occurs in a color picture display, the symptom may be difficult to interpret unless the color-intensity control is turned to minimum. With the chroma signal absent, the negative picture display becomes obvious. In addition to defective transistors, this trouble symptom can also be caused by certain kinds of short-circuits between printed-circuit board conductors. As a practical note, a video-amplifier transistor is occasionally reversed when it is replaced in its socket. This error can also produce a negative picture display.

It is instructive to consider a typical troubleshooting situation in which the trouble symptoms are a blue-tinted screen displaying two vertical color bars with red and green hues approximately 1½ inches wide at the left hand side of the screen. None of the hues in the image is correct, in this instance. The trouble condition is also intermittent. A black-and-white image is displayed normally except for a sepia-tinted screen. The first conclusion in this situation is that the Y channel is operating normally. Next, with the color-intensity control turned to maximum, the blue-tinted screen becomes a bright blue and the vertical color bars are also much more intense. Defocusing is also apparent in this display. The brightness and contrast controls operate normally. Tube replacement does not affect the trouble symptoms.

Next, DC-voltage measurements provide some preliminary trouble clues. Most of the values are out of tolerance, with some too high and others too low. Also, the DC voltages on the phase-detector diodes and the color-killer diodes are out of balance; this imbalance accounts in part for the trouble symptoms. It is verified that the 3.58-MHz oscillator is operating. Output waveforms from the R–Y demodulator are subnormal in amplitude, although the B–Y demodulator output waveforms are practically normal. Fine-tuning action is normal, and it is apparent that the bandpass amplifier is operating. In turn, suspicion falls in the area around the 3.58-MHz oscillator. The large unbalance in control voltages possibly indicates that although the oscillator is operating, it is being pulled so far off normal operating frequency that normal colors cannot be developed, and no colors can be developed at times.

Proceeding to disable the burst amplifier, the technician then determines that the subcarrier oscillator is operating normally. This finding throws suspicion onto the burst amplifier. A systematic component checkout discloses that the screen-bypass capacitor at the burst-amplifier tube (Fig. 10-12) is intermittently open-circuiting. In turn, the burst amplifier evidently develops parasitic oscillation that forces the burst output off-frequency with a ringing reaction that develops the spurious color-bar display. This kind of component defect can be pinpointed to best advantage by "bridging" suspected defective capacitors with known good capacitors, or by systematically substituting new capacitors.

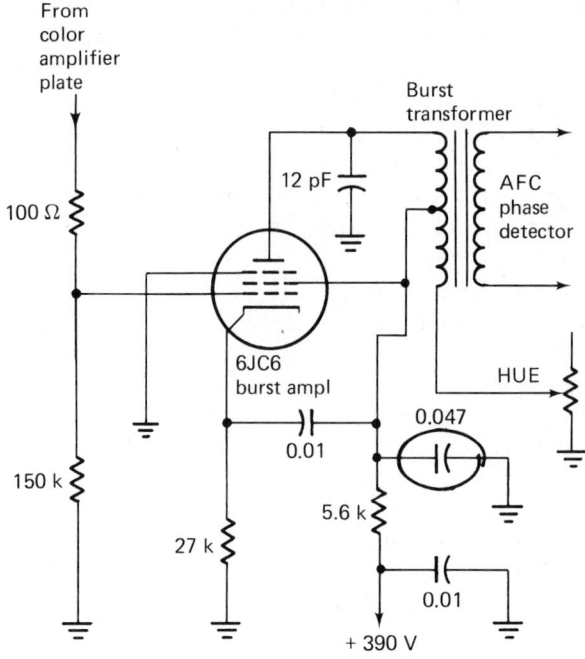

FIGURE 10-12   An open screen-bypass capacitor in this circuit permits parasitic oscillation to occur.

Next, it is helpful to study the configuration depicted in Fig. 10-13. When the deflection-plate voltages at the chroma-demodulator tubes are unequal, trouble symptoms are developed. For example, flesh tones will appear unnatural, and there may be more or less drift in hue reproduction. In this situation, the deflection-plate voltages at the chroma-demodulator tubes should be checked. The permissible tolerance is about 5 volts. For a quick-check, the two demodulator tubes may be interchanged. Then, if the voltage unbalance is reversed, the tubes should be replaced. Note that if one of the DC voltages is approximately correct, and the other is incorrect, both tubes should nevertheless be replaced.

EXERCISES

Questions

1. What is the meaning of the term *convergence*?

2. Why must the deflection yoke be precisely located on the picture-tube neck?

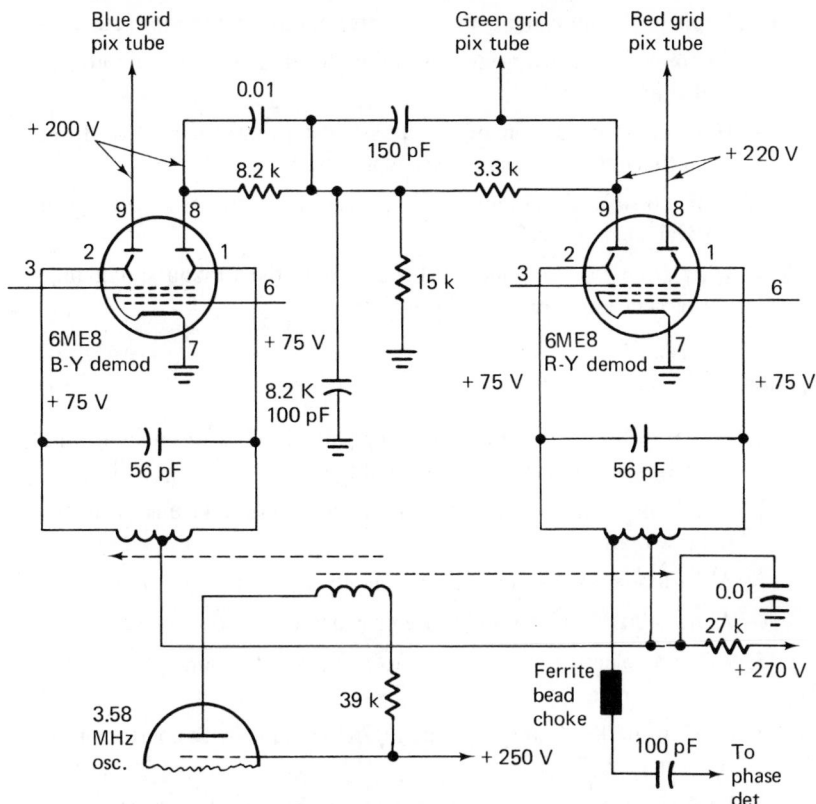

FIGURE 10-13   Unequal deflection-plate voltages at the chroma-demodulator tubes will cause distorted and drifting color reproduction.

3. How do *static convergence* controls differ from *dynamic convergence* controls from the viewpoint of function?

4. Explain why color picture tubes require dynamic convergence, in addition to static-convergence facilities.

5. Do static-convergence assemblies employ permanent magnets or electromagnets?

6. Do dynamic-convergence assemblies utilize permanent magnets or electromagnets?

7. Where are the purity rings located with respect to the convergence assembly?

8. How are parabolic convergence waveforms produced?

9. Why are diodes included in dynamic-convergence circuitry?

10. Describe the development of second-harmonic waveforms in dynamic-convergence configurations.

11. Explain why dynamic-convergence functions tend to interact.

12. Where is the convergence assembly located with respect to the deflection yoke?

13. How does the action of a blue static-convergence control differ from the action of a blue lateral corrector?

14. Is it possible to obtain good horizontal convergence with poor vertical convergence?

15. Why is the DC-component value critical in the horizontal dynamic-convergence process?

True-False

1. It is possible to manufacture color picture tubes so precisely that convergence procedures are not required.

2. The convergence point for the three electron beams is at the shadow mask.

3. A shadow mask contains approximately 1,200,000 phosphor dots.

4. About 400,000 perforations are provided on the viewing screen.

5. All three electron beams come to a focus at the same point on the viewing screen.

6. Shadow-mask picture tubes are interchangeable with aperture-grille tubes in color-TV receivers.

7. Purity adjustments are correct when all three electron beams come into focus at the same point on the viewing screen.

8. Parabolic convergence waveforms are partially rectified by a diode to produce a DC component in the waveforms.

9. Differential current flow adjustment is made to correct unsymmetrical errors on the left and the right sides of the screen.

10. Third-harmonic waveshaping is employed in the horizontal dynamic-convergence network.

11. After convergence adjustments have been properly made, the color picture tube can be replaced, and the new picture tube will be in good convergence.

12. Approximately 12 dynamic-convergence controls are provided for the shadow-mask type of color picture tube.

13. A typical dynamic-convergence configuration employs six semiconductor diodes.

14. Static convergence controls affect beam landings to a greater extent at the edges of the screen than at the center of the screen.

15. There is no interaction between the static-convergence function and the dynamic-convergence function.

Multiple Choice

1. Convergence is correct when the three electron beams in a color picture tube intersect _____ .
   (a) at the shadow mask
   (b) on the viewing screen
   (c) between the shadow mask and the screen
   (d) behind the shadow mask

2. Convergence coils function to _____ .
   (a) obtain screen purity
   (b) produce dynamic beam convergence
   (c) provide linear scanning action
   (d) correct pincushion distortion

3. Convergence magnets function to _____ .
   (a) provide edge-screen purity
   (b) produce edge-screen convergence
   (c) eliminate keystoning
   (d) provide center-screen convergence

4. A shadow-mask color picture tube utilizes a/an _____ .
   (a) aperture grille
   (b) shadow mask
   (c) single-gun assembly
   (d) two-gun assembly

5. Screen purity denotes _____ .
   (a) normal convergence
   (b) display of a pure color, such as an uncontaminated red field
   (c) precise focus of the red electron beam
   (d) precise focus of all three electron beams

6. Horizontal and vertical deflection coils are part of the _____ .
   (a) scanning system
   (b) convergence system
   (c) chroma system
   (d) purity network

7. Triad and _____ are equivalent terms.
   (a) triode
   (b) doublet
   (c) triplet
   (d) trade

8. A raster is a display of _____ .
   (a) converged white dots
   (b) a test pattern
   (c) a color-bar pattern
   (d) horizontal scanning lines

9. Diodes are utilized in dynamic-convergence circuits to _____ .
   (a) develop a DC component
   (b) rectify the chroma signal
   (c) decode the color signal
   (d) bypass flyback pulses

10. Dynamic-convergence waveshaping networks employ _____ .
    (a) second-harmonic ringing circuits
    (b) subharmonic ringing circuits
    (c) third-harmonic ringing circuits
    (d) none of the above

11. _____ waveforms are utilized in dynamic-convergence systems.
    (a) Exponential
    (b) Sinusoidal
    (c) Parabolic
    (d) Hyperbolic

12. A typical dynamic-convergence system provides _____ controls.
    (a) two
    (b) three
    (c) six
    (d) twelve

13. Dynamic-convergence controls comprise _____ .
    (a) potentiometers and slug-tuned coils
    (b) varactor diodes and trimmer capacitors
    (c) tuned lines and ferrite beads
    (d) tickler coils and rheostats

14. Dynamic-convergence controls are classed as _____ controls.
    (a) operating
    (b) maintenance
    (c) automatic
    (d) semi-automatic

15. A red pole piece is so named because _____ .
    (a) its radiation has a red wavelength
    (b) the pole piece is painted red
    (c) it is dangerous to touch
    (d) it provides red-beam adjustment

Problems

1. Two electron beams have sources one inch apart, and intersect fifteen inches from the sources. What is the angle between the electron beams?

2. A horizontal-section ringing coil has an inductance of 10 mH and develops a third-harmonic voltage. What value of capacitance is in shunt to the coil?

3. What is the reactance of the coil in problem 2 at the horizontal-scanning frequency?

4. What is the reactance of the capacitor in problem 2 at the horizontal-scanning frequency?

5. What is the reactance of the coil in problem 2 at the third harmonic of the horizontal-scanning frequency?

6. What is the reactance of the capacitance in problem 2 at the third harmonic of the horizontal-scanning frequency?

7. What is the reactance of the coil in problem 2 at the vertical-scanning frequency?

8. What is the reactance of the capacitance in problem 2 at the vertical-scanning frequency?

9. What is the reactance of the coil in problem 2 at the second harmonic of the horizontal-scanning frequency?

10. What is the reactance of the capacitance in problem 2 at the second harmonic of the horizontal-scanning frequency?

Chapter 11

Horizontal-Sweep
And
High-Voltage System

11.1   HORIZONTAL-OSCILLATOR OPERATION

Since the horizontal scanning rate must be maintained precisely in step with
the horizontal sync pulses, the horizontal oscillator functions in combination
with an automatic frequency-control (AFC) section, which is shown in Fig.
11-1. Horizontal sync pulses are fed via R552 to the base of the phase-splitter
transistor Q551. In turn, a positive pulse drops across emitter resistor R555
and is fed via R556 and C553 to the AFC detector diode D552. A negative
pulse appears across collector resistor R551, and is coupled through C552 to

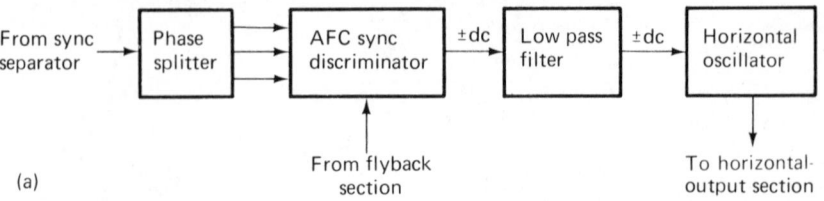

(a)

FIGURE 11-1   Horizontal oscillator and AFC section. (a) Block
diagram.

216

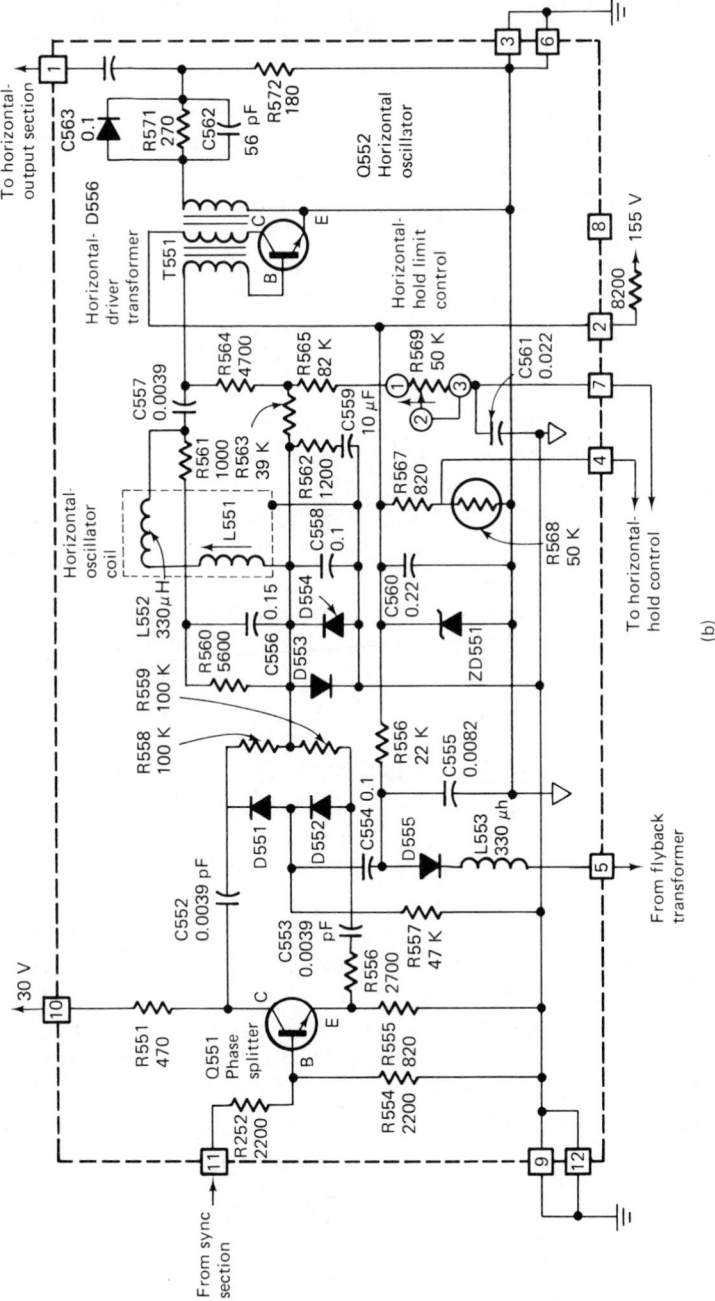

FIGURE 11-1 Horizontal oscillator and AFC section. (b) Example of a horizontal-oscillator circuit board.

(b)

detector diode D551. Pulses from the flyback transformer are conducted through L553 and D555 to the integrator capacitor C555. In turn, the integrated comparison pulse is coupled by C554 to the junction of the detector diodes.

In the discriminator, the horizontal sync pulses are compared with the flyback waveform, and a DC voltage is produced in case there is a phase difference between the two waveforms. This AFC voltage is filtered into smooth DC by C558 with R562 and C559. Diodes D553 and D554 prevent the AFC voltage from swinging more than ±0.5 volt in order to establish a proper hold-in range. The AFC voltage is then fed through resistor R563 to the base-bias network of the horizontal-oscillator transistor Q552 to hold the oscillator output in step with the horizontal sync pulses. It is desirable that one use the AFC arrangement, instead of triggering the horizontal oscillator directly, because much better noise immunity is realized with the former; that is, the low-pass filter circuit has an appreciable time-constant, which tends to average out incoming noise pulses, and thereby prevents the picture from pulling or tearing.

We observe in Fig. 11-1 that T551 provides positive feedback for Q552 and also couples the output pulses to the horizontal-output section. Pulses from the collector are fed back to the base and drive Q552 into cutoff as a result of collector saturation. During the time that Q552 remains cut off, C557 discharges to a point at which the base voltage starts going positive. In turn, regenerative amplification occurs, Q552 is promptly driven into cutoff, and the operational cycle is repeated. This is called the *blocking-oscillator* mode of operation. We observe that C557 discharges through R564, R565, the horizontal-hold limit control R569, and the horizontal-hold control (not shown). Thus, the setting of the horizontal-hold control can determine the horizontal-oscillator frequency by varying its discharge time; that is, the free-running rate of the horizontal oscillator is determined by the setting of the horizontal-hold control.

In addition to its being related to manual control of the horizontal-oscillator frequency in Fig. 11-1, note that the AFC control voltage is also combined with the hold-control voltage. Therefore, automatic control is provided for the oscillator frequency. A certain amount of the fixed DC bias voltage is provided by R567 and thermistor R568. This thermistor is included to provide temperature compensation so that the oscillator frequency is not affected by temperature variation. Observe also that oscillator stability is also increased by the provision of the ringing circuit comprising L551 and C556. The ringing voltage waveform causes the oscillator to come out of cut-off rapidly, as Fig. 11-2 depicts. Therefore, random noise pulses have much less effect on the oscillator frequency (jitter is minimized). Finally, the series resonant circuit L552-C557 is utilized to provide an output waveform with a very steep leading edge (fast rise time). This wave-shaping action is supplemented in the collector circuit of Q552 by D556, C562, and R571.

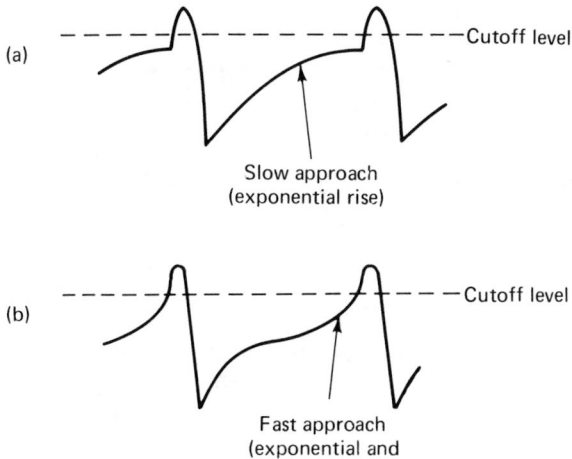

FIGURE 11-2   Effect of ringing coil on cutoff approach. (a) Without a ringing coil; (b) with a ringing coil.

Note that in Fig. 11-1 Q551 is operated from a 30-volt supply, whereas Q552 is operated from a 155-volt supply. A separate power source is employed for Q552 so that sudden current demand can be met when the drive waveform is applied to the horizontal-output section. The 155-volt source is fed through an 8200-ohm resistor to Zener diode ZD551, providing a 30-volt regulated source with adequate current capability for the horizontal oscillator. Referring to Fig. 11-3, note that the output pulse from the horizontal oscillator is coupled via C563 to the gate of the silicon controlled rectifier SCR701. Note that C563 and R572 in Fig. 11-1 also function as a differentiating circuit, narrowing the output pulse from the oscillator into a spike. As will be explained in greater detail below, this spike triggers the SCR into conduction. Immediately thereafter, D556 becomes reverse-biased, and C563 discharges via R571 and R572. This discharge action holds the waveform negative until the next positive pulse is generated. In turn, the input to the SCR is held negative during the forward scan to eliminate the possibility of false triggering.

11.2   HORIZONTAL-OUTPUT CIRCUITRY

Basically, the function of the horizontal-output configuration shown in Fig. 11-3 is to generate the current waveform depicted in Fig. 11-4. Current flow through the horizontal-deflection coils during the first half of the forward scanning interval, from the left-hand side of the center of the screen, is produced by current I_{T1}. The second half of the forward-scanning interval, from the center of the screen to the right-hand side, is produced by current I_{T2}.

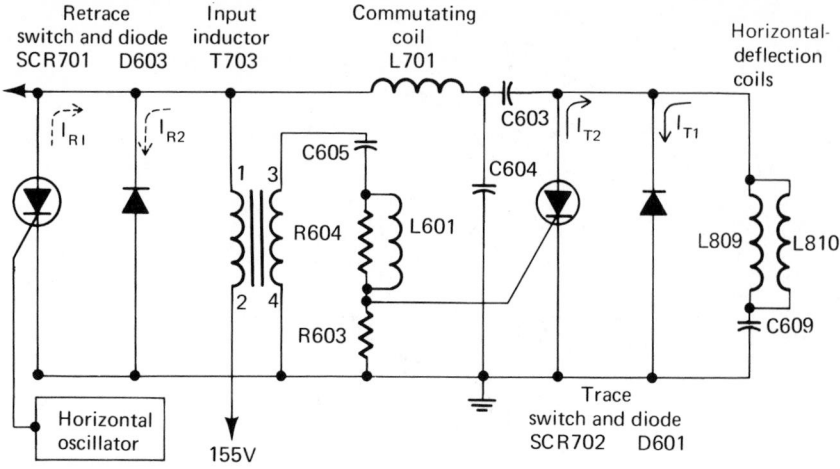

FIGURE 11-3 Simplified schematic diagram of a horizontal-output circuit.

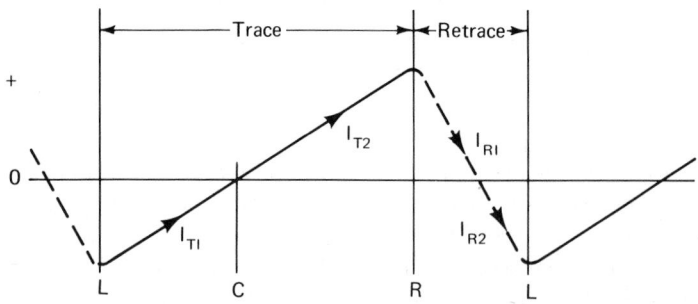

FIGURE 11-4 Detail of a forward-scanning and flyback deflection current waveform.

Retrace or *flyback* is initiated at the end of the I_{T2} interval, and the first half of the retrace period is produced by current I_{R1}; the second half of the retrace period is produced by current I_{R2}.

Observe that during the forward-scanning period (I_{T1} and I_{T2} in Fig. 11-4), diode D601 and rectifier SCR702 in Fig. 11-3 control the current flow to the horizontal-deflection coils. Diode D603 and rectifier SCR701 control the current flow in the horizontal-deflection coils during the retrace or flyback interval. Energy storage and timing of circuit action are provided by L701, C603, C604, and C609. A charge path for L701 and C603 is provided through T703. Referring to Fig. 11-4, note that induced current through L809-L810 from the collapsing magnetic field during flyback reaches a peak negative value at point L and then arrives at a zero value at point C. Observe that current flow through D601 supplies a charge to C609.

At point C in Fig. 11-4, the deflection current is zero, and C609 starts to discharge through the horizontal-deflection coils (Fig. 7-3). D601 turns off, and SCR702 is turned on at this time. In turn, the deflection current I_{T2} starts to flow and rises to a positive peak value. As noted above, the scanning beam is thereby deflected from center screen to the right-hand side at time R. The forward scan stops at point R because a pulse is applied to SCR701 from the horizontal oscillator, thereby triggering SCR701 into conduction. In turn, a charge previously stored by C603 is released into a commutating circuit comprising L701, C603, and C604. Thus, the current through SCR702 rapidly falls to zero, causing the turning off of SCR702 and the initiation of flyback.

The retrace or flyback current in Fig. 11-4 flows through a series-resonant circuit comprising L701, C603, C609, L809, and L810 in Fig. 11-3, the flyback time being determined by the L and C values of this resonant circuit. With reference to Fig. 11-4, retrace current I_{R1} comes to zero at the midway point of the flyback interval, whereupon SCR701 stops conduction. Diode D603 becomes forward-biased and conducts for the remainder of the retrace interval I_{R2}. Effectively, the energy that had been stored in C603 has been returned to the inductance of the horizontal-deflection coils; D601 thereupon becomes forward-biased, and the next forward-scanning interval starts. Note that during flyback, the primary of T703 is connected between the 155-volt source and ground via SCR701 and D603. When D603 stops conduction, this primary winding is disconnected from ground, and C603 charges through the winding from the 155-volt source to replenish energy in the deflection-coil circuit. Observe in Fig. 11-3 that the voltage developed across the primary during the charging of C603 is coupled to the gate of SCR702 through the secondary of T703 and the wave-shaping network comprising C605, R603, R604, and L601. In turn, the gate waveform that is generated possesses adequate amplitude to enable SCR702 to conduct while its anode is forward-biased.

11.3   TROUBLESHOOTING SCR HORIZONTAL-SWEEP CIRCUITS

Many defects in an SCR horizontal-sweep circuit can be analyzed by means of the resulting picture symptoms. As an illustration, the SCR sweep configuration depicted in Fig. 11-5 will produce the following picture symptoms:

1. Foldover on left of screen; retrace diode is probably open.
2. Center foldover with narrow raster; check the trace diode.
3. Straight line on left-hand side of picture; retrace diode is probably defective.

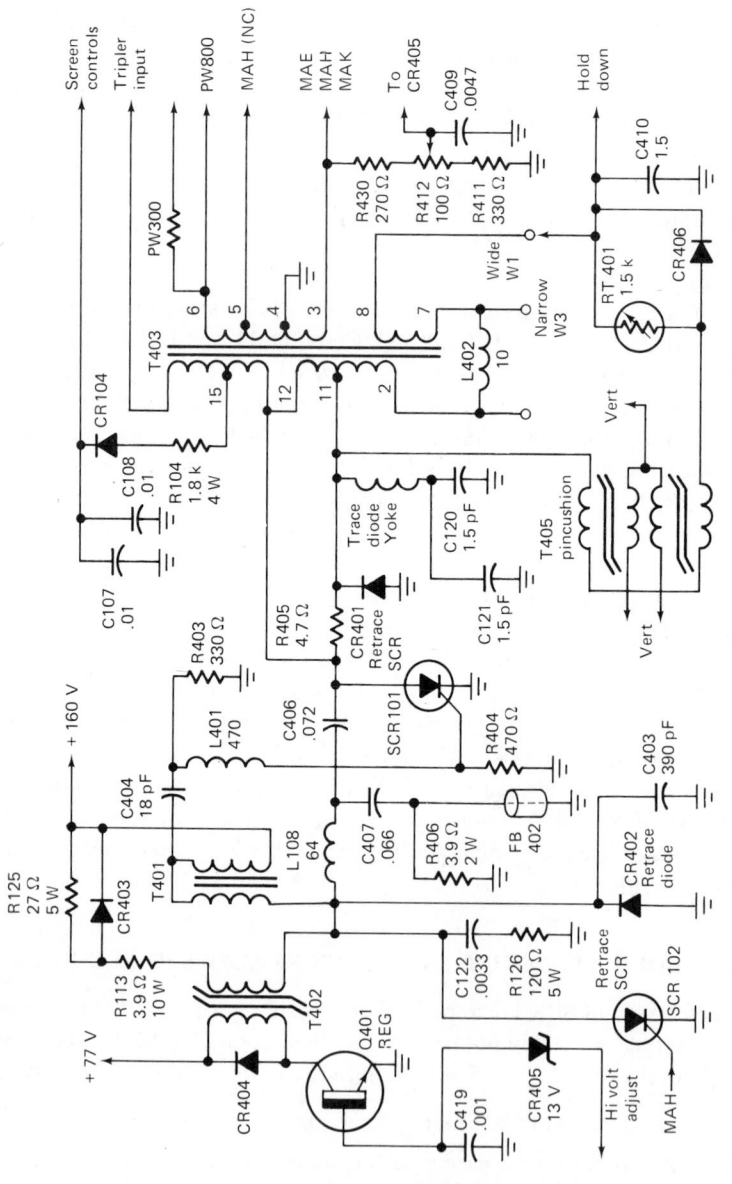

FIGURE 11-5 Detailed schematic of an SCR horizontal-sweep circuit.

222

4. Straight line in the center of the picture; trace diode is likely to be defective.

5. Wide band at the right-hand side of raster center; check trace diode.

6. Straight line at the right-hand end of the raster; the trace SCR is probably defective.

7. Straight interference line near right end of screen; check the horizontal-deflection (MAH) module.

8. Left tilted curve around the middle of the raster; regulator clamp diode is likely to be defective.

9. Partial double-banding on left side of screen; quadrupler (HV rectifier) malfunction.

10. Popping circuit breaker; generally caused by a short-circuited diode, filter capacitor, or SCR.

11.4   HIGH-VOLTAGE CIRCUITRY AND OPERATION

Accelerating voltage for the color picture tube is generated by the high-voltage section during the horizontal flyback interval. Since flyback action involves a rapid change in magnetic-field strength, a high potential is induced in the horizontal-output secondary winding. As shown in Fig. 11-6, a separate secondary winding, which is connected to a rectifier, is employed by the high-voltage section. This high-voltage rectifier may be a semiconductor diode stack in some receivers, or a high-vacuum diode in other receivers. Pulsating DC voltage from the rectifier is filtered to steady DC by high value series resistors in combination with the input capacitance of the color picture tube and the stray capacitance of the high-voltage circuitry.

Observe in Fig. 11-6 that a resistive voltage divider is utilized to obtain a suitable potential for the focus anode. This divider also serves as a bleeder to discharge the high-voltage system when the receiver is turned off. Regulator action is provided to maintain a fixed high-voltage value and to prevent screen damage from excessive beam current flow. The high-voltage value is regulated by frequency control of the energy available to the horizontal-output section. Recall that these circuits are supplied by energy stored during the forward-trace interval, particularly by the commutating capacitors C603 and C604 in Fig. 11-3. These capacitors are charged during forward scan through one winding of T703. Note that the inductance of T703 resonates with C603 and C604 at a frequency that is approximately twice as great as the horizontal-scanning frequency. Energy control is provided by adjustment of this resonant frequency.

Figure 11-7 depicts the relations between the subsections of the high-voltage regulating system. A saturable reactor, T702, is employed to

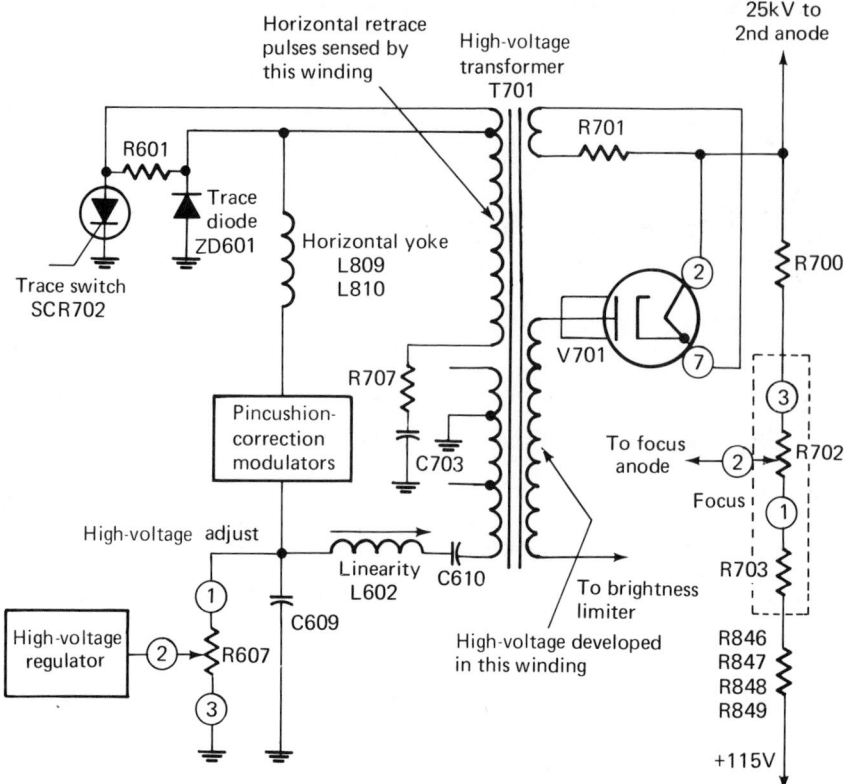

FIGURE 11-6   Example of a high-voltage configuration.

adjust the resonant frequency of T703-C603-C604. Note that the load winding of the saturable reactor has green and red leads; the control winding has white and black leads in the diagram. Current flow in the control winding determines the effective inductance of the system. This control current is provided by a high-voltage regulator transistor, Q601. Its collector current is a function of the voltage across C609, which varies with any change in high-voltage value. A reference voltage is provided by Zener diode ZD601, and the resulting difference voltage determines the amount of conduction in regulator transistor Q601.

Horizontal-scanning linearity is optimized by an adjustable resonant circuit, as depicted in Fig. 11-8. Inherent nonlinearity results from voltage drops across resistance in the forward-scanning circuitry. Referring to Fig. 11-3, note that there are voltage drops across trace diode D601 and trace switch SCR702. These drops are minimized by connecting the diode to T701 at a more negative point than the SCR, as Fig. 11-6 shows. Remaining non-

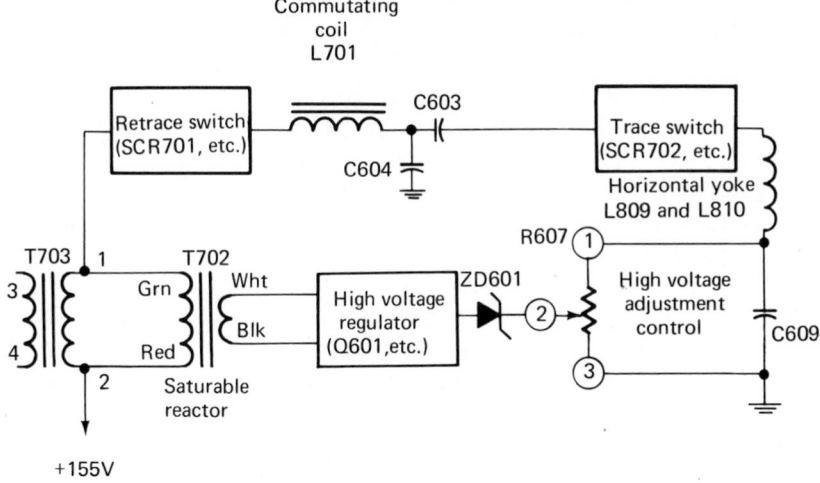

FIGURE 11-7   Subsections of a high-voltage regulating system.

linearity is corrected by a damped series-resonant circuit, such as the one in Fig. 11-8. This resonant circuit develops a damped sine wave of current, which adds or subtracts with the charge on C609. In turn, current flow through the horizontal-deflection coils during the forward-scanning interval is practically linear. Note in passing that the pincushion correction modulators indicated in Fig. 11-6 will be discussed in detail in the following chapter.

Observe the typical tube-type horizontal-output system diagrammed

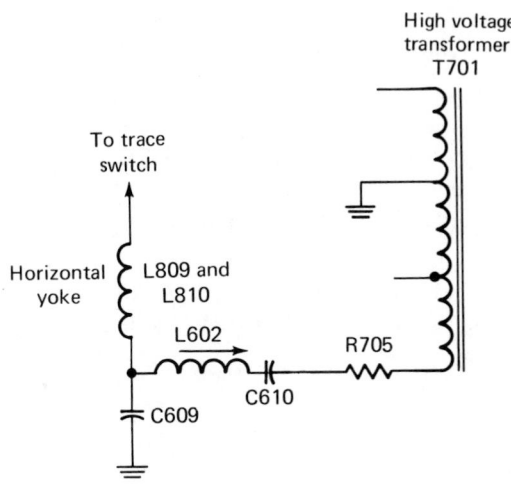

FIGURE 11-8   Example of a horizontal-linearity circuit.

in Fig. 11-9. This arrangement is found in many tube-type color receivers. Operating voltages are much higher than in corresponding solid-state circuitry. Note that DC-voltage measurements should not be attempted at the plate of the horizontal-output tube nor at the plate of the damper tube. High-voltage pulses are present at these points, which will damage a voltmeter. Similarly, waveform checks should not be made at these points unless a suitable high-voltage capacitance-divider probe is available. When malfunction occurs, tubes are checked or replaced first. Then, if the trouble symptom persists, circuit tests must be made.

Technicians often make signal-substitution tests in preliminary analysis. In Fig. 11-10, horizontal-deflection troubles can often be localized by use of the horizontal grid-drive and horizontal plate-drive outputs from a television analyzer. Horizontal grid drive tests are used to substitute for a missing or distorted driving signal to the horizontal-output tube. Note that a receiver should never be operated without a proper horizontal-driving waveform. A horizontal-output tube normally operates in class C, and excessive current drain and overheating will occur if the driving voltage waveform is absent. Troubleshooting procedure is as follows:

1. A VHF signal is applied to the antenna-input terminals of the tuner.

2. Disconnect the power source to the receiver.

3. Remove the plate cap from the high-voltage rectifier tube.

4. Connect the high-voltage indicator lamp to the insulated wire going to the plate cap disconnected above. This indicator lamp is one of the analyzer accessories.

5. Connect the red lead from the horizontal grid-drive jack on the analyzer to the grid terminal of the horizontal-output tube.

6. Reapply power to the receiver. If the indicator lamp glows, there are high-level AC pulses on the plate-cap lead of the high-voltage rectifier. On the other hand, if the indicator lamp does not glow, pulses are absent and the trouble has been isolated to the horizontal-output circuit or the flyback transformer.

7. If application of a driving pulse to the horizontal-output tube results in restoring normal operation, it is indicated that the trouble will be found in the horizontal-oscillator section.

Figure 11-11 shows that a horizontal plate-drive signal can also be applied from a television analyzer to obtain additional localization data.

Test procedure is as follows:

1. A VHF signal is applied at the antenna-input terminals of the tuner.

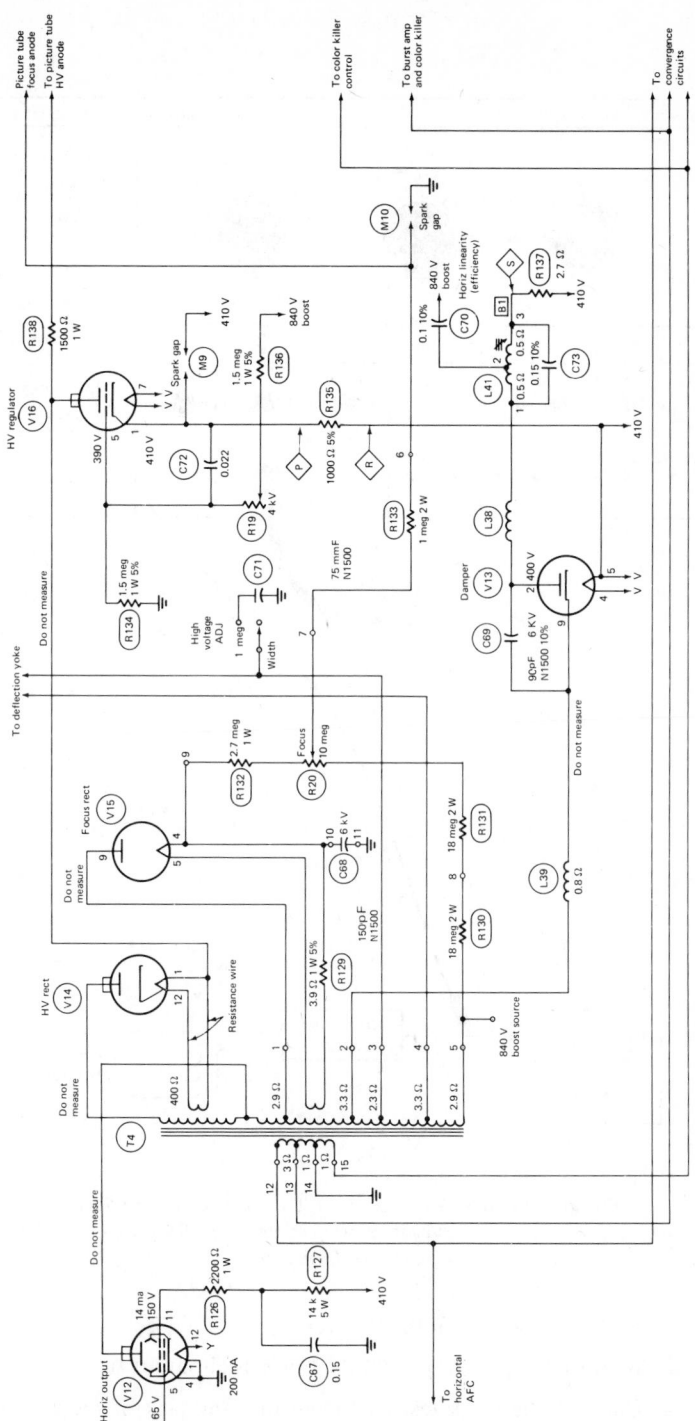

FIGURE 11-9 Typical tube-type horizontal-output system.

227

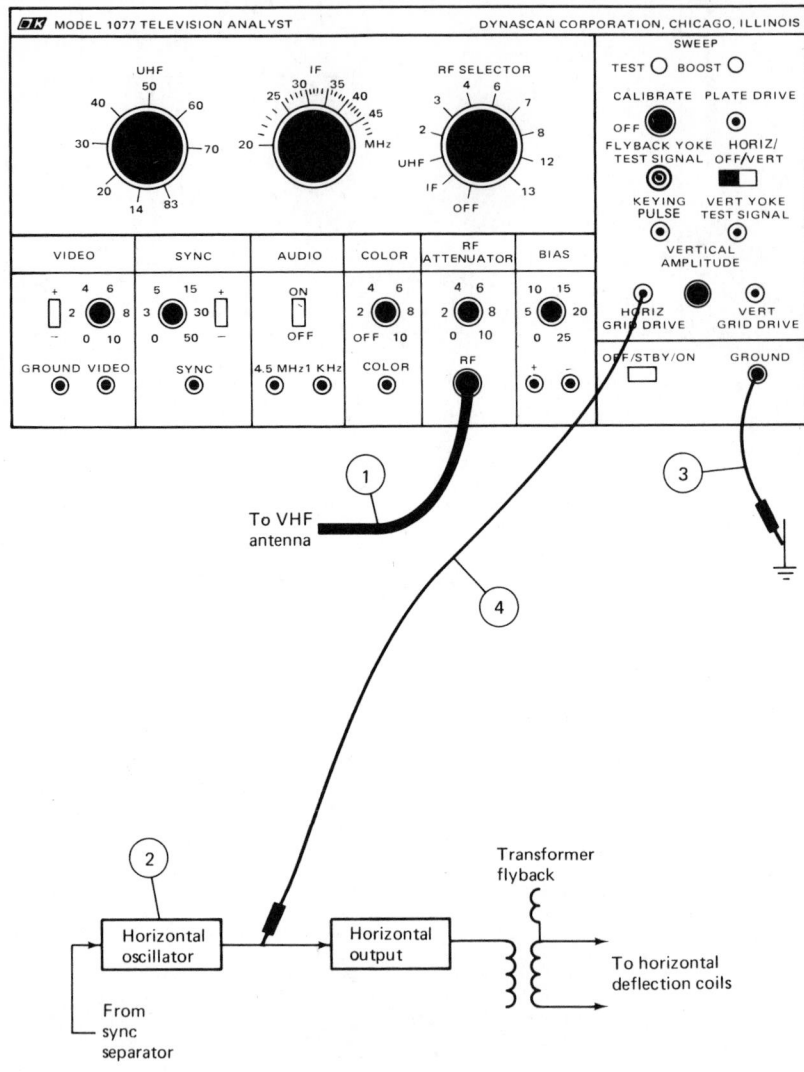

FIGURE 11-10   Application of TV analyzer for grid-drive signal substitution. *(courtesy of B&K Precision Mfg. Co., Div. of Dynascan Corp.)*

2. Remove power from the receiver.

3. Remove the plate cap from the horizontal-output tube.

4. Connect the black test lead from the Gnd jack of the analyzer to the receiver chassis.

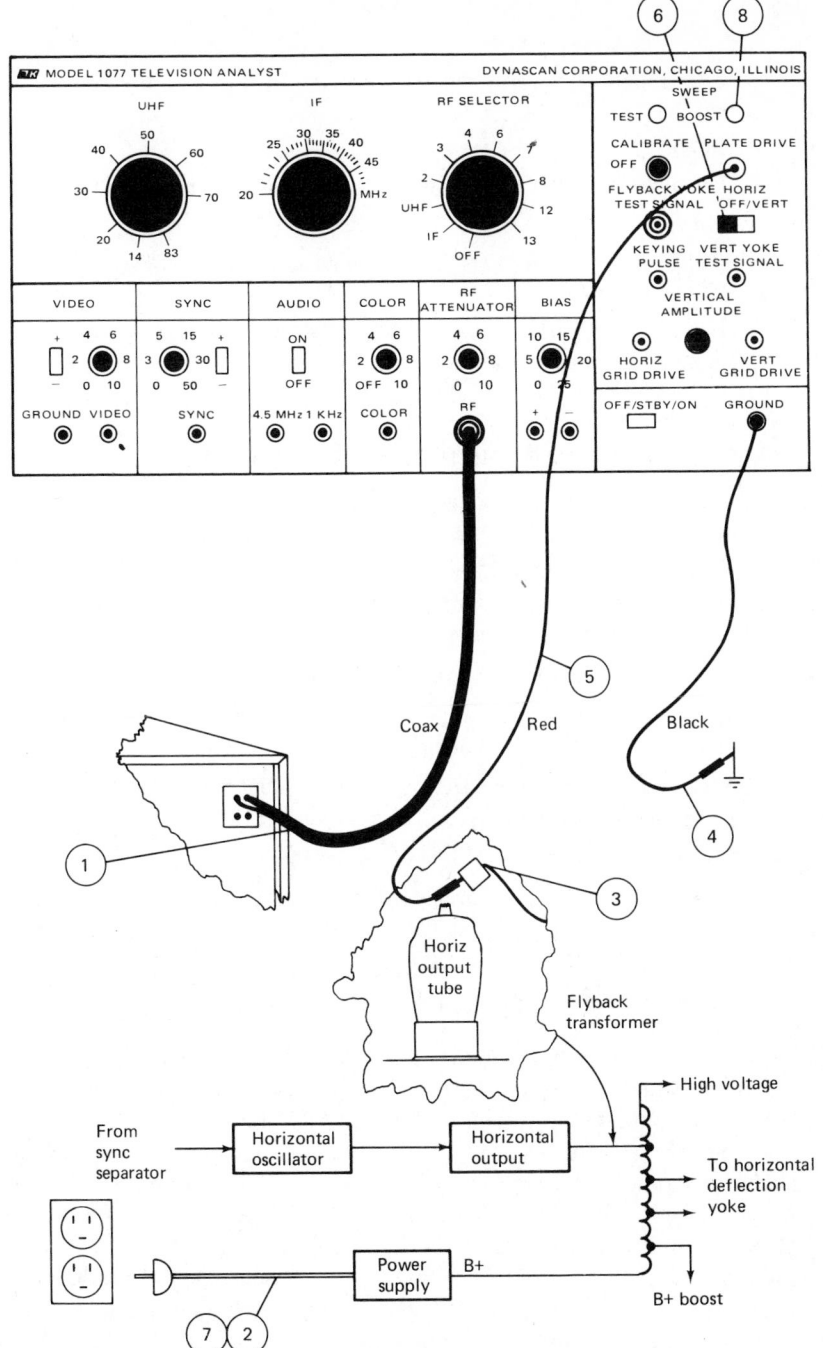

FIGURE 11-11 Connections for horizontal plate drive signal and B+ boost indicator. *(courtesy of B&K Precision Mfg. Co., Division of Dynascan Corp.)*

5. Connect the red test lead from the plate-drive jack on the analyzer to the plate lead that was removed from the horizontal-output tube.

6. Place the Horiz/Off/Vert. switch on the analyzer in its horizontal position.

7. Apply power to the receiver. Normally, horizontal sweep action will be restored.

8. In case boost B+ voltage is being generated, the boost-indicator lamp on the analyzer will glow.

11.5   TROUBLESHOOTING HORIZONTAL-SECTION MODULES

A typical horizontal-section module is illustrated in Fig. 11-12. If a module is suspected of malfunctioning, it can be easily replaced by a known good module. This type of module can also be repaired in most cases without difficulty. Transistors are plugged into sockets, and the other components can be unsoldered and replaced if necessary. More than half of the faults that occur in horizontal-section modules can be corrected as follows:

1. Look for loose connections or cold-soldered connections at the terminals where the module plugs in.

2. Make a careful visual inspection for poor connections, loose components, and physical damage.

3. Check out all of the solid-state devices.

4. Look for discolored or burned resistors, and replace any resistor that is out of tolerance.

5. Inspect the circuit board carefully for cracks, and check suspicious conductors for continuity while flexing the board slightly.

Before making circuit-action checks, measure the supply voltages to determine whether they are normal. A block diagram for a typical modularized horizontal-sweep system is shown in Fig. 11-13. The following discussion concerns the blocks to the left of the dotted line. Horizontal locking is largely the function of the AFC circuits and diodes. However, we will find that locking can occur in some cases, although the oscillator has jumped to a higher or lower frequency. Next, the driver stage functions chiefly to generate the required deflection waveform, and to provide adequate drive for the horizontal-output transistor. Drive defects usually become apparent as nonlinear horizontal sweep, and/or incorrect picture width.

Two basic waveforms are involved in operation of the AFC section. Horizontal-sync pulses are processed along with a sample of the horizontal-sweep waveform to develop a control voltage for the horizontal oscillator.

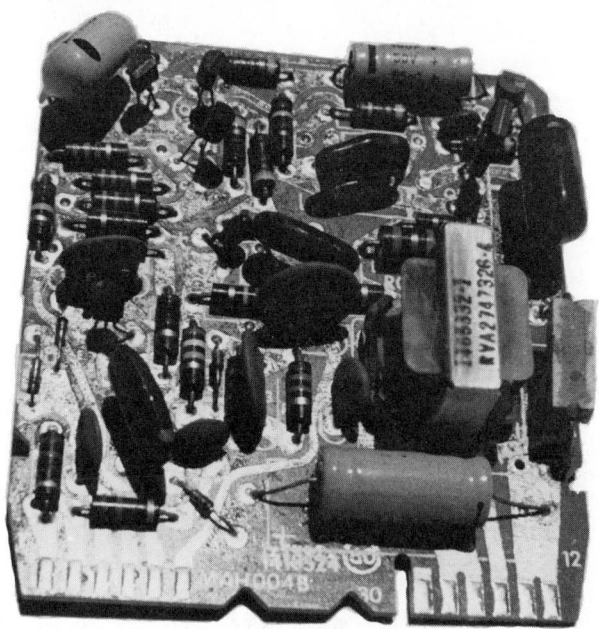

FIGURE 11-12   Appearance of a horizontal-section module. *(courtesy of Heath Co.)*

This control voltage pulls the oscillator back on-frequency, in case it tends to drift. Referring to Fig. 11-13, negative-going sync pulses with approximately 20 v p-p amplitude are applied to terminal 14. These pulses pass through R801 and C801, and are applied to the common cathodes of the germanium diodes CR801 and CR802. At the same time, negative-going pulses of approximately 65 v p-p amplitude are fed from a winding on the flyback transformer to terminal U2. A sawtooth shaper provides two steps of integration, after which the sawtooth waveform passes through C802 to the anode of CR801. The foregoing waveforms are rectified by CR801 and CR802, and the following troubleshooting points should be observed:

1. If the sync pulses are missing, but the sawtooth waveform is normal, there will be practically zero volts at the anode of CR801, and a positive voltage will be measured at the common cathodes.

2. If the sawtooth waveform is missing, but the sync pulses are normal, the DC voltages are the same as above (a scope will show definitely which waveform is missing).

3. When both waveforms are normal, a positive voltage will be measured at the common cathodes, and a DC voltage which may be zero, positive, or negative will be measured at the anode of CR801. Its polarity depends on the relative phase of the two waveforms.

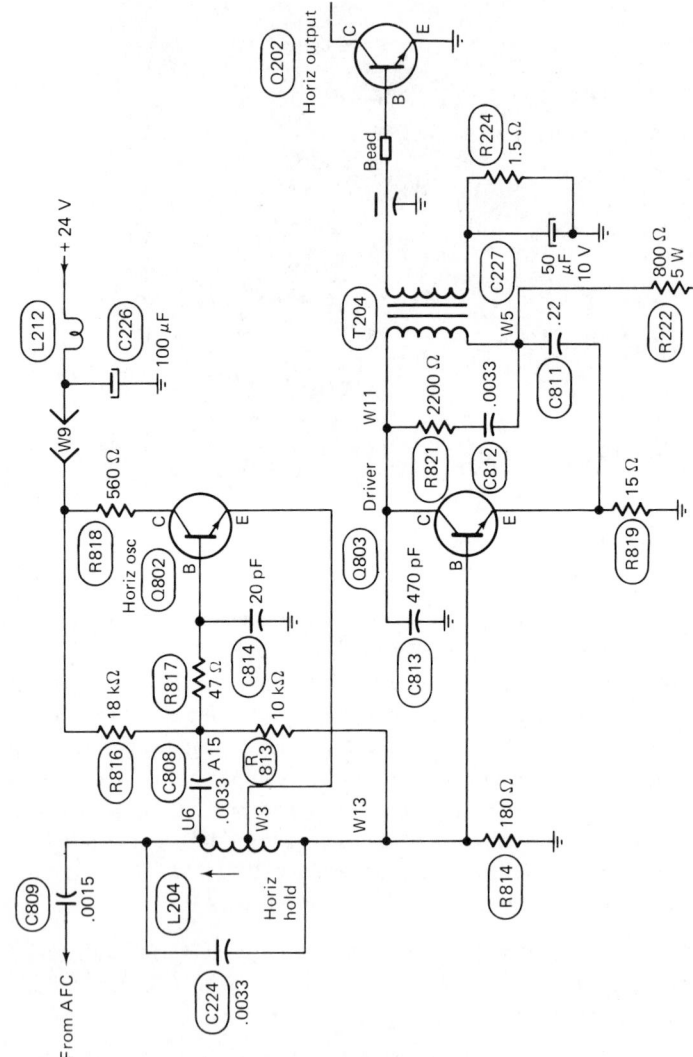

FIGURE 11-12 (continued)

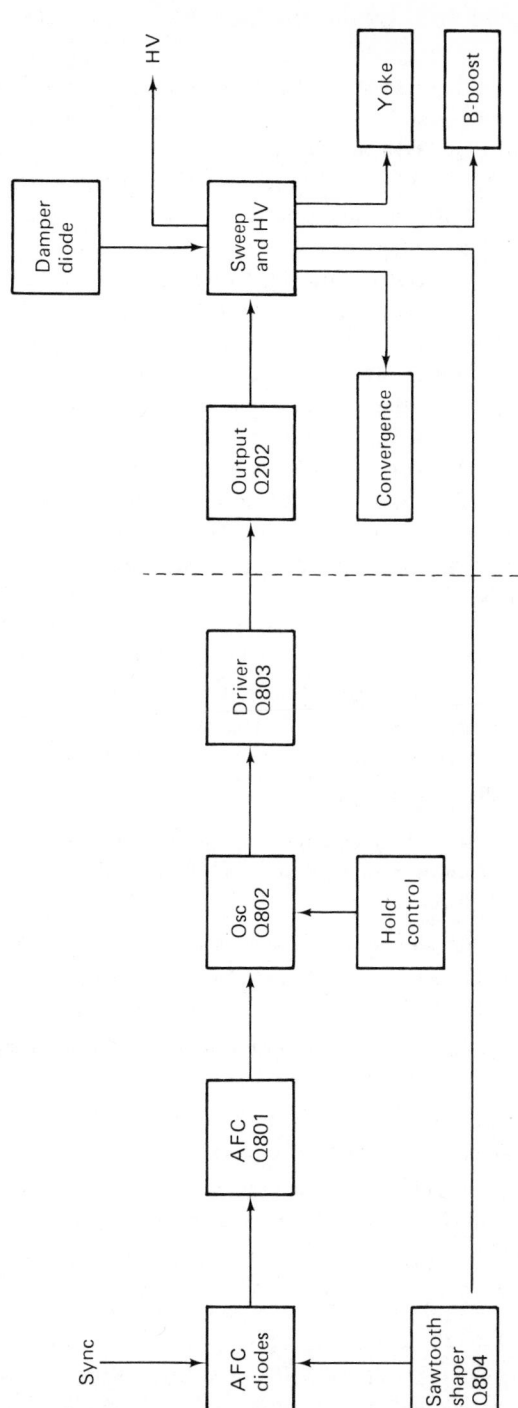

FIGURE 11-13 Block diagram of a modularized horizontal-sweep section.

233

This DC voltage that appears at the anode of CR801 is the AFC control voltage that pulls the horizontal oscillator back on-frequency when necessary. With reference to Fig. 11-14, R804 and C807 remove most of the horizontal pulse waveform (the feedthrough from the phase detector). R806 and C806 function as a stabilizing circuit to prevent a "piecrust" picture symptom. R807 isolates the bias voltage for Q801 from the AFC control voltage. Most of the forward bias for Q801 is provided via R805, although some bias is provided via R809. AFC control voltage from R807 varies the base bias voltage, and in turn changes the collector-emitter resistance of Q801. Since this collector-emitter resistance operates in series with C809, which is connected to the oscillator tank, the resonant frequency of C224 and L204 is corrected as required, as seen in Fig. 11-15.

A circuit action that may not be obvious is that the collector voltage for Q801 is not obtained entirely through R809 in Fig. 11-14. In other words, diode CR803 rectifies part of the oscillator signal that feeds back through C809 and thereby supplies approximately +5 volts to the collector. Only picture symptoms of poor sync lock or incorrect frequency can be caused by AFC circuit defects. This is the case because defects in Q804, Q801, CR801, or CR802 cannot stop Q802 from oscillating. Figure 11-16 shows normal waveforms at the AFC diodes. Loss of sync lock occurs if either of these waveforms is weak or missing. Therefore, a scope check should be made before particular components are tested. The chief defects and picture symptoms that occur are as follows:

1. In case CR801 is short- or open-circuited, the picture will be out of sync, with the oscillator running too slow. Horizontal strips (bars) will slant downhill to the left.

2. In case CR802 is short- or open-circuited, the picture will be out of sync, with the oscillator running too fast. Horizontal strips will slant downhill to the right.

3. If the horizontal-hold control can almost lock the picture, it is likely that C806 or C807 is short-circuited.

4. If 10 to 20 diagonal strips are displayed, check Q801 for an open circuit or for a collector-emitter short-circuit.

5. In case Q804 has a collector-emitter short-circuit, horizontal lock can be obtained, although not as stably as in normal operation.

6. An open- or short-circuit in CR803 causes many diagonal strips to be displayed.

Referring to Fig. 11-14, a useful test to determine whether incorrect oscillator frequency is being caused by AFC trouble or by the oscillator circuitry is to temporarily ground the junction of C807, R804, R806, and R807. This places the junction at zero volts, and the picture should freewheel

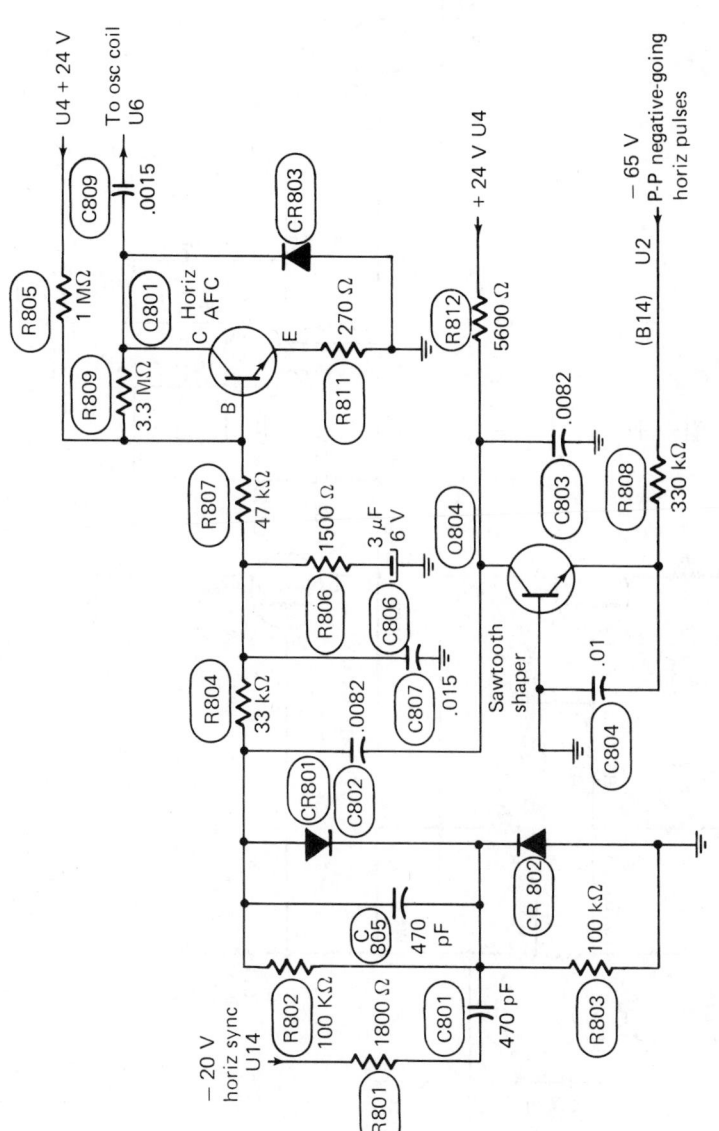

FIGURE 11-14 AFC circuitry for the example module.

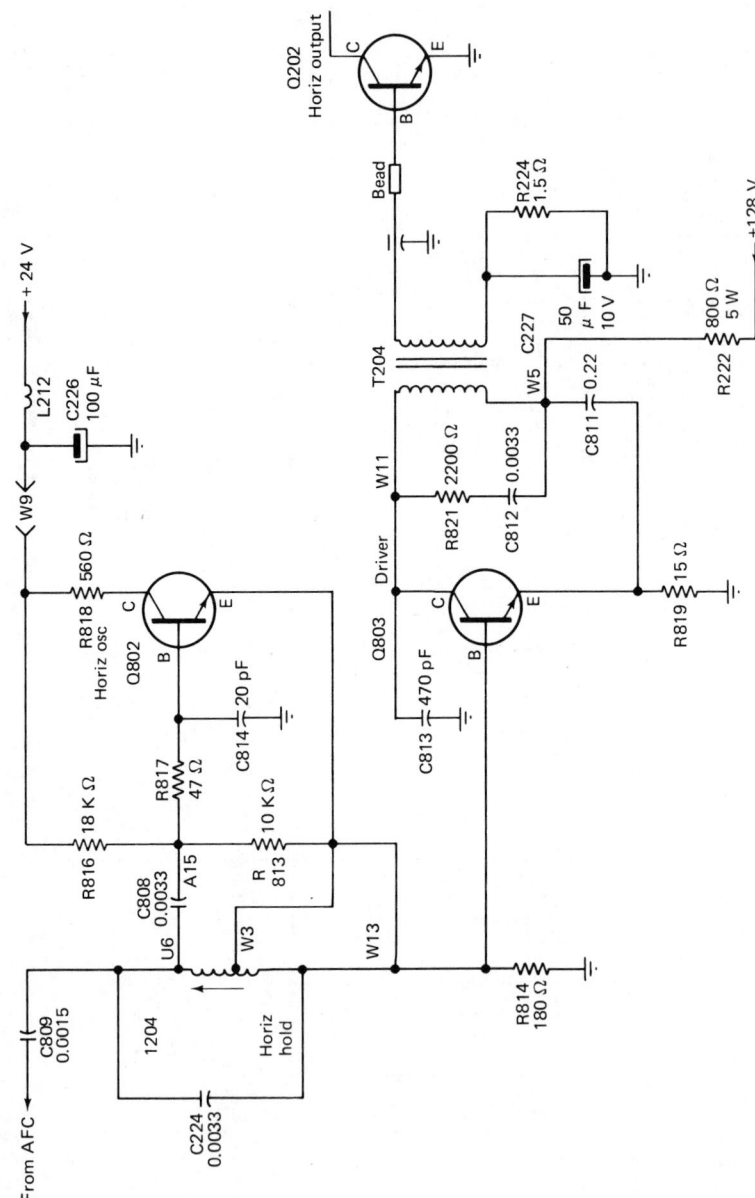

FIGURE 11-15 Schematic diagram of the example oscillator and driver stages.

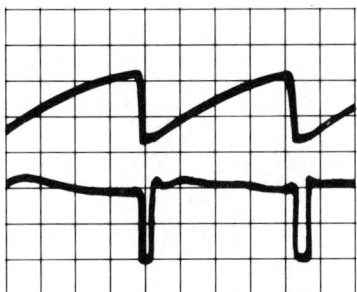

FIGURE 11-16   Normal waveforms at AFC diodes. (a) 15V p-p sawtooth at anode of CR801; (b) 9V p-p sync pulses at cathodes of CR801 and CR802 (DC voltage at cathodes is +7.5V).

approximately in sync, although it will drift left or right on the screen. In this case, look for a defective diode. On the other hand, if there are many diagonal strips displayed and the picture refuses to freewheel, the trouble will probably be found in the Q801 stage, or in the oscillator stage (Fig. 11-15). Normally, horizontal lock is sufficiently tight that the horizontal-hold control cannot throw the picture out of sync.

The horizontal oscillator is an emitter-coupled sine-wave design, in which L204 and C224 are the chief frequency-determining components. R816 and R813 provide forward bias, but base rectification of the signal normally makes the DC base voltage less positive than the emitter voltage. In other words, this transistor is normally more or less reverse-biased. The oscillator emitter current flows through L204, and thence through R814 to ground. The voltage drop across R814 provides the input signal for Q803. In turn, defects that cause Q802 to draw too much emitter current result in excessive forward bias on Q803. Therefore, Q803 runs hot, although the trouble is not in the Q803 stage. In case the defect cannot be localized to the module, remember that there are some off-module components in this example also, such as L204 and C224. Open- or short-circuits in these components will stop the oscillator. When checking DC-voltage distribution, keep in mind that transistor bias voltages are relatively critical, and be suspicious of even small bias-voltage errors.

Some practical precautions to remember when servicing modules are as follows:

1. Turn the power off before removing or plugging in a module.

2. Some chassis detour a power-supply source through one or more modules as a form of interlock; if one module is removed, the others cannot be operated.

3. When a module has burned components, find out why and correct the cause before a new module is plugged in.

4. Use only recommended cleaner solutions on plugs.

5. Be very careful not to short-circuit adjacent module terminals with meter probes; otherwise, a string of transistors or IC's may be ruined.

EXERCISES

Questions

1. What is the basic function of the automatic frequency-control section?

2. How does an AFC section operate?

3. Why are noise pulses practically rejected by an AFC network?

4. Briefly describe the horizontal-oscillator arrangement utilized in most receivers.

5. Define an *SCR*.

6. Explain a basic distinction between a tube-type and a solid-state horizontal-sweep and high-voltage system.

7. Name two devices that are employed to rectify high-voltage AC pulses.

8. Discuss the basic action of a high-voltage regulating arrangement.

9. What is the function of a phase splitter in an AFC network?

10. How does the horizontal-hold control establish the free-running horizontal-oscillator frequency?

11. Identify the source of the comparison waveform for an AFC system.

12. Why is a ringing coil utilized in a horizontal-oscillator configuration?

13. Describe the function of a thermistor in an AFC arrangement.

14. Explain how the possibility of false SCR triggering is avoided.

15. Briefly discuss the disadvantage of triggering a horizontal oscillator directly from the sync-separator output.

True-False

1. An AFC sync discriminator is located between the video amplifier and the sync phase splitter.

2. A low-pass filter is included between the sync phase splitter and the AFC sync discriminator.

3. Flyback pulses are applied to the AFC sync-discriminator section.

4. Output from the sync phase splitter is applied to the horizontal-output section.

5. Most horizontal oscillators employ a basic Colpitts configuration.

6. A typical AFC range is from +0.5 volt to −0.5 volt.

7. Comparison waveforms are produced by integrating flyback pulses.

8. Typical AFC discriminators employ semiconductor diodes.

9. An AFC ringing coil has a resonant frequency of 15,750 Hz.

10. A silicon-controlled rectifier is a three-junction semiconductor device that is normally an open circuit until an appropriate gate pulse is applied, whereupon it switches to its conducting state.

11. A configuration that employs both tubes and solid-state devices is called a *hybrid circuit.*

12. Typical high-voltage configurations provide an output of +115 volts.

13. The focus anode requires a higher DC potential than the accelerating anode in a color picture tube.

14. An SCR trace switch is associated with a trace diode.

15. Zener diodes are generally employed in AFC discriminator circuits.

Multiple Choice

1. An AFC sync discriminator is basically a _____ circuit.
   - (a) single-phase
   - (b) two-phase
   - (c) three-phase
   - (d) none of the above

2. If an AC pulse is passed through a semiconductor diode, the output is _____ .
   - (a) sinusoidal
   - (b) parabolic
   - (c) exponential
   - (d) a DC pulse

3. Thermistors are employed in AFC networks to _____ .
   - (a) filter the AFC output
   - (b) rectify the AFC output
   - (c) stabilize AFC operation
   - (d) control the AGC voltage

4. An AFC ringing coil operates to ———————— .
   - (a) minimize picture jitter
   - (b) suppress parasitic oscillation
   - (c) optimize high-voltage regulation
   - (d) resonate the deflection coils

5. Tube-type horizontal-sweep and high-voltage systems operate at ————————————— DC voltages as their solid-state counterparts.
   - (a) the same
   - (b) lower
   - (c) higher
   - (d) none of the above

6. Tube-type AFC discriminators generally employ ———————— .
   - (a) tubes
   - (b) semiconductor diodes
   - (c) silicon controlled rectifiers
   - (d) varicaps

7. Trace switches and trace diodes conduct in ———————— .
   - (a) the same direction
   - (b) opposite directions
   - (c) both directions
   - (d) resonance

8. Retrace switches and retrace diodes conduct in ———————— .
   - (a) the same direction
   - (b) opposite directions
   - (c) both directions
   - (d) anti-resonance

9. A commutating coil is located between the ———————— .
   - (a) retrace switch/diode and the trace switch/diode
   - (b) sync phase splitter and the blocking oscillator
   - (c) the high-voltage section and the picture tube
   - (d) ringing coil and the deflection yoke

10. Horizontal-scanning linearity is optimized by———————— .
    - (a) ferrite beads
    - (b) an adjustable resonant circuit
    - (c) thermistors
    - (d) light-dependent diodes

11. A hybrid configuration employs both ———————— .
    - (a) tubes and semiconductors
    - (b) series circuits and parallel circuits
    - (c) printed circuits and point-to-point wiring
    - (d) modules and regulator diodes

12. The horizontal retrace interval is ⎯⎯⎯⎯⎯⎯ than the forward-trace interval.
    (a) longer
    (b) shorter
    (c) higher
    (d) lower

13. Retrace and flyback are ⎯⎯⎯⎯⎯⎯ .
    (a) opposite circuit actions
    (b) equivalent terms
    (c) regulatory functions
    (d) none of the above

14. A charged capacitor in an RC circuit decays in a/an ⎯⎯⎯⎯⎯⎯ manner.
    (a) sinusoidal
    (b) parabolic
    (c) hyperbolic
    (d) exponential

15. Horizontal sync pulses are compared with flyback pulses in the automatic ⎯⎯⎯⎯⎯⎯ section.
    (a) frequency control
    (b) regulator
    (c) speed
    (d) damper

Problems

1. A horizontal-output system is resonant at 70 kHz, and the flyback interval comprises one-half of a sine wave. What is the duration of the flyback interval?

2. If the peak voltage across the horizontal-deflection coils is 100V, the peak current is 10A, and the voltage and current are $90°$ out of phase, what is the peak power value?

3. If the peak voltage across the deflection coils is 100V, the peak current is 10A, and the voltage and current are in phase, what is the peak power value?

4. If the peak voltage across the deflection coils is 100V, the peak current is 10A, and the voltage and current are $45°$ out of phase, what is the peak power value?

5. A pair of horizontal-deflection coils have an inductance of 650 μH. The ringing frequency during flyback is 70 kHz. What is the capacitance value across the deflection coils?

6. What is the reactance of 650 μH at 70 kHz?

7. What is the reactance of 650 μH at 15,750 Hz?

8. What is the reactance of 0.007 μF at 70 kHz?

9. What is the reactance of 0.007 μF at 15,750 Hz?

10. If a sine-wave current of 2.83 amperes p-p flows through a 4-ohm resistor, how much power does the resistor dissipate?

Chapter 12

Vertical-Sweep And Raster-Geometry Circuitry

12.1   VERTICAL-SWEEP REQUIREMENTS

Vertical-deflection coils have significant resistance in addition to their inductance. Accordingly, the vertical-deflection coils present an inductive impedance load to the vertical-sweep section. Since the vertical-deflection current waveform has a sawtooth shape, the corresponding voltage waveform has a shape that is a function of the inductive and resistive load components. As depicted in Fig. 12-1, a pulse-voltage waveform is required to drive a sawtooth-current waveform through a pure inductance. On the other hand, a sawtooth-voltage waveform is required to drive a sawtooth-current waveform through a pure resistance. It follows from the superposition theorem that a peaked-sawtooth voltage waveform will be required to drive a sawtooth-current waveform through an inductive impedance. A *vertical-sweep section* functions to generate the necessary peaked-sawtooth output-voltage waveform.

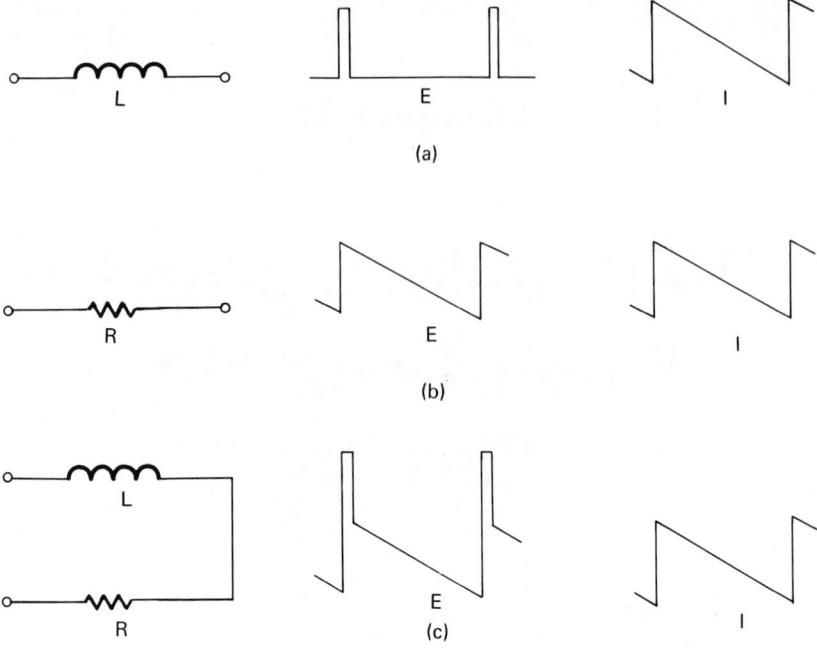

FIGURE 12-1   Vertical-sweep voltage and current waveforms. (a) Voltage and current waveforms for pure inductance; (b) voltage and current waveforms for pure resistance; (c) voltage and current waveforms for an RL load.

12.2   VERTICAL-SWEEP CIRCUITRY AND OPERATION

A block diagram for a typical vertical-sweep section is shown in Fig. 12-2. This is basically a free-running sweep-oscillator arrangement, which is synchronized by trigger pulses from the vertical-sync integrator. Figure 12-3 depicts the formation of a vertical trigger pulse from the vertical sync-pulse sequence. As noted previously, the time-constant of the vertical integrator is sufficiently long that little output voltage results from application of horizontal sync pulses that are comparatively narrow. Since the equalizing pulses are only half the width of horizontal sync pulses, the equalizing pulses produce even less output. On the other hand, the comparatively wide vertical sync pulses charge up the integrator capacitor substantially and, thus, produce an output trigger waveform.

Note that the equalizing pulses are provided in the sync-pulse sequence to obtain good interlacing, since successive horizontal-sync trains end with a half-line interval from one vertical pulse and a full-line interval from the next vertical pulse. The vertical-sync control level indicated in Fig.

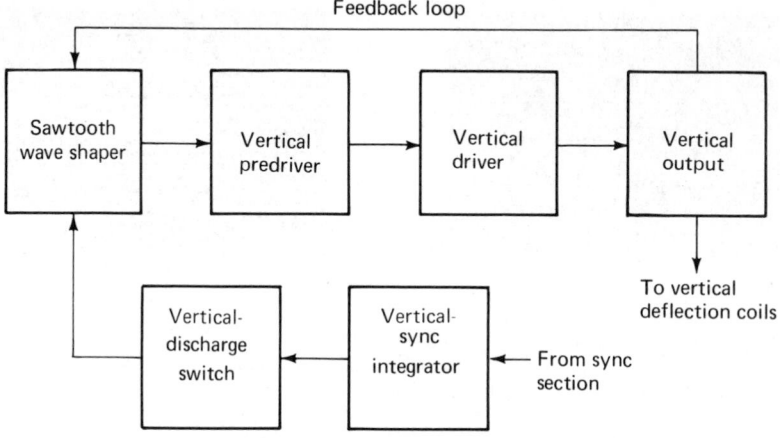

FIGURE 12-2   Block diagram of a vertical-sweep section.

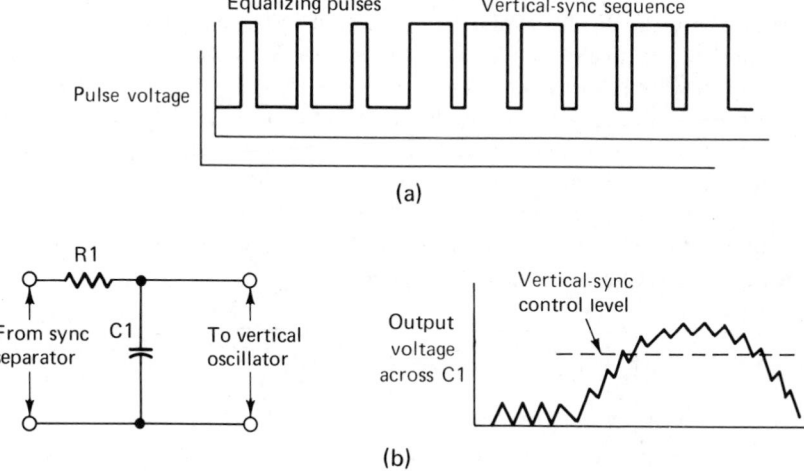

FIGURE 12-3   Operation of a vertical-sync integrator. (a) Vertical sync-pulse sequence; (b) voltage build up across integrating capacitor.

12-3(b) would shift somewhat from one field to the next, unless equalizing pulses were present. Inasmuch as the equalizing pulses are quite narrow, the integrator has time to discharge virtually to zero from the end of a horizontal-sync train to the start of the next vertical-sync sequence. It follows from the foregoing principles that the vertical oscillator must not be exposed to fly-back-pulse or horizontal-sync pulse fields, and that the integrator capacitor must not be defective. Otherwise, complete or partial loss of interlacing will result, as illustrated in Fig. 12-4. Poor interlacing results in loss of picture detail.

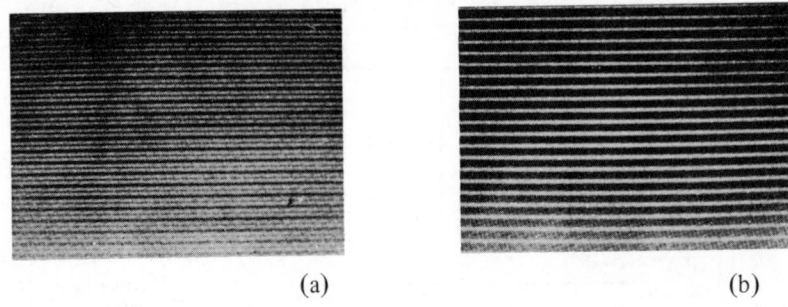

(a)   (b)

FIGURE 12-4   Examples of poor interlacing. (a) Partial loss of
interlace (line pairing); (b) complete loss of interlace (line overlay).

Referring to Fig. 12-2, note that the output pulse from the vertical-
sync integrator is applied to the *vertical-discharge switch*. This is an elec-
tronic-switch arrangement that causes the sawtooth wave shaper to start its
retrace interval slightly earlier than it would in its free-running mode, thus
locking the vertical-sweep rate to the vertical sync-pulse rate. Basically, a
sawtooth wave shaper operates by charging a capacitor through a resistance.
Since the charge rises exponentially, as depicted in Fig. 12-5, suitable circuit
means must be employed to linearize the charge rate. In some receivers, a
constant-current source is utilized; in others a negative-feedback loop is used.
Additionally, in general practice a vertical-linearity control is provided so that
scanning linearity can be optimized when required because of component
aging, or replacement component tolerances.

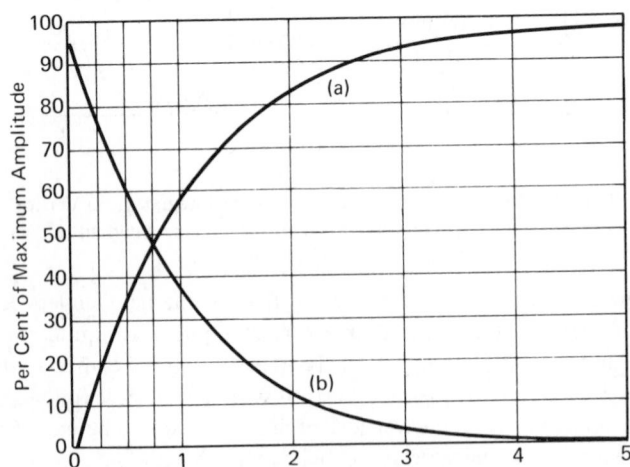

FIGURE 12-5   Universal RC time-constant chart.

To generate a peaked-sawtooth waveform, a peaking circuit like the one shown in Fig. 12-6 is often employed. Note that the transistor is biased to cut-off and is driven into conduction for brief intervals by pulses applied to the gate. During the comparatively long time that the transistor is cut off, capacitor C charges through R and generates an approximate sawtooth waveform. Then when a pulse is applied to the gate, the transistor suddenly shunts a low value of resistance across the capacitor C and the peaking resistor. Observe that if the peaking resistor were short-circuited, an approximate sawtooth waveform would be obtained. On the other hand, if the capacitor were short-circuited, an amplified pulse waveform would be obtained. Therefore, with both C and the peaking resistor in the circuit, a combination sawtooth and pulse waveform is obtained. The amplitude of the peaking pulse depends upon the value of the peaking resistor.

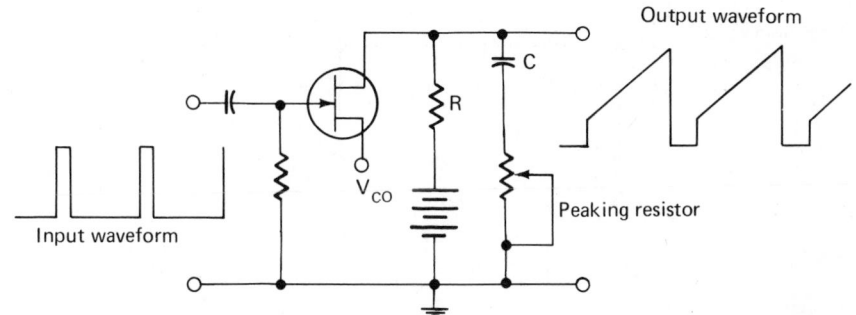

FIGURE 12-6   A peaked-sawtooth waveform generator.

Most receivers utilize vertical-blanking pulses. These pulses are usually obtained by shaping the peaked-sawtooth waveform, as depicted in Fig. 12-7. A peaked-sawtooth waveform has a low-frequency region and a high-frequency region in its sawtooth interval and pulse interval, respectively. When a peaked-sawtooth waveform is passed through a differentiating circuit with a suitable time-constant, the input waveform is changed into a pulse output waveform; that is, a differentiating circuit is a form of a high-pass filter that operates to reject the low-frequency component and to pass the high-frequency component of the peaked-sawtooth waveform.

12.3   CONFIGURATION OF A VERTICAL-SWEEP SYSTEM

A typical vertical-sweep system is depicted in Fig. 12-8. Three amplifier stages are utilized, comprising vertical predriver Q503, vertical driver Q504, and vertical-output transistor Q802. These stages operate in a relaxation-oscillator configuration to generate a peaked-sawtooth waveform, which is

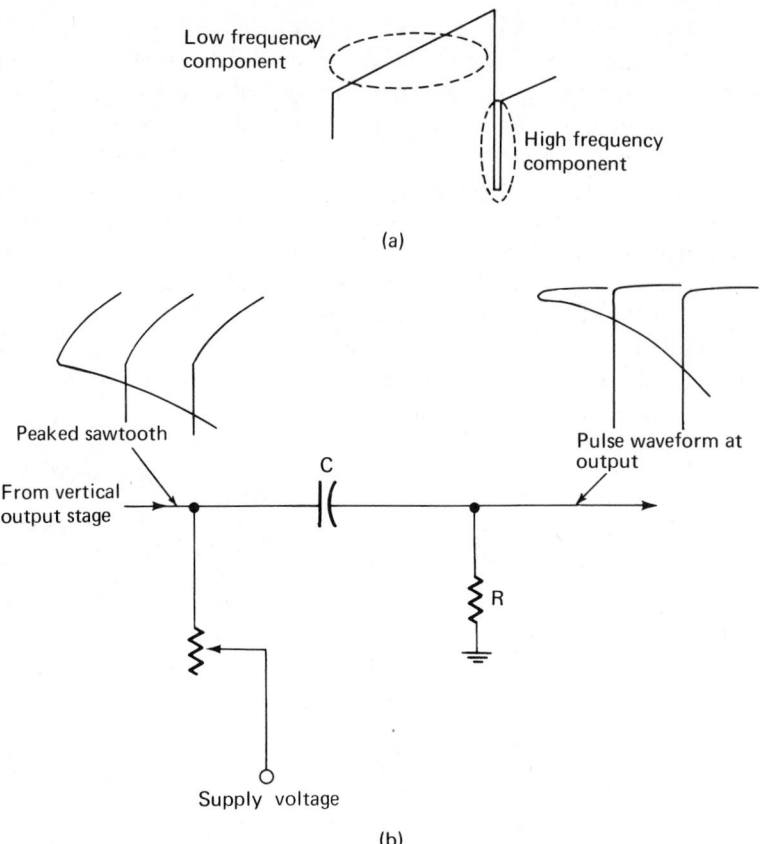

(a)

(b)

FIGURE 12-7   Waveshaping circuit for producing blanking pulses. (a) Frequency components of a peaked-sawtooth waveform; (b) differentiating circuit produces blanking-pulse waveform.

applied to the vertical-deflection coils. At the start of a vertical scanning interval, C502 begins to charge, and Q503 goes into conduction, followed by Q504 and Q802, causing current to start flowing through the vertical-deflection coils. Note that feedback resistor R506 operates in series with the deflection coils through a secondary winding on the vertical-output transformer T802, dropping a voltage, which is proportional to the deflection-coil current across this resistor. This feedback to the integrating capacitor C502 tends to keep the charging rate constant, thereby linearizing the sawtooth deflection waveform.

Observe that the vertical switch Q501 in Fig. 12-8 provides a discharge path for C502. Vertical-sync pulses via R511 and Zener diode ZD501 are coupled through C506 to trigger the vertical switch, Q501 causing the

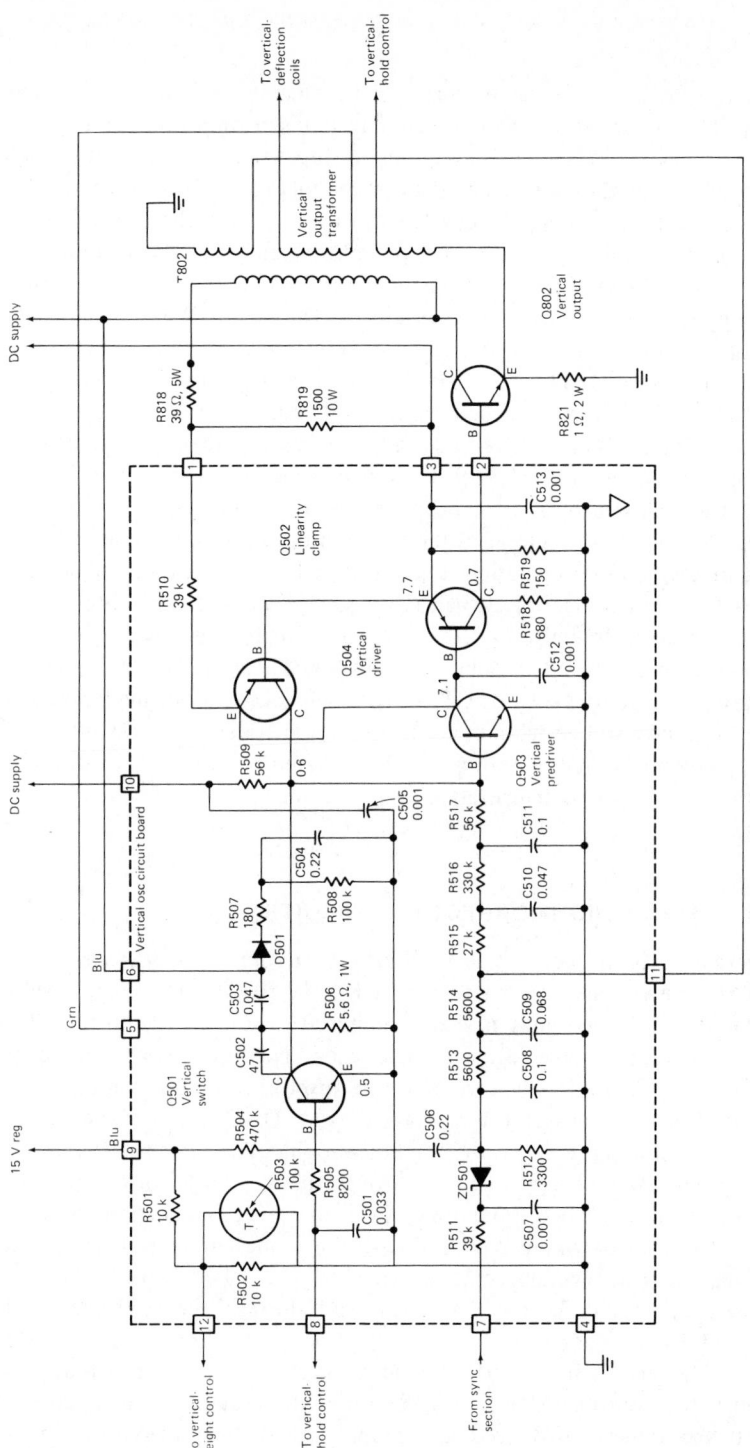

FIGURE 12-8 Configuration of a typical vertical-sweep system.

249

relaxation oscillator to operate slightly faster than it would in its free-running mode. The generated waveform is maintained in synchronism with the incoming vertical-sync pulses. This free-running repetition rate is adjustable by means of a potentiometer called the *vertical-hold control* (not shown), which sets the base-bias voltage on Q501. Note also that the charging voltage for C502 is adjustable by means of a potentiometer called the *vertical-height control* (not shown). The vertical-height control allows the user to determine the amplitude of the generated waveform. Thermistor R503 stabilizes this amplitude over an appreciable range of temperature variations. Finally, the clamp transistor Q502 establishes the initial charging current level so that the raster is fixed in position and cannot shift vertically.

Here, it is instructive to consider the typical tube-type vertical-sweep system diagrammed in Fig. 12-9. This arrangement employs a triode blocking oscillator, followed by a triode output stage. As in all tube-type receivers, the tubes are checked or replaced at the outset in the event of malfunction. However, in case normal operation is not restored, circuit tests must be made. Apart from tubes, electrolytic capacitors such as C2 and C3 in Fig. 12-9 are prime suspects. Note also that the internal resistances of the circuits are extremely high compared to solid-state designs. Therefore, slight amounts of leakage in capacitors such as C72 or C70 will cause serious malfunctioning. Keep in mind also that worn or erratic potentiometers such as R3 and R4 can cause puzzling trouble symptoms. A deteriorated potentiometer often tends to drift in value as its temperature changes gradually owing to circuit current flow.

12.4   PINCUSHION-CORRECTION CIRCUITRY

Large-screen color picture tubes require *pincushion-correction circuitry* to eliminate raster curvature, as shown in Fig. 12-10. This is accomplished by driving the deflection coils at greater amplitude as the central portion of the scan at the edge of the raster is approached. Top-bottom pincushioning is corrected by modifying the vertical-sweep waveform. Note that in Fig. 12-11, the vertical-sweep waveform is coupled from Q802 through the vertical-output transformer T802 to the top-bottom pincushion circuit. Observe that the vertical-sawtooth current flows through the two center or control windings of reactor T553. Current flows from the vertical-deflection coil L812, through a control winding of T553, through one winding of L802, back through the other winding of L802, through the other control winding of T553, to vertical-deflection coil L811, and through the vertical-centering control R766 to ground.

Observe that the vertical-deflection current in Fig. 12-11 that passes through the control winding of T553 varies the reluctance of the cores on which the outer (load) coils are wound. Also, the horizontal-deflection

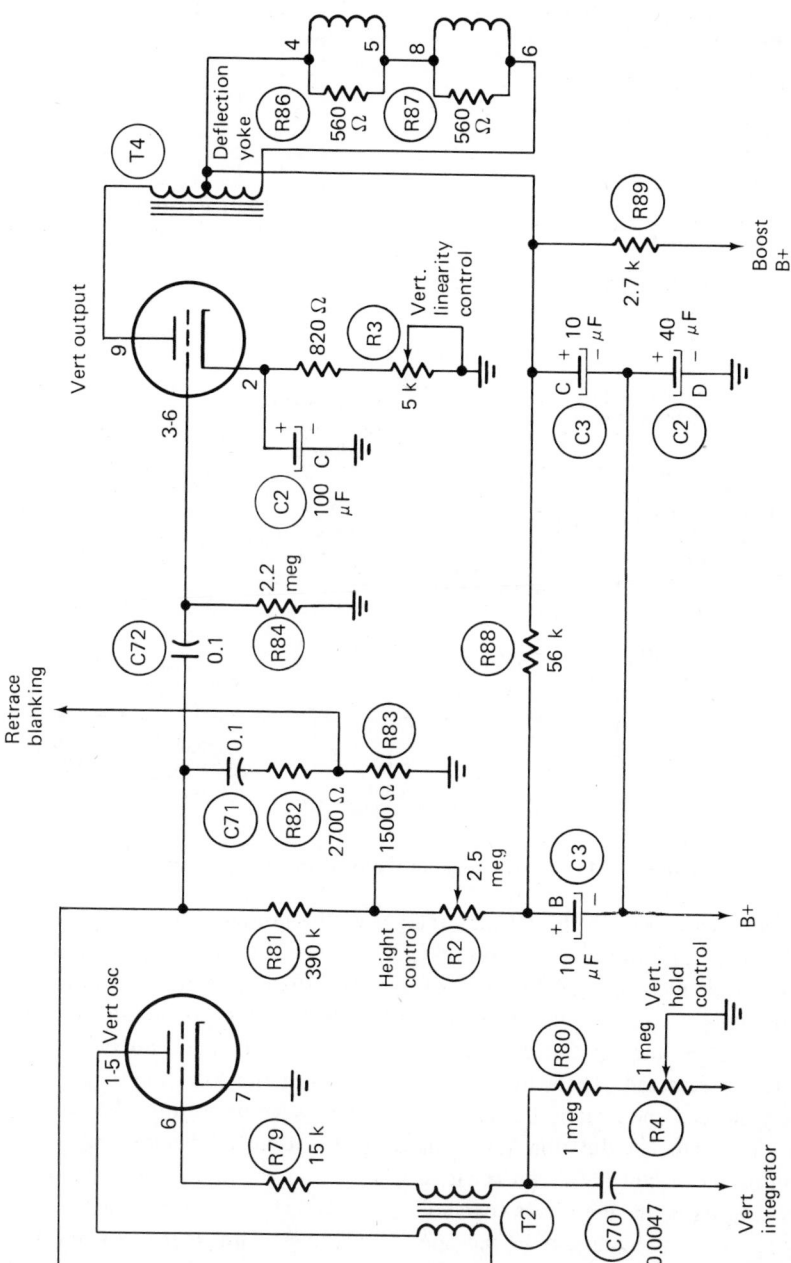

FIGURE 12-9   Typical tube-type vertical-sweep system.

251

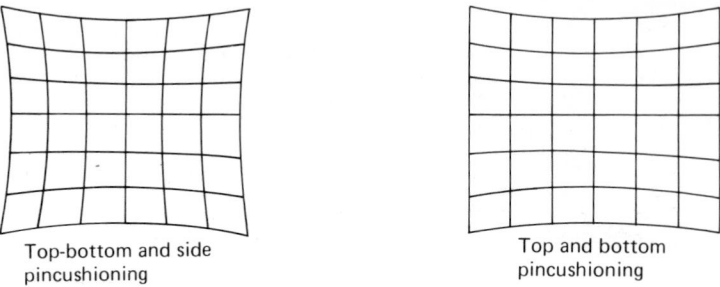

Top-bottom and side pincushioning

Top and bottom pincushioning

FIGURE 12-10   Appearance of raster pincushion distortion.

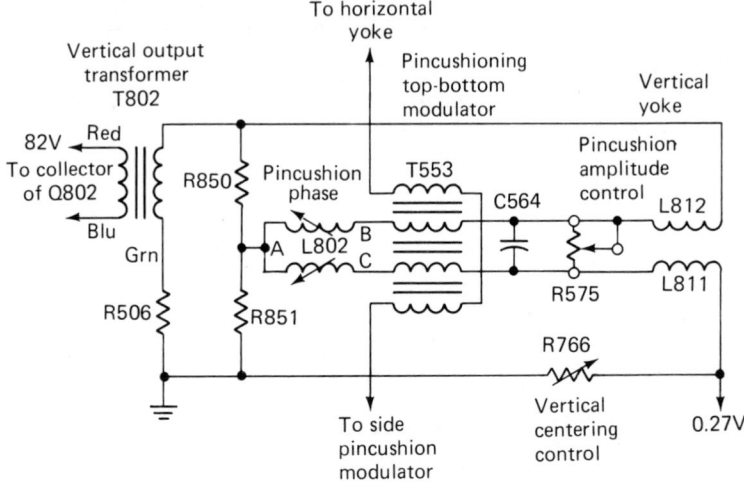

FIGURE 12-11   Top-bottom pincushion correction circuitry.

current flows through these load windings to the horizontal-deflection coils. Reluctance variation in the cores of T553 causes a corresponding variation in load inductance. During one-half of the vertical-deflection waveform, one load winding increases in inductance with respect to the other load winding. On the other half of the vertical-deflection waveform, the inductance relations are reversed. These inductance variations in the load windings effectively induce modulated horizontal-deflection waveforms into the control windings. This modulation takes place at the vertical-deflection rate. Pincushion phase coil L802 and capacitor C564 filter the deflection waveforms into approximate sine waves (parabolic waves), which are then added to the vertical-deflection current in the control windings. Thus is top-bottom pincushion correction effected.

It is evident that pincushion-correction sine waves must be correctly phased with respect to the vertical-deflection currents in Fig. 12-11, so that the added amplitude variation is introduced at the correct time. Phase is established by adjustment of the inductance in L802. It is also necessary that the amplitude of the correction waveform has a value that provides adequate redirection of the scanning beam, without either over-correction or under-correction. Accordingly, R575 is provided for pincushion amplitude control. These maintenance adjustments are attended to in case of component replacement to compensate for manufacturing tolerances.

Side pincushion correction is accomplished by amplitude-modulating the horizontal-deflection current at the vertical-scanning rate. The result of this modulation is an increase of the horizontal scan width in the center of the raster as compared with the scan width at the top and bottom of the raster. A saturable-reactor arrangement is employed, as shown in Fig. 12-12. Observe that a parabolic wave with the vertical-frequency repetition rate is generated by the control winding of T552, capacitor C565, and resistors R577 and R574. In turn, this parabolic waveform, which is coupled to the horizontal deflection-coil circuit by T552, modulates the amplitude of the horizontal-deflection current and effects the required variation in horizontal scanning width.

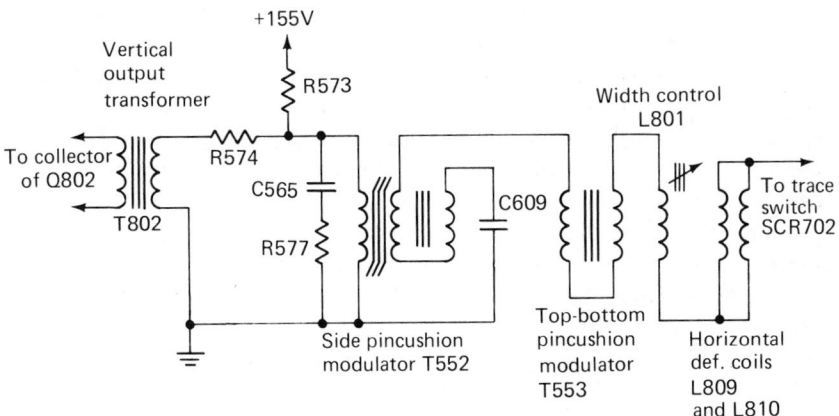

FIGURE 12-12   Side pincushion correction circuitry.

12.5   TROUBLESHOOTING VERTICAL-SWEEP MODULES

Low-level components of the vertical-deflection circuit are mounted on a module, but the output transistors, control circuits, and load are found at various points on the main chassis in one type of solid-state receiver. In turn, when trouble is suspected in the low-level section, the technician need only replace the module to determine whether normal operation is restored. Referring to Fig. 12-13, vertical retrace starts when Q1 is driven into conduction either by a sync pulse or by a signal from a feedback loop B. In turn, this conduction discharges capacitor C7, and removes the forward bias from Q2 and Q3 so that they cut off. Also, the resulting maximum positive collector voltage on Q3 saturates Q101, the NPN output transistor, while cutting off the PNP output transistor Q102.

Maximum electron flow occurs through R13 on the module, through the yoke and pincushion circuitry, and continues through the NPN output transistor to the positive-source voltage. This current deflects the scanning beam from the center of the picture-tube screen to the top (half of the retrace interval). During this half of the retrace interval, a positive-going retrace pulse is generated by the inductance of the yoke windings, and part of this retrace voltage is fed back to the module via feedback loop C. In turn, the leading edge of this pulse drives Q1 hard and into saturation (to decrease its turn-on time), and also saturating Q4 so that its collector resistance grounds out the sync. In other words, grounding the sync prevents possible false triggering from any noise spikes that might be present with the sync. Next, the trailing edge of the retrace pulse drives both Q1 and Q4 back into cutoff.

At the same time that Q1 is cut off (Fig. 12-13), C7 begins to charge through R8 and the height control towards the positive source voltage. Also, Q2 and Q3 begin to conduct, and the voltage at the collector of Q3 slowly decreases. Inasmuch as the time constant of C7 and R8 (including the height control) is long compared to the time interval involved, this voltage decrease is quite linear. Feedback loops A and E improve the linearity enough that a linearity control is not required. Since the oscillator must continue to operate even when no vertical-sync pulse is applied, a turn-on pulse for Q1 must be provided from the yoke circuit. As the scan approaches the bottom of the raster, conduction of the PNP output transistor is increasing toward maximum, and the voltage in feedback loop B is also increasing. When this voltage reaches a critical value, Q1 is turned on and retrace begins. In this manner, the vertical-sweep system can operate as a free-running oscillator.

The resistance of the height control determines the amount of charge that C7 will take during the allotted time. In a pulse circuit, the voltage value determines the pulse amplitude. An increase in resistance value reduces the voltage across C7 and reduces the raster height. Conversely, a decrease in resistance values increases the height. Figure 12-14 shows the vertical-output stage and the yoke circuit for the modular chassis. Q101 and Q102 operate as a complementary-symmetry push-pull output stage. The base drive to the transistors is a sawtooth waveform that rises to a maximum positive value at the end of retrace. In turn, Q101 conducts while scanning the upper half of the raster, and Q102 conducts while scanning the lower half of the raster. Diode CR102 provides a voltage offset between the two bases to

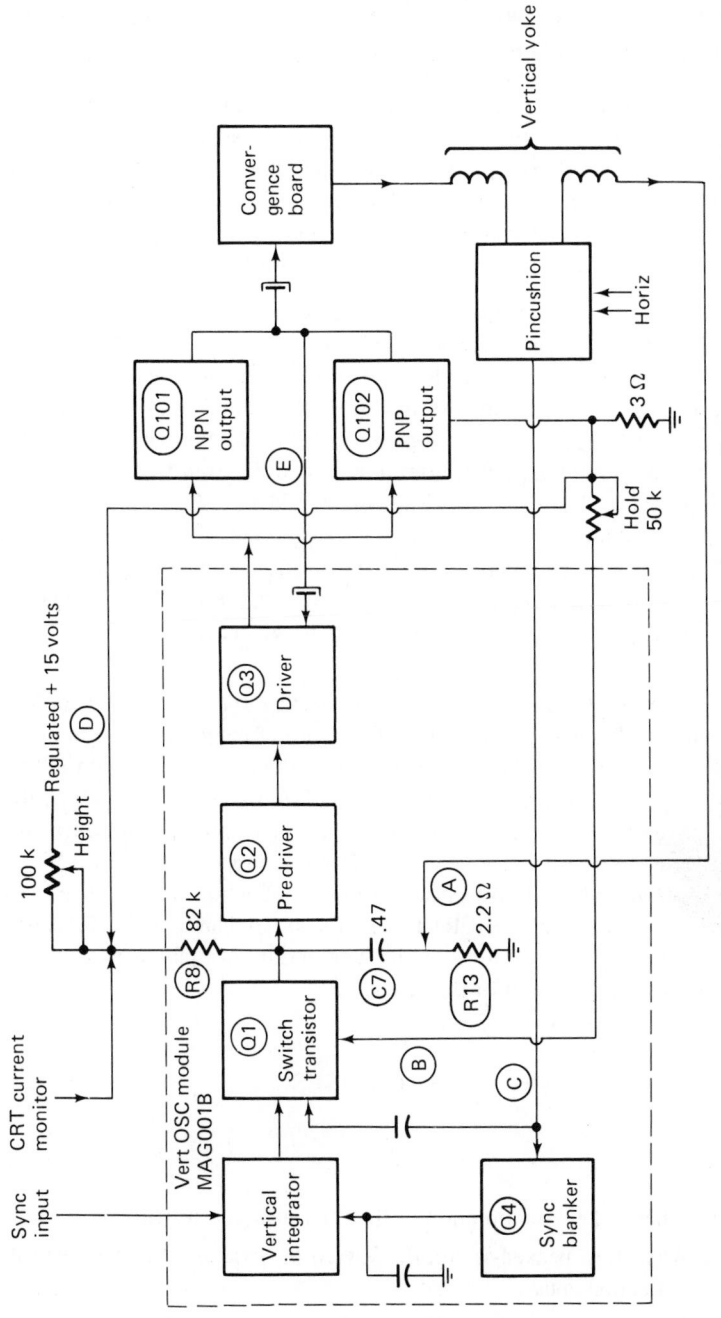

FIGURE 12-13 Block diagram of vertical-sweep section (module enclosed by dotted lines).

minimize crossover distortion that might occur while scanning the center of the raster. The Zener voltage regulator, CR4 (located on the module) limits the maximum positive drive to the base of Q101. This 65-volt limit prevents excessive current flow during retrace.

R128 and R129 (Fig. 12-14) limit the emitter current to Q101 and Q102. R4 serves as a fuse to protect the power supply in case of a short-circuit. It also develops a pulse voltage which is fed back through loop B in Fig. 12-13, to trigger oscillation of Q1, and is also fed to the height control via loop D to improve linearity at the bottom of the raster. Troubleshooting the output stage usually involves the following procedures:

1. The most likely fault in the output stage is shorted transistors. If one of the transistors short-circuits, it is likely that the other transistor will also short-circuit because of the overload. In turn, R4 burns out, which removes the overload from the power supply. Because R4 is used as a fuse, an exact replacement should be provided. When the transistors are replaced, attention should be given to the insulating spacers, and ample silicone grease should be used to provide good heat sinking. Short-circuits in the output transistors can cause other failures on the module. As an illustration, a base-collector short-circuit in Q101 applies 74 volts directly to CR4, the Zener regulator, causing it to burn out. Unless CR4 is replaced, the new transistor is apt to be short-lived.

2. The output transistors should be checked for short-circuits before installing a new module. Also, CR4 should be checked for an open-circuit before replacing the output transistors. In case both of the output transistors are short-circuited from collector-to-emitter, the driver transistor is likely to be found open-circuited. Although this fault will not damage new replacement transistors, there will be no vertical deflection. Absence of vertical deflection can also be caused by an open-circuited CR102. On the other hand, if CR102 short-circuits, a barely visible white bar about one-half inch wide will appear in the center of the raster.

EXERCISES

Questions

1. What is the basic function of the vertical-sweep system?
2. Why is a peaked-sawtooth waveform required by the vertical-deflection coils?
3. How does the vertical integrator develop a trigger pulse?

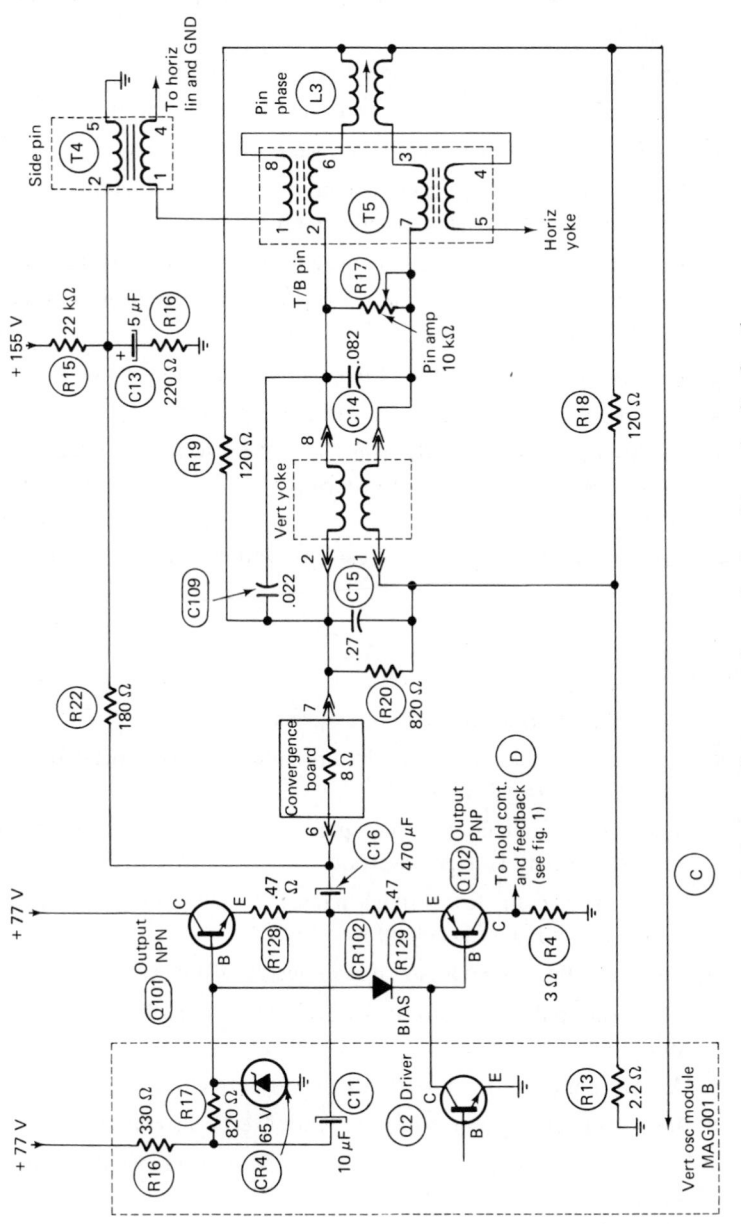

FIGURE 12-14 Output, yoke and pincushion circuits for the vertical-sweep module.

257

4. When does the output waveform from the vertical integrator rise to its maximum amplitude?

5. Where are the equalizing pulses located in the composite sync signal? The serrated pulses?

6. Explain the basic operation of a sawtooth wave shaper.

7. Describe a method of generating a peaked-sawtooth waveform.

8. Briefly discuss the malfunction termed *poor interlacing.*

9. Identify the low-frequency component and the high-frequency component in a peaked-sawtooth waveform.

10. What is the result of differentiating a peaked-sawtooth waveform?

11. How does a tube-type vertical-sweep system differ from its solid-state counterpart?

12. Explain the basic operation of *pincushion-correction circuitry.*

13. Describe two types of raster pincushion distortion.

14. Discuss the exponential rise and fall of the waveform produced by an RC differentiating circuit in terms of its first and second time-constants.

15. What is the result of processing a sine wave through an integrating circuit?

True-False

1. Vertical-deflection coils have inductive and resistive components.

2. Sawtooth current flow through pure inductance is associated with a sawtooth voltage waveform.

3. Sawtooth current flow through pure resistance is associated with a peaked-sawtooth voltage waveform.

4. Sawtooth current flow through series inductance and resistance is associated with a pulse voltage waveform.

5. Vertical-deflection coils have distributed capacitance.

6. Equalizing pulses are provided in the sync-pulse sequence to provide good interlacing.

7. A vertical integrator has very little response to equalizing pulses.

8. Negative feedback is employed in typical vertical-sweep systems.

9. A constant-current charging source may be utilized instead of negative feedback in a vertical-sweep system.

10. To develop a peaked-sawtooth waveform, a pulsed transistor may include a series RC load-circuit component.

11. Differentiation of a peaked-sawtooth waveform results in an output pulse waveform.

12. The low-frequency component of a peaked-sawtooth waveform is contained in the peaking pulse.

13. A peaked-sawtooth waveform has its high-frequency component in the ramp portion of the waveform.

14. After one time-constant, a capacitor is almost completely discharged through its load resistor in an RC circuit.

15. After five time-constants, a capacitor is discharged to 37% of its initial value through its load resistor in an RC circuit.

Multiple Choice

1. A vertical-deflection current waveform has a/an _____ waveshape.
   (a) pulse
   (b) exponential
   (c) sinusoidal
   (d) sawtooth

2. A vertical-deflection voltage waveform has a/an _____ waveshape.
   (a) pulse
   (b) exponential
   (c) sawtooth
   (d) peaked-sawtooth

3. Equalizing pulses have _____ the width of horizontal-sync pulses.
   (a) 0.1
   (b) 0.2
   (c) 0.5
   (d) 0.75

4. Equalizing pulses are provided to obtain _____ .
   (a) noise rejection
   (b) sync-buzz rejection
   (c) good interlacing
   (d) good scanning linearity

5. Negative feedback is employed in vertical-sweep systems to _____ .
   (a) linearize the scanning action
   (b) provide good sound fidelity
   (c) cancel noise pulses
   (d) none of the above

6. If a capacitor is charged from a constant-current source, the charging waveform is _____ .
   - (a) exponential
   - (b) sinusoidal
   - (c) trapezoidal
   - (d) a linear ramp

7. *Line pairing* is also termed _____.
   - (a) poor interlace
   - (b) retrace blanking
   - (c) foldover
   - (d) pincushion distortion

8. When a capacitor discharges through a resistor, its initial voltage decays to 37% after _____ time-constant(s).
   - (a) 0.1
   - (b) 1
   - (c) 10
   - (d) 100

9. When a capacitor charges through a resistor, its terminal voltage is nearly maximum after _____ time-constants.
   - (a) 0.1
   - (b) 0.5
   - (d) 5
   - (d) 50

10. A peaked-sawtooth waveform has its high-frequency component associated with its _____ .
    - (a) peaking pulse
    - (b) ramp
    - (c) ramp and peaking pulse
    - (d) none of the above

11. A peaked-sawtooth waveform has its low-frequency component associated with its _____ .
    - (a) peaking pulse
    - (b) ramp
    - (c) ramp and peaking pulse
    - (d) none of the above

12. When a peaked-sawtooth waveform is differentiated by an RC circuit with a suitable time-constant, the output waveform is a

    _____.
    - (a) sinusoid
    - (b) paraboloid
    - (c) peaked sawtooth
    - (d) pulse

13. Tube-type vertical-deflection systems operate at a _____ than solid-state systems.
    (a) higher speed
    (b) lower speed
    (c) higher voltage
    (d) lower profile

14. *Pincushion distortion* denotes _____ .
    (a) pinpoint spots in the image
    (b) a "soft" image
    (c) curved raster edges
    (d) none of the above

15. Pincushion correction circuitry may employ _____ .
    (a) reluctance variation
    (b) inductance variation
    (c) neutralizing circuits
    (d) ferrite beads

Problems

1. Fields are scanned at the rate of 60 per second, and frames are scanned at the rate of 30 per second. What is the field period, and what is the frame period?

2. The vertical-deflection coils in a typical receiver have a winding resistance of 42.6 ohms. If the vertical-deflection current has a peak-to-peak amplitude of 400 mA, what is the peak-to-peak power value dissipated by the deflection coils?

3. A vertical decoupling circuit employs a 10 μF capacitor in series with a 40 μF capacitor. What is the reactance of this series combination at 60 Hz?

4. What is the reactance of the series combination of capacitors in the foregoing problem at the fifth harmonic of the vertical-deflection waveform?

5. A 10 μF bypass capacitor is shunted by a 260-ohm resistor. What is the impedance of this parallel combination at 60 Hz?

6. The vertical-deflection coils in a receiver have an inductance of 48 mH. What value of inductive reactance do the coils develop at 60 Hz?

7. What is the inductive reactance of the deflection coils in the foregoing problem at the fifth harmonic of the vertical-deflection waveform?

8. A vertical blocking oscillator employs a 0.0047 μF grid capacitor

with a 2-megohm discharge resistance. What is the time-constant of this RC combination?

9. If the vertical-deflection coils have an inductance of 48 mH and a winding resistance of 42.6 ohms, what is the Q value of the coils at 60 Hz?

10. A 100-ohm decoupling resistor has an average DC-current flow of 0.1 mA. What wattage rating should the resistor have?

Chapter 13

Intercarrier Sound
And
Audio Sections

13.1 FUNCTION OF THE INTERCARRIER SOUND AND AUDIO SECTIONS

A conventional television receiver is composed of a *picture-signal channel* and a *sound-signal channel*. The first processing circuits are common to both the picture and sound signals. After the signals have been stepped up to an appreciable level, branch circuits are employed to process the picture and sound signals separately. The plan of a typical signal system is depicted in Fig. 13-1. Observe that the picture signal is processed through an ordinary superheterodyne arrangement, whereas the sound signal is processed through a *double-conversion* superheterodyne system. It is instructive to follow the progress of the sound signal through the various receiver sections.

The sound signal at the antenna-input terminals falls in the range of 55.25-885.25 MHz. After amplification through the RF amplifier, the sound signal is converted to a typical intermediate frequency of 41.25 MHz. The 41.25 MHz signal is then stepped up through the IF amplifier and fed to the picture detector. Note that the sound signal undergoes a second conversion in the picture detector where it is heterodyned with the picture carrier to

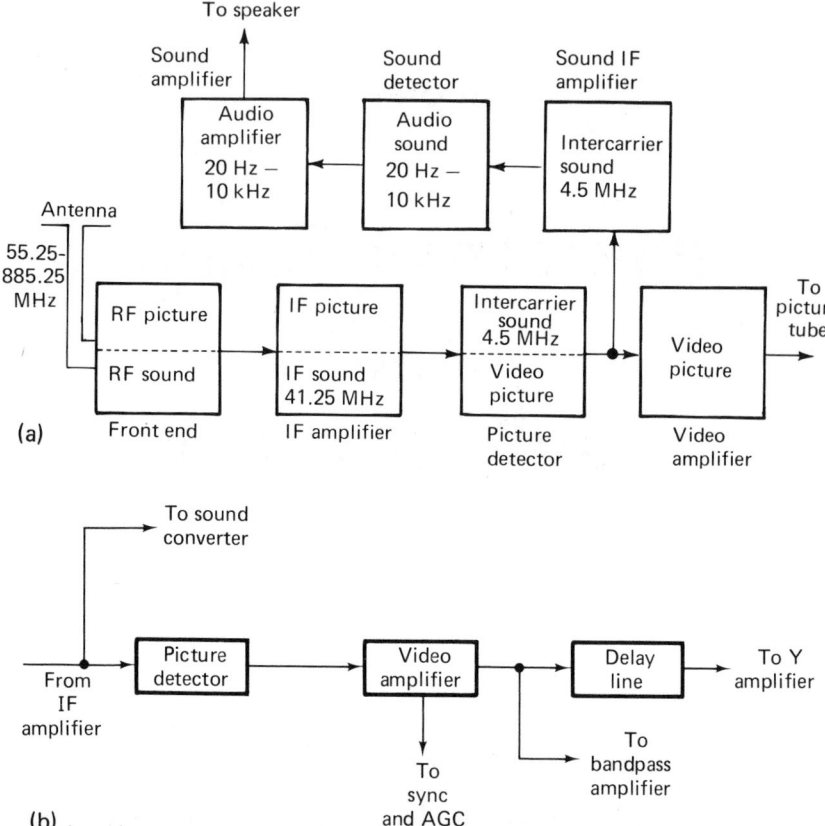

FIGURE 13-1 (a) Plan of a typical black-and-white television picture and sound-signal system; (b) sound take-off point in color receiver.

produce the 4.5-MHz intercarrier sound signal. Although picture-sideband frequencies are also present in the picture-detector circuit, the picture-carrier frequency is the strongest component, so that in effect the IF sound signal is being heterodyned with the picture carrier signal.

In Fig. 13-1, the picture and sound signals branch from the picture detector into separate channels. A 4.5-MHz tuned circuit at the output of the picture detector "picks off" the intercarrier sound signal and feeds it to the sound detector. The intercarrier sound signal is then demodulated by the sound detector and applied to the audio amplifier. This audio amplifier is conventional and has a frequency response from approximately 20 Hz to 10 kHz. The TV sound signal is a frequency-modulated transmission. Therefore, the sound detector is an FM configuration, as explained in greater detail below. It is instructive to note that the 4.5-MHz intercarrier sound signal is actually first available at the mixer output in the front end. In other words, this signal is one of the heterodyne conversion products. However, since the 4.5-MHz signal has a very low level at this point, it is not "taken off" here, but rather rejected by the IF tuned circuits. Thus, the 41.25-MHz conversion product is stepped up through the IF amplifier, and the 4.5-MHz intercarrier sound signal becomes available next as a conversion product from the picture detector.

It was noted previously that the sound-IF signal must be maintained at a comparatively low level through the IF amplifier to avoid objectionable amplitude modulation of the FM sound signal by the AM picture signal. This means that unless the sound IF level is restricted to 10 percent or less of the picture IF level, the vertical sync pulses are likely to modulate the FM sound signal to such an extent that the limiter in the sound IF section would not be able to remove the resulting 60-Hz "notches" from the FM signal. The practical consequence of this difficulty will be a rasping 60-Hz "buzz sound" from the speaker when picture signals with white backgrounds are being transmitted. Since the IF amplifier can provide only limited gain for the sound IF signal, a 4.5-MHz sound IF amplifier is employed prior to the FM sound detector, as is shown in Fig. 13-1.

13.2   INTERCARRIER SOUND AND AUDIO CIRCUITRY

Either a tuned coil or a tuned transformer may be utilized in the sound-takeoff circuit, as shown in Fig. 13-2. Sound takeoff may be located at the picture-detector output or the video-driver output. The video-driver stage provides some additional amplification for the intercarrier sound signal. Although sound takeoff could be located at the output of the video section, this is seldom done because of the problem of controlling intercarrier buzz. Suppression of buzz requires that the intercarrier signal be amplified by Class A amplifiers. Since the video-output stage would need an excessive dynamic range to avoid occasional momentary overload, it is impractical to pass the intercarrier signal through the video-output stage.

Observe in Fig. 13-2 that the sound-takeoff coil (or transformer) serves more than one purpose. In addition to picking off the 4.5-MHz sound signal, it also operates as a sound trap. A takeoff coil or the primary of a takeoff transformer inserts a substantial 4.5-MHz impedance in series with the video signal path, thus preventing the 4.5-MHz signal from passing into the video-output stage. If the sound-takeoff circuit is mistuned, the intercarrier sound signal can gain entry into the output stage resulting in the appearance of a 4.5-MHz dot (grain) interference in the picture. Note that since the bandwidth of the intercarrier sound signal is approximately 50 kHz, the sound-takeoff circuit has a fairly high Q value, typically around 80.

Sound IF amplifiers operate with much less bandwidth than video IF

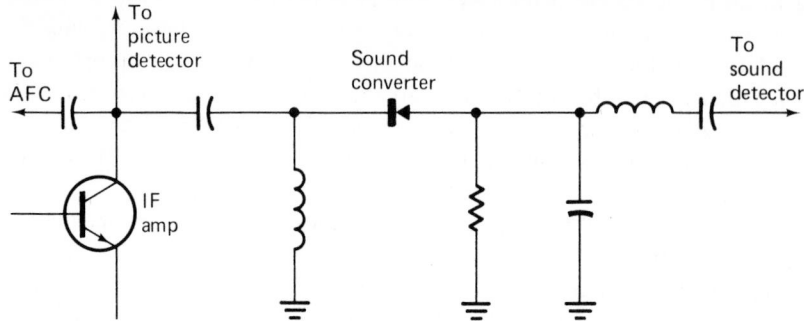

FIGURE 13-2   Typical sound take-off configuration.

amplifiers. Otherwise, the circuitry is much the same in both receiver sections. A typical intercarrier sound IF configuration is depicted in Fig. 13-3. Two stages are employed in the common-emitter configuration, followed by an FM detector. Bifilar tuned transformers are used in this example, although single-tuned coupling circuits are utilized in some receivers. To obtain a 50-kHz bandwidth the primary of T1 is loaded by R2, a 20-kilohm resistor. Neutralization of Q1 and Q2 is required to prevent regeneration and possible oscillation. This is the function of neutralizing capacitors C3 and C7. Regeneration would result in greatly subnormal bandwidth and distorted sound output, whereas oscillation would block the signal passage and produce a no-sound symptom.

Next, observe the typical tube-type intercarrier sound system shown in Fig. 13-3. This arrangement is typical of many tube-type color receivers. If a defective tube is not responsible for trouble symptoms, it is advisable to signal-trace the network from test points A through H with an oscilloscope. An incoming 4.5-MHz signal must be available, either from a TV station or from a suitable generator. These waveform checks serve to localize the faulty section in most situations. Note that if A31 or A32 are seriously misaligned, the sound output will be weak and distorted, or zero. Both tuned circuits are normally peaked at 4.5 MHz. Apart from tubes, electrolytic capacitors such as C1 and C3 are prime suspects. The operating voltages in this tube network are much higher than in its solid-state counterpart. However, the circuit actions are essentially the same.

After preliminary localization of the trouble area by means of waveform checks, it is often possible to pinpoint the defective component by DC-voltage and resistance measurements. Note that an open capacitor such as C55 or C56 in Fig. 13-3 is comparatively difficult to pinpoint on the basis of electrical measurements. An open bypass capacitor results in low stage gain and weak output. In most cases, a technician will "bridge" a suspected open capacitor with a known good capacitor to determine whether normal operation is resumed. Beginners are often puzzled by persistent sync buzz. In this

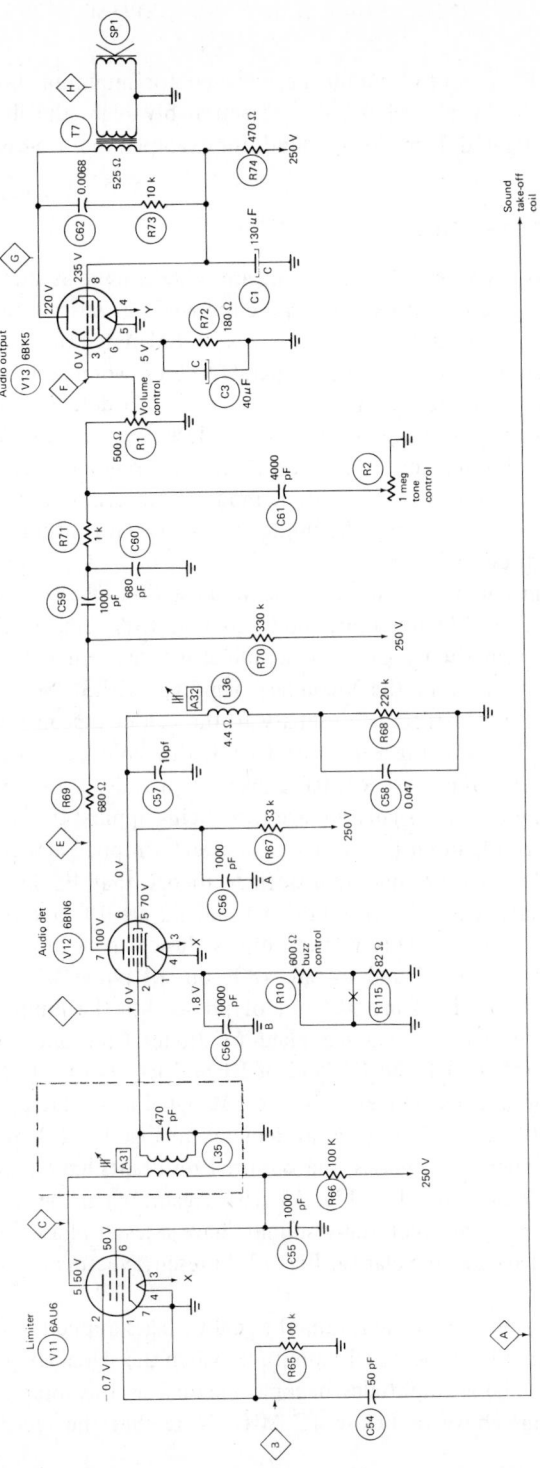

FIGURE 13-3 A typical tube-type intercarrier-sound system, with test points indicated.

267

situation, the buzz control should be adjusted for minimum buzz interference. In case the buzz level remains objectionably high, the limiter action should be investigated. It might be found, for example, that C54 is leaky.

13.3   FM DETECTION

Ratio detectors, as shown in Fig. 13-4, are widely used as intercarrier FM demodulators. Discriminators are also employed in various receivers. Although the basic operations of both types of FM detectors are essentially the same, a discriminator requires a separate limiter, whereas a ratio detector has inherent limiting action. However, since a ratio detector cannot completely limit high percentages of AM signal, designers occasionally also provide a separate limiter for a ratio detector. The fundamental action of a limiter is shown in Fig. 13-5. Amplitude-modulation variations are clipped by a limiter stage by overdriving the limiter transistor into saturation and into collector-current cut-off.

It is instructive to analyze the circuit action of a conventional discriminator type of FM detector, shown in Fig. 13-6. Signal voltage E1 is applied to the primary L1 of the discriminator transformer. In turn, signal voltage E2 is induced in the secondary winding L2-L3. Note that signal voltage is also coupled from the primary to the center tap on the secondary by capacitor C2. From the center of L2-L3, this coupled signal voltage is applied to diodes CR1 and CR2. RF choke L4 provides a DC return circuit and prevents the coupled signal voltage from being applied to the junction of C4 and C5. Thus, the input voltage E1 is, in effect, dropped across L4.

It follows from tuned-transformer theory that E2 in Fig. 13-6 is 180° out of phase with E1. Also, both E1 and E2 are 90° out of phase with the primary current I1. This relationship is shown in vector form in Fig. 13-5(b). Observe that E3 and E4 are 180° out of phase with each other, and that both E3 and E4 are 90° out of phase with the input voltage E1. Accordingly, the signal voltages applied to diodes CR1 and CR2 are the vectorial resultants of E1 and E3, and of E1 and E4, as Fig. 13-6(b) shows. When the instantaneous frequency of the FM signal is 4.5 MHz, these resultants have a 90° phase difference, as depicted in Fig. 13-7(a). When the instantaneous frequency decreases, the vectorial resultants have a greater phase difference, as demonstrated in Fig. 13-7(b). Again, when the instantaneous frequency increases, the vectorial resultants have a lesser phase difference, as shown in Fig. 13-7(c). For clarity, Fig. 13-7 presents vectors E6, E7, and ER only.

Observe that the total resultant signal voltages applied to diodes CR1 and CR2 in Fig. 13-6 have equal amplitudes when the signal frequency is 4.5 MHz. However, these amplitudes become unequal as the intercarrier signal frequency swings above or below 4.5 MHz. Note that the rectified output

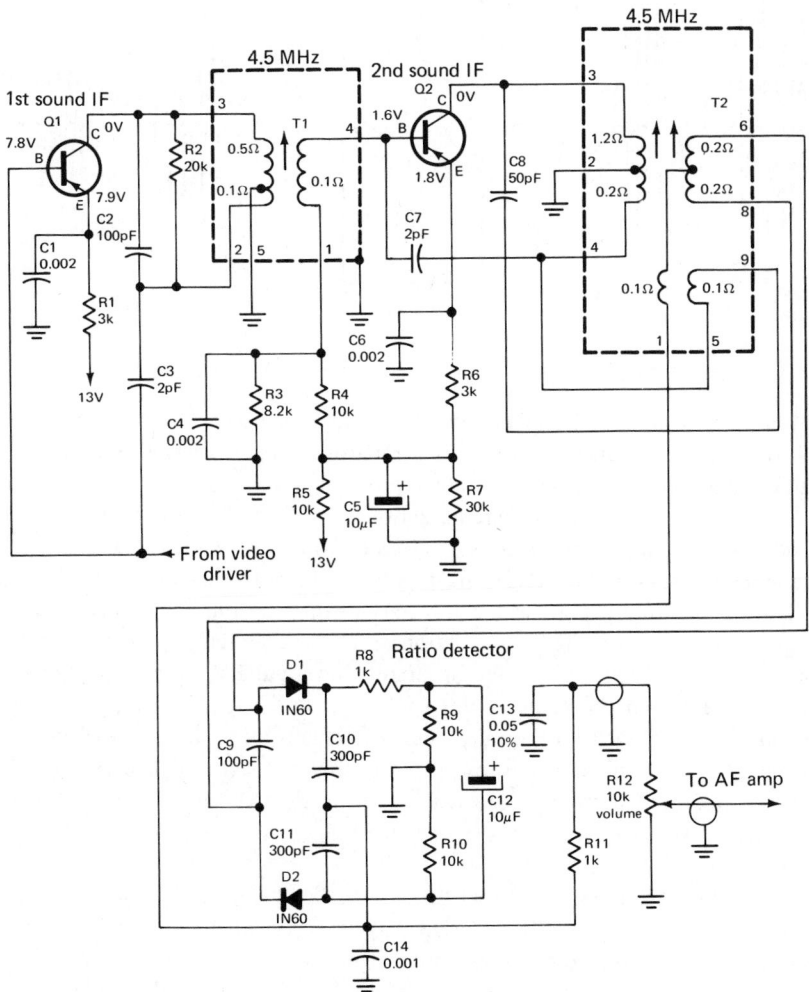

FIGURE 13-4   Example of an intercarrier sound-IF configuration with ratio detection.

voltages E8 and E9 oppose each other and cancel at 4.5 MHz. When E8 is greater than E9, the discriminator output voltage is positive, and when E9 is greater than E8, the output voltage is negative. The amplitude of this audio output voltage becomes greater as the intercarrier signal swings further from the 4.5-MHz center frequency. To repeat a basic point, the input signal to the

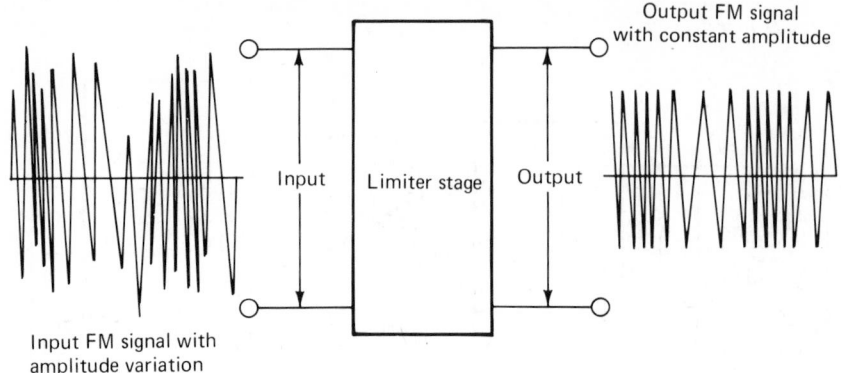

Output FM signal with constant amplitude

Input

Limiter stage

Output

Input FM signal with amplitude variation

FIGURE 13-5   Fundamental action of a limiter stage.

discriminator has constant amplitude, which is ensured by prior passage through a limiter stage. Thus a discriminator operates solely to recover the audio-modulating component from the FM signal.

Now consider the circuit action of a ratio detector in which the audio-modulating component is recovered from the FM signal with simultaneous limiting action. Referring to Fig. 13-8, note that diodes CR1 and CR2 are oppositely polarized in contrast to the previous discriminator configuration. However, the demodulation process is essentially the same in both arrangements. In the ratio-detector circuit, the input FM signal is not capacitively coupled to the center point of the diode branch; instead, a tertiary winding is employed on the ratio-detector transformer. Thus, mutual inductance M2 serves the same purpose as the coupling capacitor in a discriminator circuit. Inherent limiting action is provided in Fig. 13-8 by C3 by means of capacitive storage. It is instructive to follow the sequence of limiting operation.

Observe in Fig. 13-8 that the demodulated voltages E1 and E2 across R1 and R2 (and across C1 and C2) will be equal when the FM signal frequency is 4.5 MHz. The audio-output signal will be zero since it is developed in a bridge circuit that is balanced at 4.5 MHz. Then, if the FM signal swings more or less away from 4.5 MHz, the demodulated voltages E1 and E2 will become unequal, as was depicted in Fig. 13-7. For example, E1 might become greater than E2, and the audio-output voltage will then be equal to half the difference between E1 and E2. Note that since C3 in Fig. 13-8 has a large capacitance value, its terminal voltage can change neither appreciably nor rapidly when the FM signal amplitude changes. This is just another way of saying that any envelope variation in the applied FM signal will be effectively limited by capacitive storage action. Under most reception conditions, this inherent limiting action of the ratio detector would be satisfactory. However, in the event that the incoming FM signal had been amplitude-modulated 80%, for example, by the vertical sync pulses, the limiting action

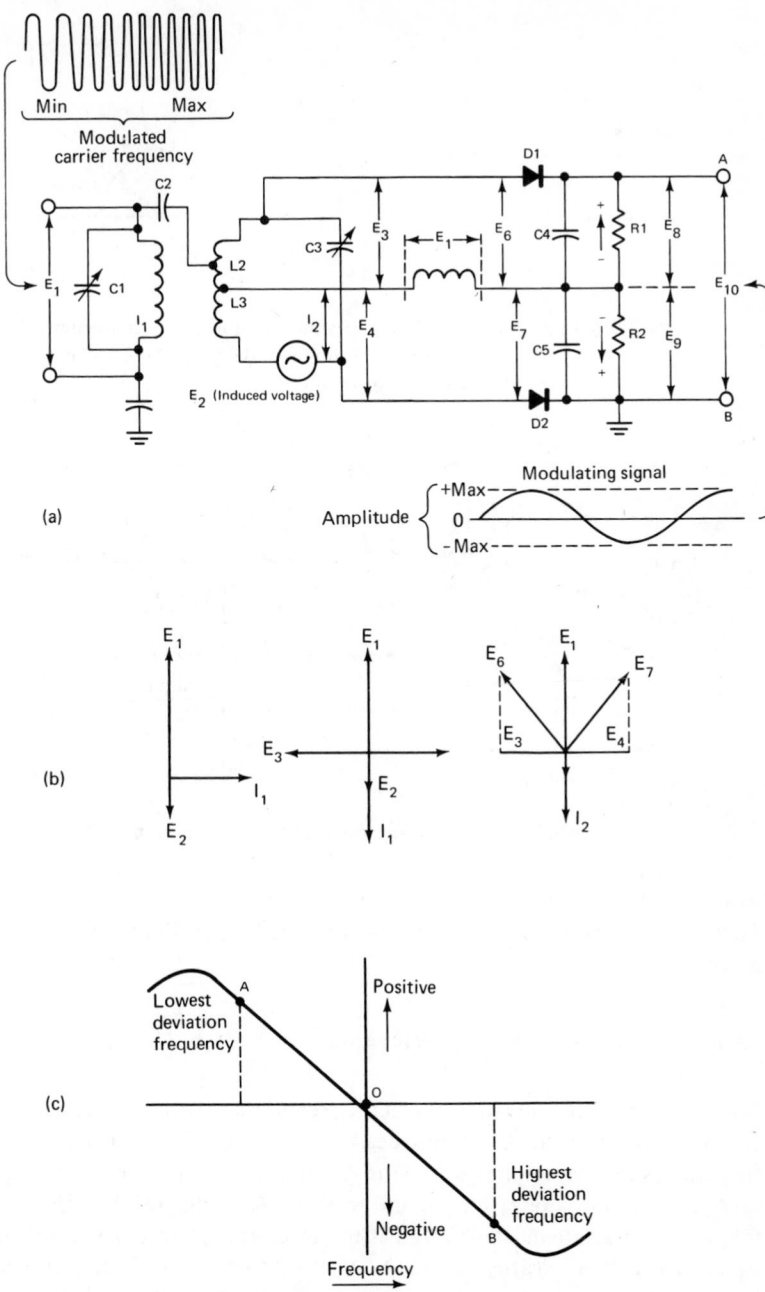

FIGURE 13-6 Discriminator type of FM detector. (a) Basic circuitry; (b) reference voltage and current phase relations; (c) discriminator-frequency response curve.

271

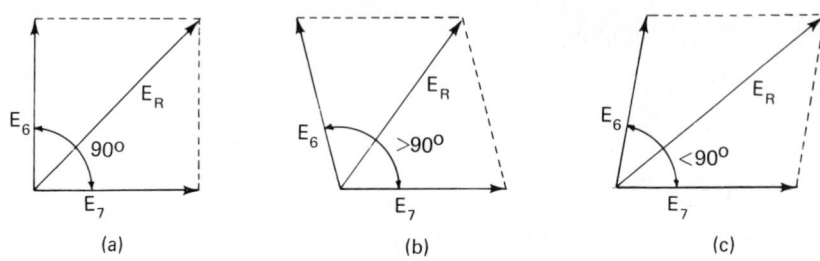

FIGURE 13-7 Vector phase variation produced by frequency swing above or below 4.5-MHz. (a) At 4.5 MHz; (b) below 4.5 MHz; (c) above 4.5 MHz.

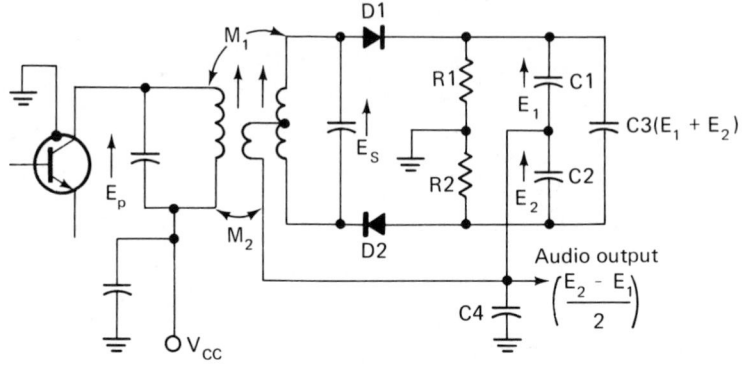

FIGURE 13-8 Basic ratio-detector configuration.

would be incomplete and sync buzz would be audible in the sound output. Therefore, a ratio detector may be preceded by a limiter in a "deluxe receiver."

13.4 AUDIO SIGNAL DE-EMPHASIS

Both discriminators and ratio detectors require a de-emphasis circuit in the audio-output branch. As in broadcast FM transmission, the higher audio frequencies are pre-emphasized at the TV transmitter in order to provide an optimum signal-to-noise ratio at the receiver. Accordingly, the higher audio frequencies must be de-emphasized at the receiver to restore the normal tonal balance. Figure 13-9(a) shows the standard pre-emphasis frequency characteristic, with the corresponding de-emphasis characteristic in Fig. 13-9(b). It is evident that the combination of these responses results in a uniform audio-frequency characteristic. An RC integrating circuit with a time constant of 75 microseconds is generally employed for de-emphasis, as shown in Fig. 13-10. This arrangement consists of three RC sections, the first consisting of the internal resistance of the FM detector and C1, followed by R2 and C2, and completed by R3 and C3 (where C3 is the capacitance of the shielded audio lead). The effective time constant of this network is approximately 75 microseconds.

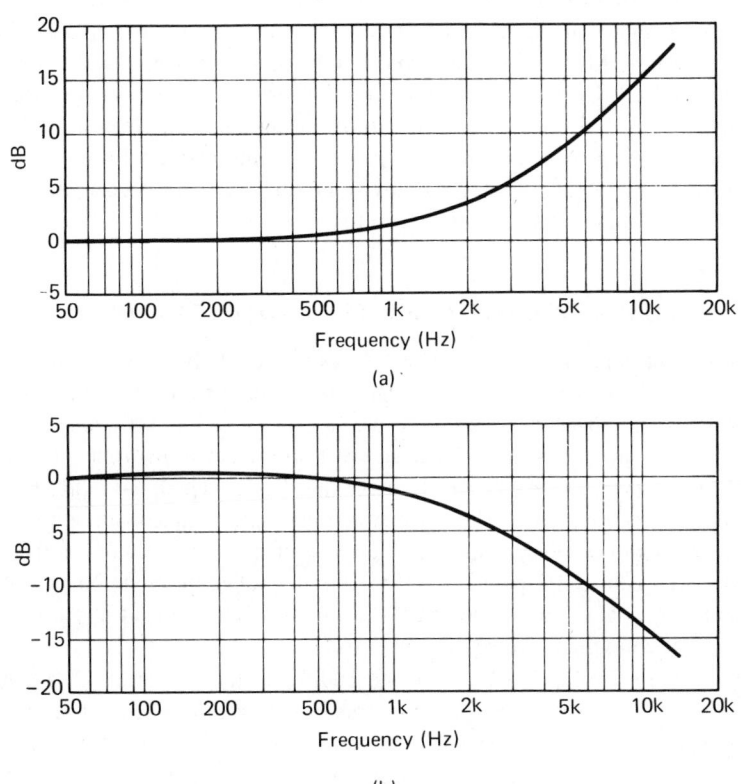

(a)

(b)

FIGURE 13-9   Pre-emphasis and de-emphasis frequency characteristics. (a) Pre-emphasis curve used at FM transmitter; (b) de-emphasis curve used at receiver.

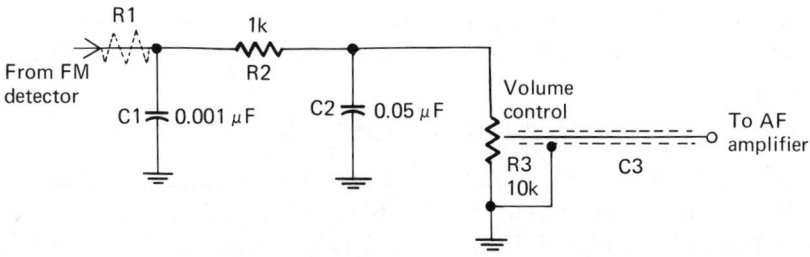

FIGURE 13-10   Typical de-emphasis network.

When limiting is employed prior to an FM detector, a transistor may be supplemented by a diode for improved limiting action, as shown in Fig. 13-11. Note that Q2 operates with a comparatively low collector-to-emitter potential of 1.4 volts. Also, the collector load for Q2 includes the 2200-ohm resistor R9. In turn, Q2 is driven into saturation on negative peaks of the base-input signal. Thus, limiting is provided on negative peaks. To obtain limiting on positive signal peaks, diode CR1 is shunted across the primary winding of T2. This diode starts to conduct on a positive-peak potential of approximately 0.3 volt. Thereafter, its shunting action increases rapidly so that limiting is provided on positive peaks. In combination with the inherent limiting of the ratio detector employed in this system, intercarrier buzz is never audible unless a circuit defect occurs.

13.5   AUDIO-AMPLIFIER CIRCUITRY AND OPERATION

Audio amplifiers in most television receivers are of the conventional type, although deluxe receivers employ high-fidelity amplifiers. The configuration shown in Fig. 13-12 is typical of those in general use. It utilizes two stages that operate in single-ended Class A, with an output transformer to couple the output stage to the speaker. Observe that C1 is a deemphasis capacitor in the ratio-detector section. C2 is a small compensating capacitor that provides improved high-frequency response at lower settings of the volume control R1. Driver transistor Q1 operates in the common-collector mode to provide a suitable impedance match between the ratio-detector output and transistor Q2. Direct coupling is utilized from Q1 to Q2.

Note in Fig. 13-12 that the audio-output transistor Q2 is a low-current, high-voltage type, similar to transistors used in video-output stages. A substantial power output with economical circuit design is thus realized. Potentiometer R6 operates as a tone control, providing a treble cut (bass boost) action. Optimum overall audio-frequency response is provided by the equalizing network composed of R6, C6 and C7. Observe that the speaker plug P1 has four terminals, two of which are arranged as a supply-voltage interlock. Accordingly, if the speaker is disconnected from the amplifier, the collector supply voltage for Q2 is removed. This is an essential safety precaution since operation of the output stage without a normal speaker load would result in an excessive peak-inverse collector voltage that would damage Q2.

13.6   INTEGRATED SOUND-SECTION CIRCUITRY

There has been a marked trend toward utilization of integrated circuitry in the intercarrier sound system. Figure 13-13 depicts a configuration in which the sound IF amplifier, FM detector, and audio amplifier are built around a single IC chip. A block diagram is shown in Fig. 13-13(a), and the external-internal plan in Fig. 13-13(b). Observe that the 4.5-MHz signal is coupled into

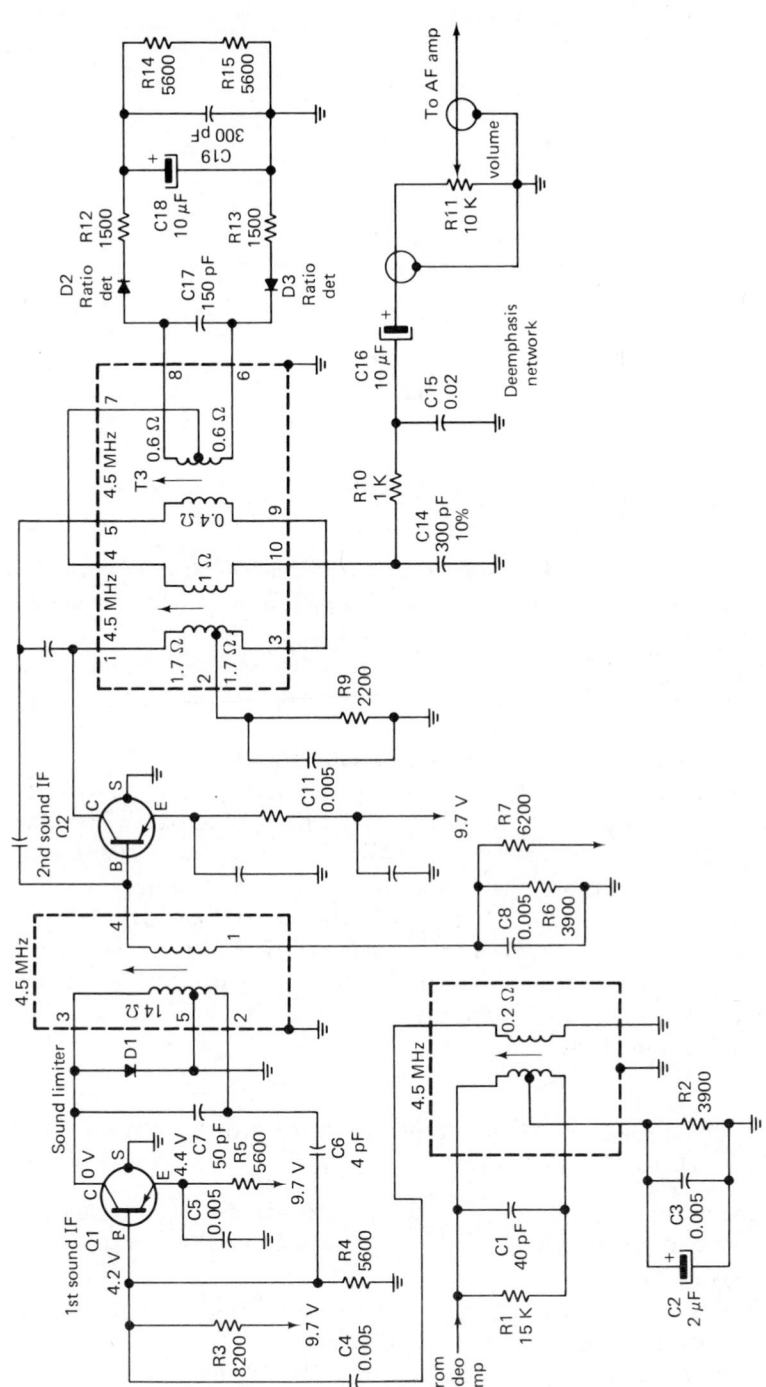

FIGURE 13-11 Transistor-and-diode limiting configuration.

275

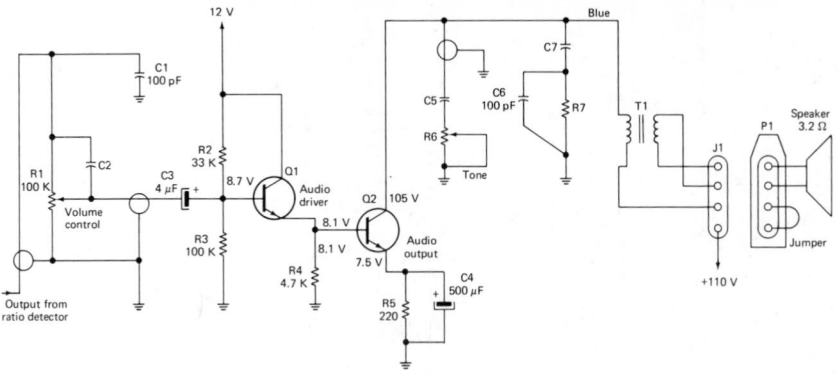

FIGURE 13-12   Typical audio-amplifier section for a television receiver.

the chip via external tuned transformer T1. The first IF stage comprises Q1, Q2, and Q3 and is directly coupled to the second IF stage. This second stage consists of Q4, Q5, and Q6. The amplified IF signal is fed to the limiter stage comprising Q7 and Q8. The limiter output in turn energizes the primary of external ratio-detector transformer T2. Note that all of the IF selectivity is provided by these two tuned transformers.

Observe in Fig. 13-13 that the ratio-detector section consists of detector diodes CR3 and CR4, with an unconventional load-resistance arrangement composed of resistors R11, R12, and reverse-biased diodes CR5, CR6, and CR7. These diodes simply operate as capacitors to filter out the 4.5-MHz component from the demodulated audio signal. An audio preamplifier circuit is also provided—consisting of transistors Q11 and Q12. As would be anticipated, a deemphasis network is required. This includes the external resistor R17 and capacitor C4. Finally, note that diodes CR1 and CR2 together with transistors Q9 and Q10 operate as a voltage regulator to stabilize the IC operation.

13.7   TROUBLESHOOTING THE INTERCARRIER SOUND AND AUDIO SECTIONS

Intercarrier sound or audio defects seldom produce any picture symptoms. Sound output symptoms include no sound, weak sound with or without buzz or background noise, normal sound with objectionable buzz level, and various forms of distorted audio output. Preliminary sectionalization is directed to location of the faulty receiver section. For example, if sync buzz is accompanied by a negative or partially negative picture, it indicates that a component defect has occurred prior to the sound takeoff point. On the other

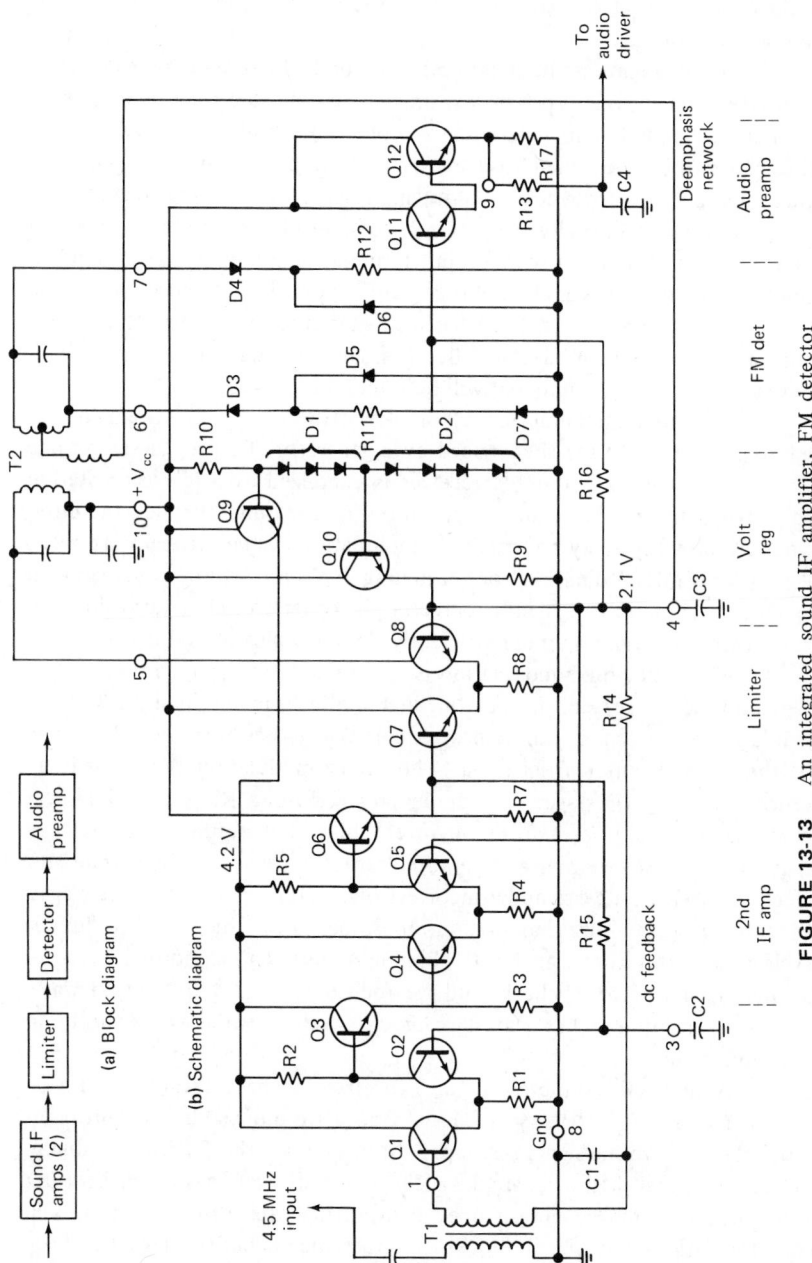

FIGURE 13-13 An integrated sound IF amplifier, FM detector and audio preamplifier. *(courtesy of RCA)*

277

hand, if picture reproduction is normal and there is no sound output, we would conclude that the trouble will be found between the sound takeoff point and the speaker.

Signal tracing is the most useful method of closing in on a defective stage. Although various types of instruments can be used for this purpose, a wide-band scope with a low-capacitance probe is most informative. Of course, the receiver must be tuned to a television station, or a suitable generator signal must be provided when signal-tracing tests are made. In normal receiver operation, a 4.5-MHz sine wave pattern will be displayed on the scope screen when the probe is applied to the input terminal of the intercarrier sound IF amplifier. The normal signal amplitude at this point is on the order of 0.2 V. This 4.5-MHz sine wave can be checked at each stage up to the FM detector. If the signal is absent or substantially attenuated at a particular stage, it indicates that a defective component will be found in this area.

Note that the amplitude of the 4.5-MHz sine wave will remain constant only if the receiver is energized by a generator. That is, the amplitude will vary to some extent if the receiver is energized by a television station signal. When there is substantial sync buzz in the sound channel, the intercarrier signal will display an amplitude modulation by the vertical sync pulse. Deep downward modulation, or "notching," places maximum demand on the limiting function of a ratio detector. In other words, a ratio detector limits upward modulation more efficiently than downward modulation.

When sync-buzz modulation is excessive, look for a stage (usually a video-amplifier stage) that is operating in a nonlinear mode. In any case, buzz modulation will occur in a nonlinear circuit that passes both the video signal and the 4.5-MHz intercarrier signal. Thus, buzz modulation can occur in an overloaded video IF stage, or even in an overloaded RF stage. Nonlinear operation is usually caused by incorrect bias that results from a leaky capacitor. However, nonlinear operation can also be caused by a transistor defect. An AGC fault can apply incorrect bias voltage to one or more stages, and thus produce buzz modulation. In some cases, the sound-IF limiter develops a defect that impairs its clipping action. This condition becomes apparent in a waveform check, and we then observe that the output waveform from the limiter has almost as much envelope variation as the input waveform in these cases.

A no-sound symptom is also localized to best advantage with an oscilloscope. The 4.5-MHz signal can be traced from the sound-takeoff point to the stage where the signal is stopped. In most cases, the difficulty is due to a defective capacitor. For example, a leaky capacitor can bias a transistor out of its operating range, and an open neutralizing capacitor often results in stage oscillation and signal blocking. After the defective stage has been localized, DC voltage measurements are generally most effective to close in on the faulty component. As noted previously in the discussion of the video IF section, oscillation results in an abnormal DC voltage output from the detector. Thus, if an intercarrier IF stage is oscillating, a high DC voltage output from the FM detector will be found.

Note that a no-sound symptom can result from a defect in the audio-amplifier section, up to and including the speaker. Again, an oscilloscope is the most useful signal-tracing instrument. Weak-sound symptoms are generally caused by marginal defects in the same components that are responsible for a no-sound symptom. For example, if a coupling capacitor has marginal leakage, the operating point of the following transistor will be shifted with resulting attenuation of the output signal. In high-level stages, particularly the output stage or the driver stage, such signal attenuation is likely to be accompanied by audio distortion. If a capacitor defect is not the cause of attenuated and/or distorted audio signals, a transistor is usually suspect. Resistors seldom cause trouble symptoms. However, a volume control, which may become excessively worn after extended service, is a notable exception.

Noisy sound output is commonly caused by insufficient limiting. Of course, the sound system develops considerable noise when no input signal is present since the signal-to-noise ratio is then zero. When noise becomes objectionable under normal reception conditions, the cause is most likely to be an open electrolytic capacitor at the output of the ratio-detector circuit. As noted previously, electrolytic capacitors may lose a substantial portion of their rated capacitance after extended use. A similar sound symptom can be caused by a defective limiter stage that fails to clip the incoming 4.5-MHz intercarrier signal. Defective transistors sometimes develop excessive noise. This condition is generally associated with low gain and abnormal or subnormal DC terminal voltages.

EXERCISES

Questions

1. Are the picture-signal channel and the sound-signal channel completely separate?
2. How does a double-conversion superheterodyne system operate?
3. What is the difference between a sound converter and a video detector?
4. Is the sound signal heterodyned with the picture carrier in a sound converter?
5. Where is the sound converter located in a color receiver?
6. Why does the intercarrier-sound IF section have a 4.5-MHz center frequency?

7. State the approximate bandwidth of the intercarrier-sound signal.

8. Discuss the requirement for neutralization in a sound-IF amplifier.

9. Describe the chief distinction between tube-type intercarrier-sound systems and their solid-state counterparts.

10. Compare ratio-detector action with discriminator action.

11. Explain preemphasis and deemphasis processes.

12. Why should an audio-output stage always be operated with a speaker load, or equivalent?

13. How many transistors does a typical integrated sound-IF amplifier, detector, and audio preamplifier include?

14. Explain the cause of sync buzz in the sound output.

15. What is a common cause of noisy sound output?

True-False

1. A color-TV receiver utilizes the same sound-takeoff point as a black-and-white receiver.

2. Intercarrier sound detection occurs in the sound-converter circuit of a color receiver.

3. Double conversion and synchronous detection are equivalent terms.

4. An audio amplifier in a color receiver has a typical frequency response from 20 Hz to 10 kHz.

5. To avoid sound interference, the sound carrier is restricted to 10 percent or less of the picture-IF level.

6. An intercarrier sound-IF amplifier has a bandwidth of approximately 4 MHz.

7. A Q value of approximately 80 is employed by typical sound-takeoff circuits.

8. Regeneration in the sound-IF amplifier results in greatly subnormal bandwidth and distorted sound output.

9. Oscillation in the sound-IF amplifier blocks signal passage.

10. Discriminators have inherent limiting action.

11. Ratio detectors have no inherent limiting action.

12. Both discriminators and ratio detectors have the same form of frequency-response curve.

13. Sync buzz is less likely to occur if a ratio detector is preceded by a limiter.

14. Preemphasis denotes progressive attenuation of the higher audio frequencies.

15. Deemphasis denotes progressive boost of the higher audio frequencies.

Multiple Choice

1. An intercarrier sound-IF amplifier has a center frequency of _____.
   - (a) 455 kHz
   - (b) 10.7 MHz
   - (c) 4.5 MHz
   - (d) 41.25 MHz

2. A TV audio amplifier has a frequency response of approximately _____.
   - (a) 20 Hz to 10 kHz
   - (b) 20 Hz to 20 kHz
   - (c) 200 to 2000 Hz
   - (d) 88 to 108 kHz

3. A sync-buzz waveform has the shape of a _____ .
   - (a) notch or pulse on the 4.5-MHz FM signal
   - (b) ripple on the FM signal
   - (c) sawtooth baseline in the FM signal
   - (d) ringing sequence in the FM signal

4. Sync-buzz pulses have a repetition rate of _____ .
   - (a) 15,750 Hz
   - (b) 4.5 kHz
   - (c) 60 Hz
   - (d) 30 Hz

5. Sound-takeoff coils in color receivers are tuned to _____ .
   - (a) 60 Hz
   - (b) 15,750 Hz
   - (c) 4.5 MHz
   - (d) 41.25 MHz

6. An intercarrier sound-IF channel has a bandwidth of approximately _____ .
   - (a) 4.5 MHz
   - (b) 3.58 MHz
   - (c) 15,750 Hz
   - (d) 50 kHz

7. Sound-takeoff circuits have a typical Q value of _____ .
   (a) 20
   (b) 50
   (c) 80
   (d) 100

8. Ratio detectors cannot _____ .
   (a) completely limit high percentages of AM signal
   (b) demodulate FM signals
   (c) operate as phase detectors
   (d) limit small percentages of AM signal

9. Clipping is accomplished by driving a limiter transistor _____ .
   (a) in class A
   (b) in class B
   (c) into saturation and into collector-current cutoff
   (d) none of the above

10. Deemphasis entails _____ .
    (a) attenuation of the lower audio frequencies
    (b) attenuation of the higher audio frequencies
    (c) attenuation of the midband audio frequencies
    (d) suppression of all audio frequencies

11. Limiter transistors operate with _____ .
    (a) high collector potential
    (b) low collector potential
    (c) injected local-oscillator voltage
    (d) sinusoidal base-emitter voltage

12. Substantial misalignment in the intercarrier-sound channel can cause _____ .
    (a) motorboating
    (b) poor picture quality
    (c) adjacent-channel interference
    (d) weak or no sound output

13. A discriminator frequency-response curve is called a/an _____ .
    (a) bandpass filter
    (b) S curve
    (d) sound takeoff circuit
    (d) deemphasis network

14. Transistor limiters are sometimes supplemented by _____ .
    (a) diode limiters
    (b) diode rectifiers
    (c) adjacent-channel traps
    (d) regulator circuits

15. If an audio-output transistor is operated under no load,_____
    is likely.
    (a) IF oscillation
    (b) RF regeneration
    (c) poor selectivity
    (d) output transistor damage

Problems

1. The Q value of a tuned circuit is approximately equal to its center
   frequency divided by its bandwidth. Calculate the Q value of a
   sound-takeoff coil that has a bandwidth of 50 kHz.

2. If a sound-takeoff coil has a Q value of 80, what is its approximate
   bandwidth?

3. An intercarrier-sound signal has a maximum deviation of ±25 kHz.
   What are the frequency limits of the signal in an IF amplifier?

4. What are the frequency limits of an intercarrier-sound signal in a
   video amplifier?

5. A deemphasis network employs a simple integrating circuit and has
   a time-constant of 75 μs. If a 0.00025 μF capacitor is utilized,
   what value of resistor will be required?

6. What is the reactance of a 0.00025 μF capacitor at 10 kHz?

7. What is the reactance of a 0.00025 μF capacitor at 1 kHz?

8. The heterodyne process generates sum and difference frequencies.
   What is the sum frequency of the sound and picture carriers generated in the sound converter?

9. Calculate the value of inductance that will resonate at 4.5 MHz
   with 0.00025 μF of capacitance.

10. What is the reactance of the inductance value found in problem
    9 at 4.5 MHz?

Chapter 14

Color-Television Accessories And Optional Features

14.1 SURVEY OF ACCESSORIES AND OPTIONAL FEATURES

Various types of accessories and optional features are available for the color-TV viewer. Some of today's standard equipment was once optional. For example, receivers were formerly offered with or without UHF tuners. However, all modern receivers are now required to have UHF tuning facilities by the Federal Communications Commission. Recently one of the most popular accessories has been some form of remote-control unit. A typical remote-control box provides pushbutton facilities for switching channels, adjusting sound volume, adjusting hue, and adjusting color intensity. Since the receiver in the previous example employs automatic fine-tuning (AFT) control, the remote-control box does not include fine-tuning facilities. Color receivers that do not employ AFT may provide a tuning-eye function to serve as a guide in adjustment of the fine-tuning control. A few receivers offer a special video-amplifier output stage to drive a video tape recorder (VTR). Figure 14-1 illustrates a typical VTR.

There is also a marked trend toward automatic tint control facilities. These arrangements differ greatly in design and in effectiveness. However, all

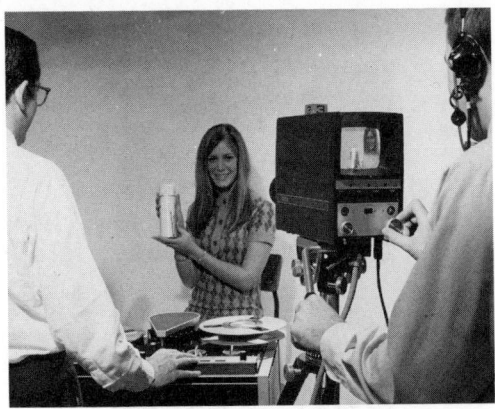

FIGURE 14-1   A typical video tape recorder. *(courtesy of Ampex)*

have the purpose of effectively stabilizing the reproduction of flesh tones so that the viewer need not readjust the hue control every time that channels are switched, or studio programming changes from live to pre-recorded transmission. To anticipate subsequent discussion, automatic tint control (ATC) is something of a misnomer, in that the ATC function entails judicious distortion of the reproduced color spectrum. For example, the hues that are adjacent to orange in the color spectrum may be modified by blending with orange, or these adjacent hues may have been eliminated entirely and replaced by orange hues. Thus, a trade-off is involved in ATC function, where the distorted reproduction of hues on either side of orange is accepted in return for greater stability of flesh-tone reproduction. ATC action is subsequently explained in greater detail. Operation of a typical ATC arrangement is summarized to good advantage by the vectorgrams shown in Fig. 14-2.

Optional RF tuners for community-antenna TV installations (CATV or cable TV) are offered by some manufacturers. When television receivers were first introduced, 13 VHF channels were utilized. However, Channel 1 was eventually abandoned, and programming has continued on the 12 VHF channels. When interference between stations became a problem, 70 UHF channels were allocated by the FCC. UHF converters were thereupon provided as optional equipment. However, as noted previously, UHF tuners have been made standard equipment by the FCC. With the growth of CATV, optional CATV tuners have been introduced, providing a total of 31 channels for cable reception, in addition to the UHF channels. In this tuner, 8 channels are inserted in the frequency gap between the 12 VHF channels, viz., between channels 6 and 7, and 11 channels are inserted above VHF Channel 13. Details of CATV are covered in Chapter 18.

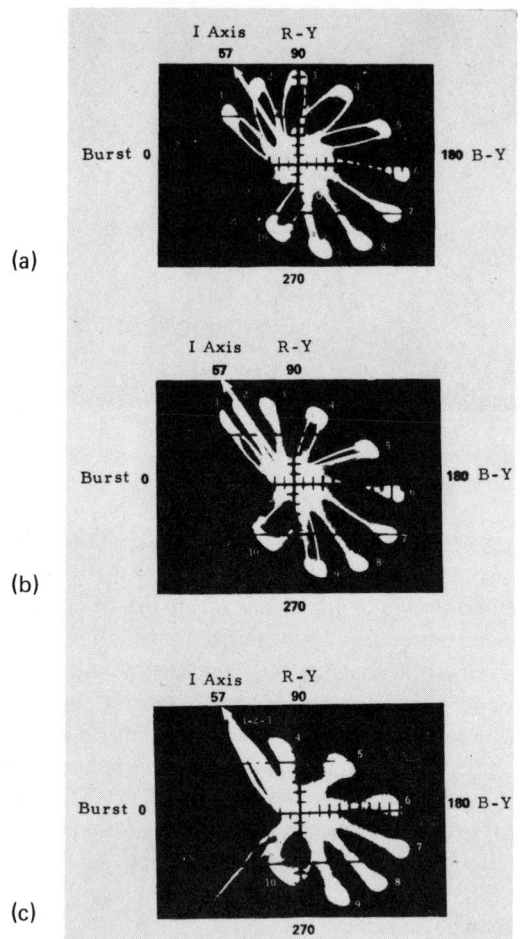

FIGURE 14-2   Vectorgrams showing the effect of ATC action. (a) ATC switched off; chroma demodulation phases normal; (b) ATC function set to partial; chroma demodulation phases in the vicinity of the I axis "pulled in"; (c) ATC function set to full; chroma demodulation phases in the vicinity of the I axis merged with the I phase.

14.2   REMOTE CONTROL FOR COLOR-TELEVISION RECEIVERS

Remote-control units enable the viewer to adjust the operating controls of a receiver from a distance. Most remote-control units employ ultrasonic sound waves to actuate the control mechanism. Figure 14-3 depicts a remote transmitter unit that generates seven ultrasonic sound-wave frequencies in the

35.5-44.5 Khz range. It utilizes a transistor oscillator (actuated when a push-button is depressed), which radiates ultrasonic sound waves from a transducer in the transmitter unit. Figure 14-4 shows the oscillator circuitry that is used in this design. The transmitter is effective at distances of up to approximately 50 feet.

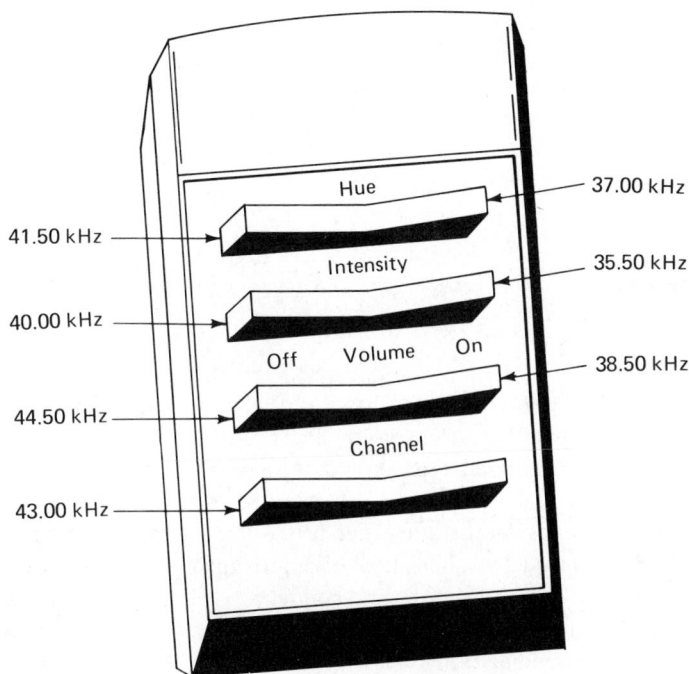

FIGURE 14-3   A remote-control transmitter unit.

At the receiver, a built-in receiver and control arrangement operates to change the settings of the controls as required. Referring to Fig. 14-5, note that a transducer (microphone) picks up the ultrasonic sound energy and converts it to a signal voltage, which is stepped up by a three-stage preamplifier. From the preamplifier, the signal voltage is fed to a driver stage, depicted in Fig. 14-6. Note that the driver energizes a broad-band, function-output transformer and a sharply-tuned, channel-change coil. This coil passes only 43-kHz signals and actuates the channel-change relay. Contacts 1 and 3 close when the relay is actuated, and power is applied to the tuner motor. Other remote-control functions, such as actuation of the on-off switch, require retention of the control signal in an electronic "memory" circuit. This circuit action is accomplished by a capacitor-charging arrangement like the one shown in Fig. 14-7. Its operation is as follows: Note that the gate of the MOSFET, depicted in Fig. 14-7, has a very high insulation resistance. Since

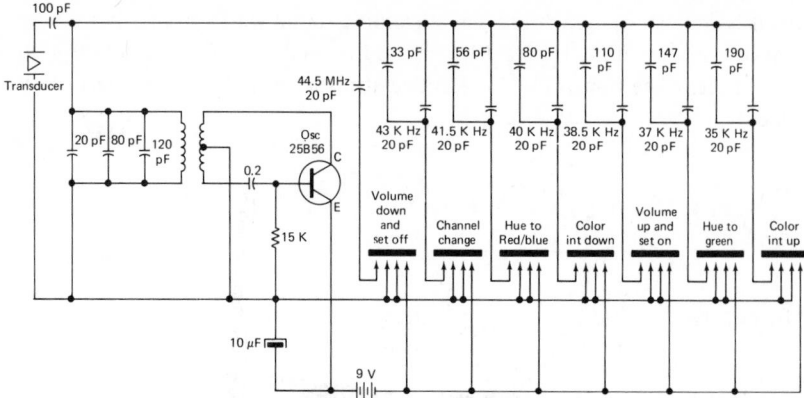

FIGURE 14-4   Configuration of a remote-control transmitter unit.

the neon bulb in the circuit also has a very high insulation resistance in its nonconducting condition, a charge on the 1.5-μF capacitor can be retained for a long period of time, typically 1000 hours. Since the MOSFET device conducts at zero bias, its output can be either increased or decreased by application of a positive or a negative bias voltage, respectively, to its gate. This bias is obtained by application of approximately 80 volts to the neon bulb, which then becomes ionized and conducts. The clamp diodes permit the capacitor to charge to any chosen potential in the range of –2 to +20 volts. In turn, a corresponding potential drop occurs across the 10-k load resistor, and this voltage is fed to the control amplifiers.

In the circuit of Fig. 14-8, the audio memory module is utilized to turn the receiver on or off. This module supplies an output of 2 volts to turn the receiver on, and 1.5 volts to turn the receiver off. This output voltage is fed to the base of Q3 in Fig. 14-6. When the transistor is turned on, its collector voltage decreases, removing the cutoff bias from the base of Q4 and turning this transistor on. Collector current actuates relay E2, closing contacts 2-3 and 6-7. Current is thus applied to the picture-tube filament transformer instant-on switch and the channel-change relay E1.

Observe in Fig. 14-8 that six sharply-tuned control coils T1 through T6 comprise the collector load for Q1. Each control coil feeds its output to a rectifier function diode. Note that if a 38.5-kHz "volume on" signal arrives, coil T2 develops an output voltage, and subsequently the neon bulb E13 is ionized. Capacitor C7 will charge to a potential determined by the length of

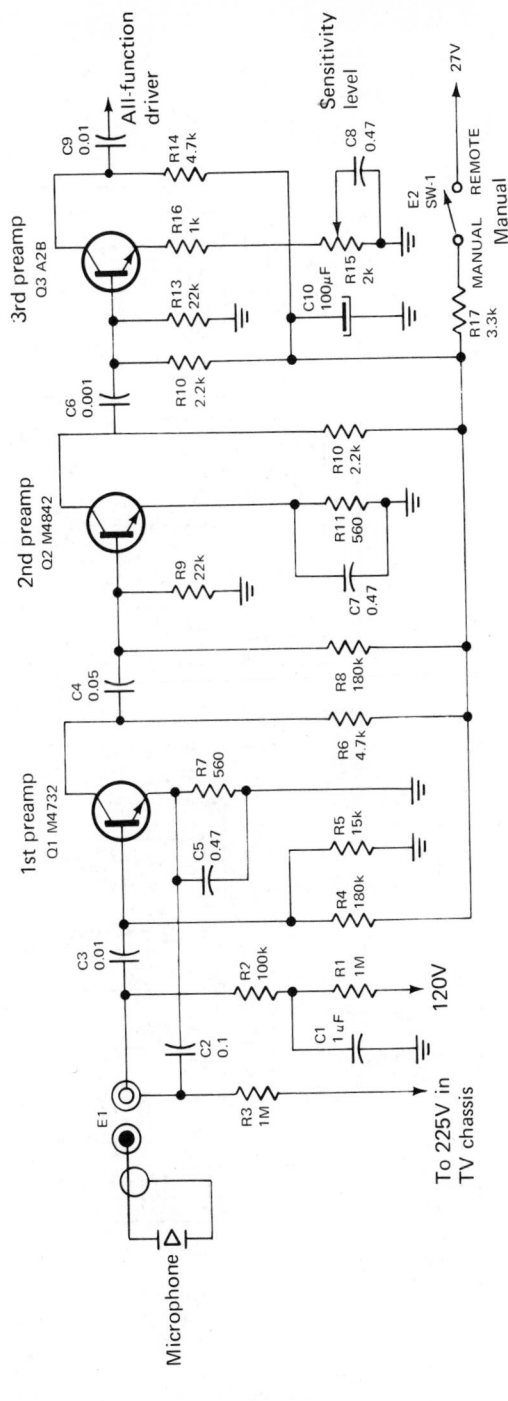

FIGURE 14-5 Input section of a remote-control receiver unit.

289

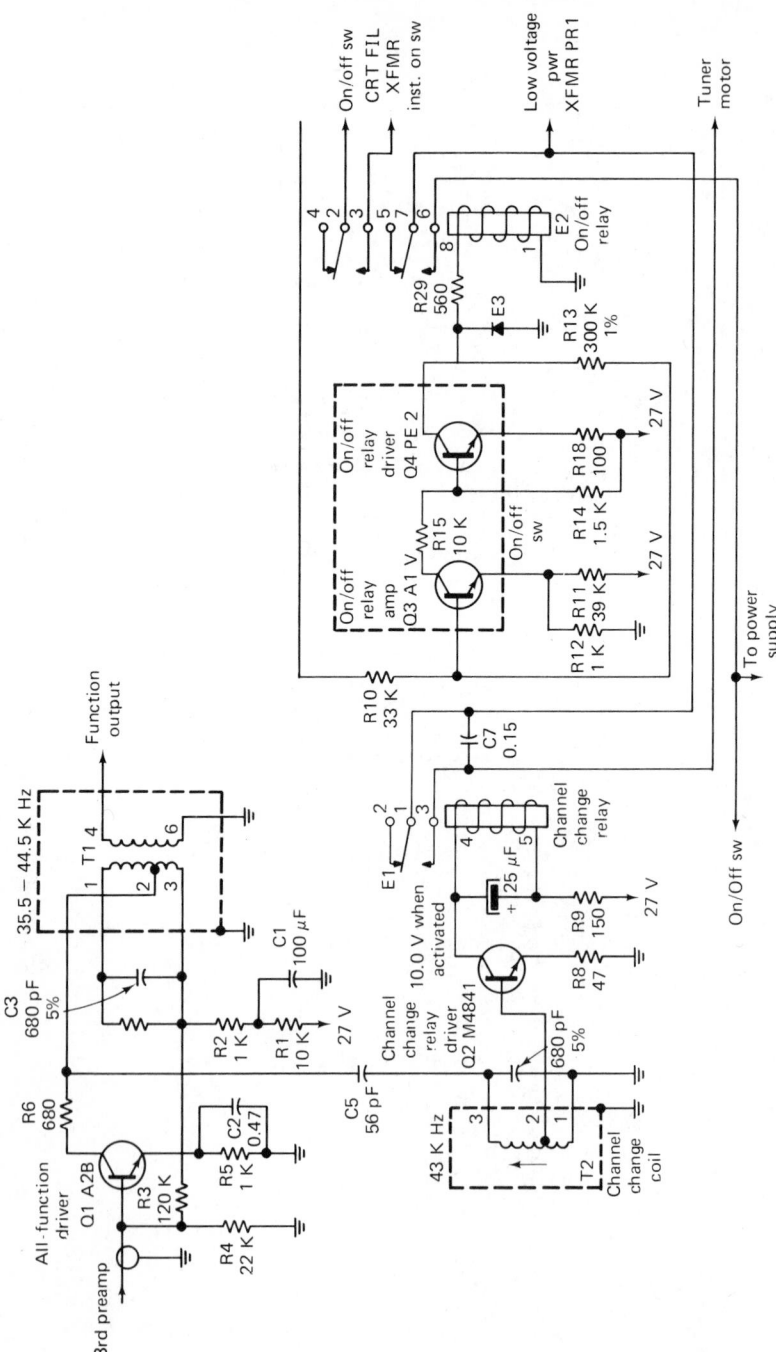

FIGURE 14-6 Driver and channel-change section configuration.

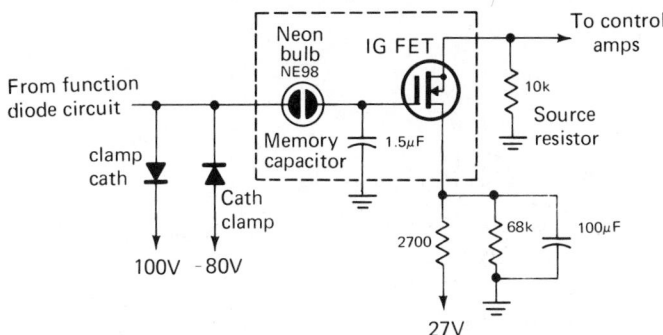

FIGURE 14-7   Electronic "memory" circuit for a remote-control
receiver.

time that the control signal persists, thus establishing the conduction level of
E14 or the base bias on Q2 and Q3. Since the audio signal is passed through
Q2 and Q3, the volume of the audio signal is controlled by the bias voltage on
Q2 and Q3. In the same general manner, the color-intensity level is varied,
and the hue is controlled by outputs T3 through T6.

14.3   AUTOMATIC TINT CONTROL

Numerous variations of automatic tint control arrangements are used to effectively stabilize the reproduction of flesh tones by judicious distortion of the
color spectrum. One of the simpler methods is depicted in Fig. 14-9. This
configuration automatically changes the raster to a sepia hue when the
receiver is tuned to a color-TV program. In turn, all of the reproduced colors
are blended with sepia, which tends to mask errors in flesh tones. Note that
the color killer produces a negative control voltage when a color signal is
present. One path for this negative voltage goes to the gate of the B–Y amplifier. This negative bias voltage causes a corresponding change in drain
potential, with the result that the raster is operated with a sepia hue. In contrast, when the receiver is tuned to a black-and-white signal, the color killer
applies no bias to the gate of the B–Y amplifier, and the raster has its normal
neutral gray hue.

Another method of obtaining an ATC function is depicted in Fig.
14-10. True colors are reproduced when chroma demodulation takes place
along the R–Y and B–Y axes. To obtain a greater "spread" of reproduced
orange hues, SW204 is closed, producing a phase shift in the 3.58-MHz subcarrier voltages and separating the demodulation axes by $124°$ instead of by
the normal $90°$. The end result is production of an abnormal phase "spread"
between orange and blue hues, and between orange and green hues. That is,
the hues adjacent to orange in the spectrum are changed into orange or near-
orange hues.

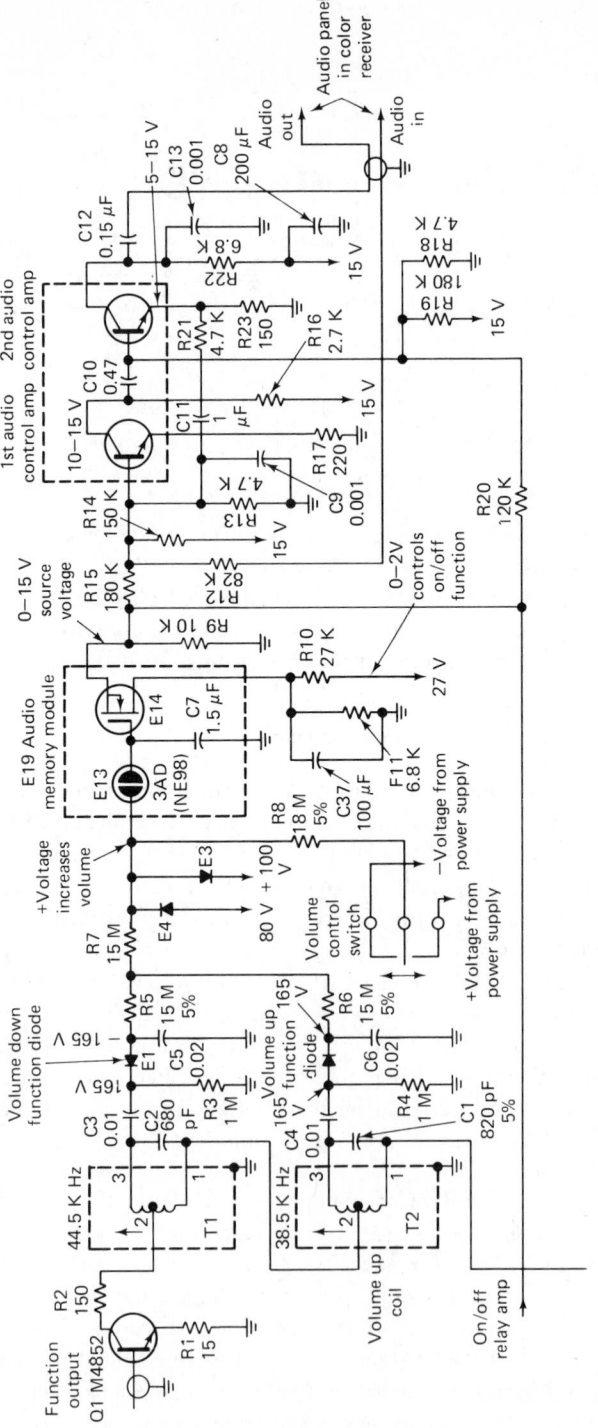

FIGURE 14-8 Remote control operational circuitry.

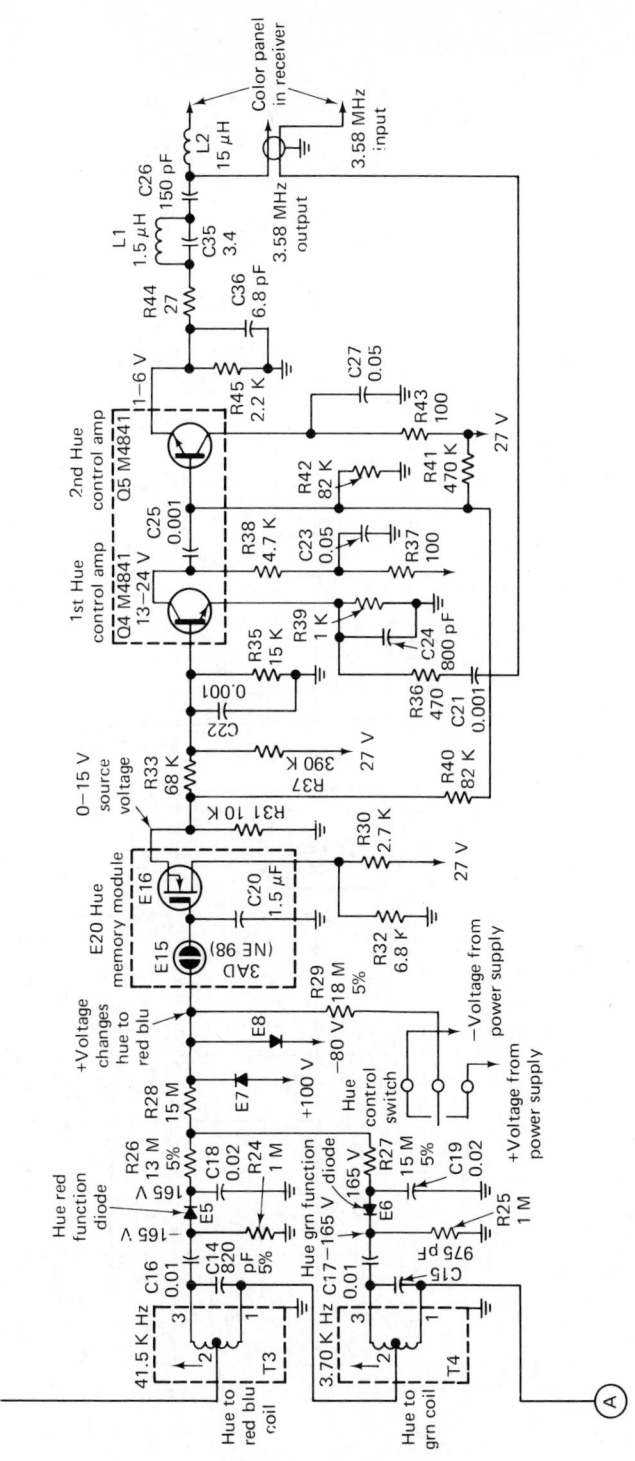

FIGURE 14-8 (Continued.)

293

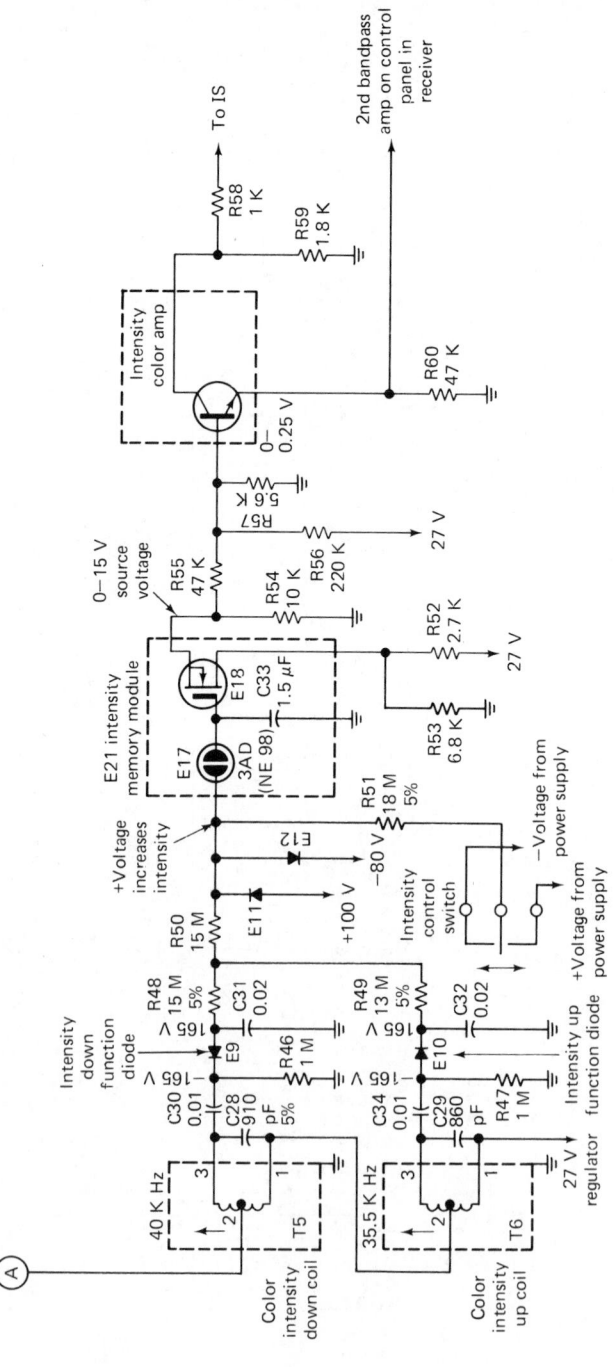

FIGURE 14-8 (Continued.)

294

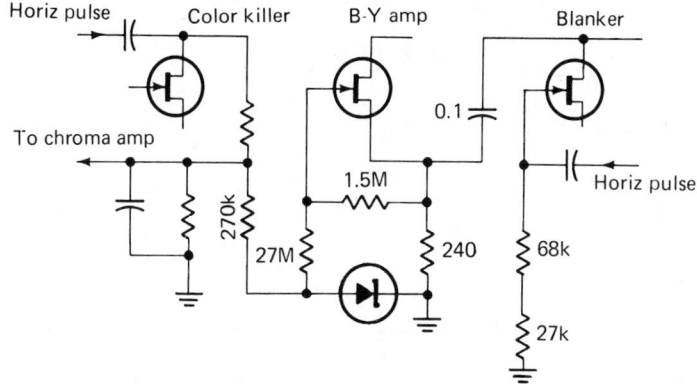

FIGURE 14-9 Color-killer action changes the raster hue to sepia when a color signal is present.

A somewhat more complex ATC arrangement is shown in Fig. 14-11. In effect, this method exploits the advantages of the demodulation phase-shift design without introducing its unnecessary disadvantages. In other words, when simple subcarrier phase shifting is employed, the "spread" of orange hues is increased, but the "spread" of blue hues is simultaneously decreased. Consequently, more color distortion is introduced than is actually necessary to obtain effective stabilization of orange hues. In the arrangement of Fig. 14-11, only positive yellow and red chroma signals are gated into the phase-shifting networks. Negative yellow and red chroma signals are processed normally without any ATC phase shift. The end result is that all hues, except those that occur adjacent to orange, are reproduced normally, as shown in Fig. 14-12. Note that the preference control in Fig. 14-11 enables the viewer to adjust the ATC system for a chosen "spread" of orange hues.

14.4 X-RAY DETECTORS

Early designs of color-TV receivers occasionally produced soft X-rays in excess of tolerable levels in accordance with standards established by the U.S. Bureau of Radiological Health. This X-radiation was produced in the high-voltage regulator tube. The National Council on Radiation Protection and Measurement (NCRP) has stipulated that X-radiation levels not exceed 0.5 milliroentgens per hour at a distance of two inches from a receiver. TV technicians have been advised to adjust the high-voltage value to the manufacturer's specification and to make certain that covers are properly installed in the high-voltage section of the receiver. Although technicians are not required to measure the X-radiation of a color-TV receiver, suitable sensing devices can be constructed without undue difficulty, as explained next.

The arrangement depicted in Fig. 14-13 employs a Geiger-Mueller

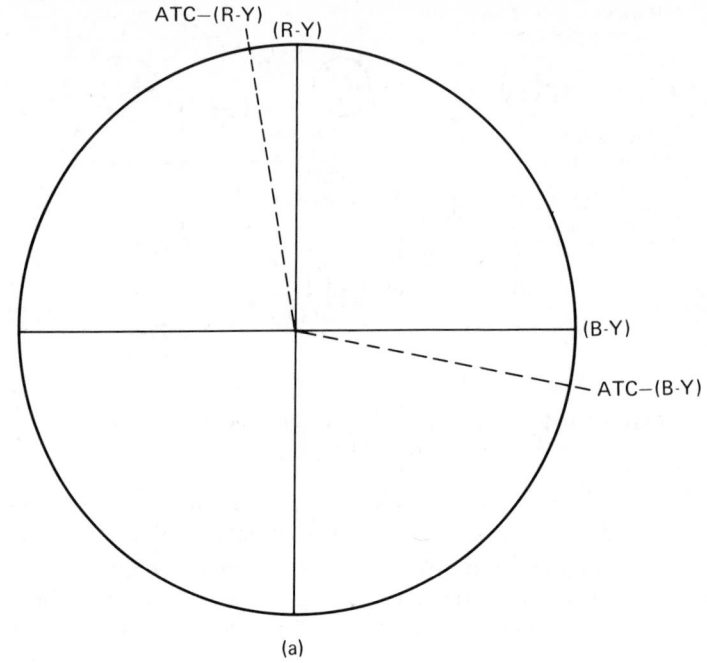

(a)

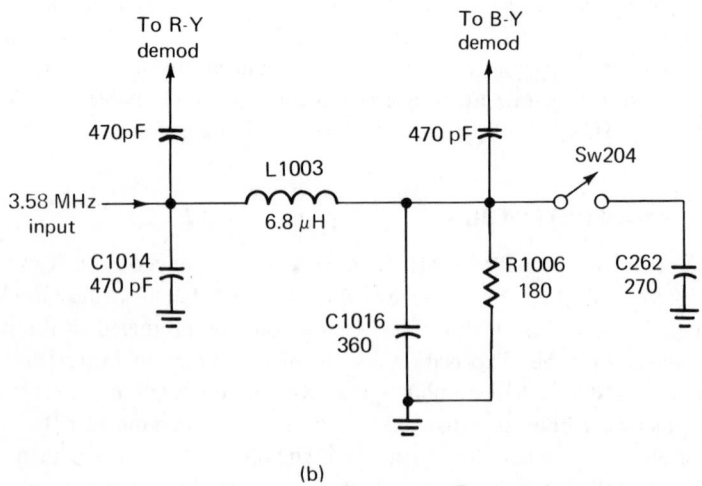

(b)

FIGURE 14-10 ATC action by shift of chroma-demodulation axis. (a) Normal R−Y and B−Y axes, with shifted ATC demodulation axes; (b) phase-shift capacitor is switched in for ATC function.

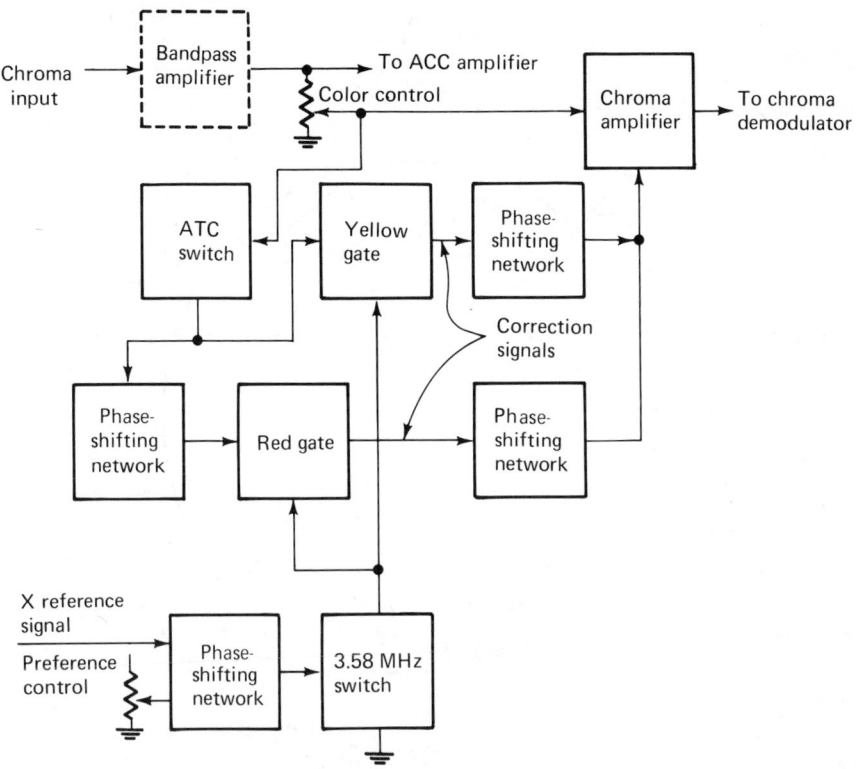

FIGURE 14-11   An ATC arrangement that shifts the phase of
positive red and yellow chroma signals only.

tube. It includes three functional sections: the sensor, the high-voltage supply, and the read-out arrangement. A Geiger-Mueller tube comprises a
cathode, anode, and the gas content of the G-M tube. When the anode is positive (Fig. 14-14), the passage of radiation will cause an avalanche of electrons
to be produced, with the formation of a pulse current. The G-M tube employed in Fig. 14-13 is a Victoreen IB85. The high-voltage supply is in the
range from 800 to 1,000 volts DC.

To protect the G-M tube against possible overload and damage, neon
lamps are utilized as voltage regulators in Fig. 14-13. Note that high voltage
is generated for charging C1 by repetitively operating switch SW1. Once that
C1 is charged adequately, it will operate the radiation detector for approximately an hour. Q1 is an amplifier that also inverts the pulses from the G-M
tube for driving the monostable multivibrator Q2 and Q3. A neon lamp is
employed for read-out, or, an earphone may be plugged into J1. Normal
background radiation will produce an average count rate in the range from 40
to 150 counts per minute. On the other hand, an X-ray field of 0.5 mR/hr

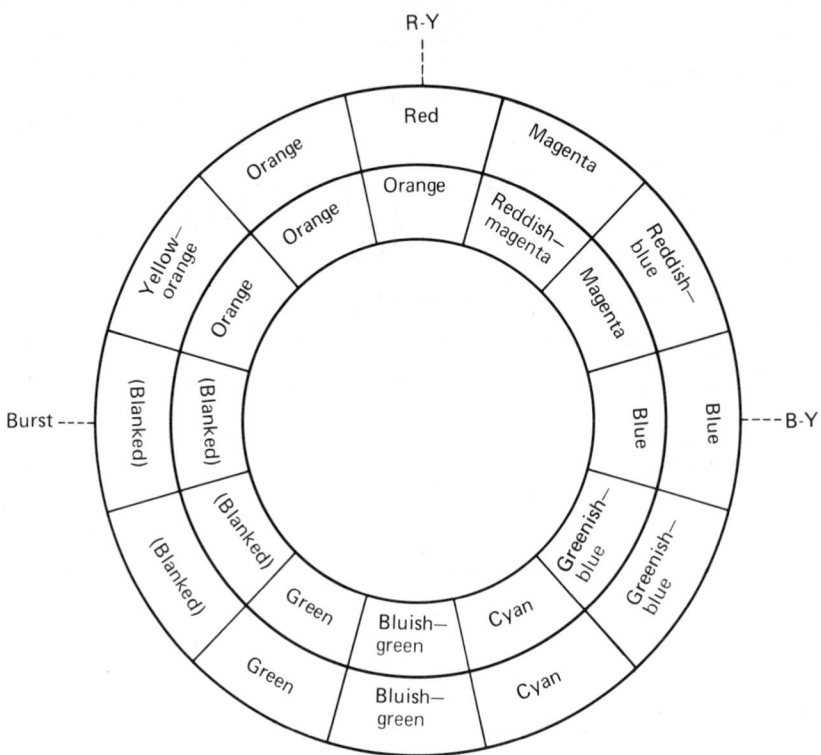

FIGURE 14-12   Phase relations of hues with and without ATC action. Outer circle: normal phase-hue relationship; inner circle: ATC phase-hue relations.

(maximum tolerable level) will produce a count rate of about 4,000 counts per minute.

14.5   TROUBLESHOOTING INTERMITTENTS

Troubleshooting intermittent color reproduction often demands patience and persistence. Many "tough-dog" service jobs are in this category. It is helpful to keep those circuit actions in mind that can prevent color display. Refer to the block diagram in Fig. 14-15. If the bias produced in the ACC stage becomes abnormal, it cuts off the bandpass amplifier and blocks passage of the chroma signal. Again, in case the killer/ACC detector stage becomes defective, the chroma signal may be stopped. Although the blanker cuts the bandpass amplifier off only for the duration of the flyback pulse (in normal operation), it is possible for a fault condition to bias the bandpass amplifier past cutoff continuously, with the result that the chroma signal is blocked.

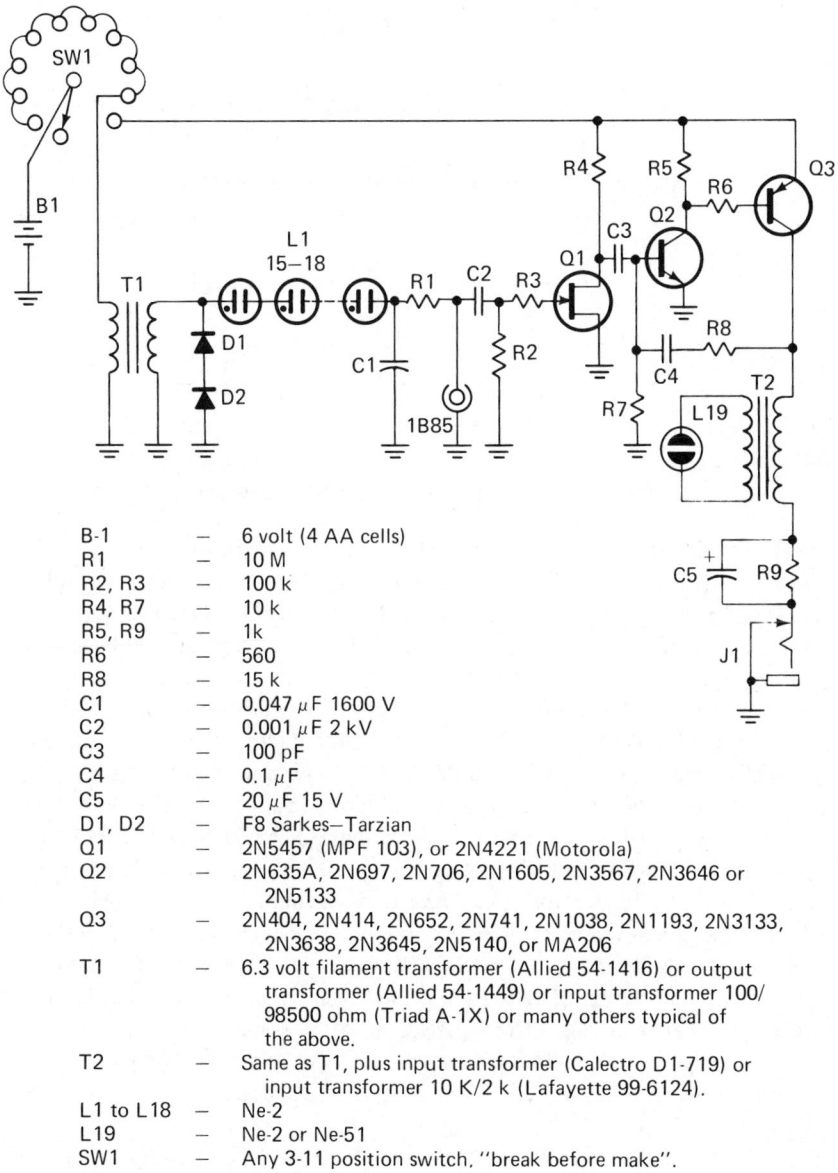

B-1	—	6 volt (4 AA cells)
R1	—	10 M
R2, R3	—	100 k
R4, R7	—	10 k
R5, R9	—	1k
R6	—	560
R8	—	15 k
C1	—	0.047 μF 1600 V
C2	—	0.001 μF 2 kV
C3	—	100 pF
C4	—	0.1 μF
C5	—	20 μF 15 V
D1, D2	—	F8 Sarkes–Tarzian
Q1	—	2N5457 (MPF 103), or 2N4221 (Motorola)
Q2	—	2N635A, 2N697, 2N706, 2N1605, 2N3567, 2N3646 or 2N5133
Q3	—	2N404, 2N414, 2N652, 2N741, 2N1038, 2N1193, 2N3133, 2N3638, 2N3645, 2N5140, or MA206
T1	—	6.3 volt filament transformer (Allied 54-1416) or output transformer (Allied 54-1449) or input transformer 100/ 98500 ohm (Triad A-1X) or many others typical of the above.
T2	—	Same as T1, plus input transformer (Calectro D1-719) or input transformer 10 K/2 k (Lafayette 99-6124).
L1 to L18	—	Ne-2
L19	—	Ne-2 or Ne-51
SW1	—	Any 3-11 position switch, "break before make".

FIGURE 14-13   Circuit diagram of NCRP circuit number 1, a simple X-ray detection instrument.

Preliminary analysis of intermittent color reproduction is often facilitated by clamping tests. In other words, consider a situation in which it is known that the bandpass amplifier is workable, although the chroma signal

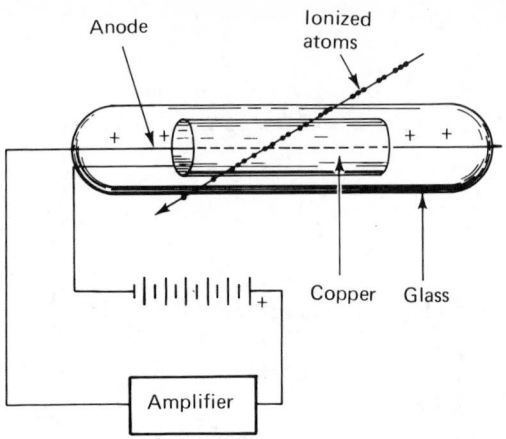

FIGURE 14-14   Geiger-Mueller tube arrangement.

disappears in this section. The troubleshooter must determine whether the fault is in the blanker, ACC, or color-killer circuitry. In turn, it is helpful to apply a DC-test voltage from an external power supply. This test voltage is applied first at the blanker connection, then at the color-killer voltage line, and finally at the input from the ACC stage. In Fig. 14-16, the DC supply is set to 4 volts. This test voltage is applied between cathode and ground, thereby clamping the cathode at 4 volts. Then, if color reproduction resumes, the technician concludes that the trouble will be found in the blanker circuitry.

On the other hand, consider a situation in which the blanker section is operating normally. It is logical to check the ACC input point in Fig. 14-16. The clamp voltage would be set to zero or a slightly positive voltage in this example and applied across C6. Then if the bandpass amplifier passes the chroma signal and color reproduction resumes, the technician concludes that the fault will be found in the color-killer section or the ACC section. In turn, DC-voltage and resistance values are measured in these sections. Oscilloscope waveform checks are less informative than in the bandpass-amplifier section because only the incoming 3.58-MHz waveform is present. However, the troubleshooter should verify that the killer/ACC detector is being driven normally.

It is instructive to review various component defects responsible for intermittent color reproduction in various receivers. For example, a color-intensity control (potentiometer) can cause this symptom. Loose transistor sockets, or a poor connection between the socket and the circuit board are other causes of intermittents. Sometimes a resistor lead may be lightly touching a capacitor lead on the component side of a circuit board, for example. The color takeoff coil occasionally becomes intermittent. Look for connections that have not been soldered; this error happens occasionally when a

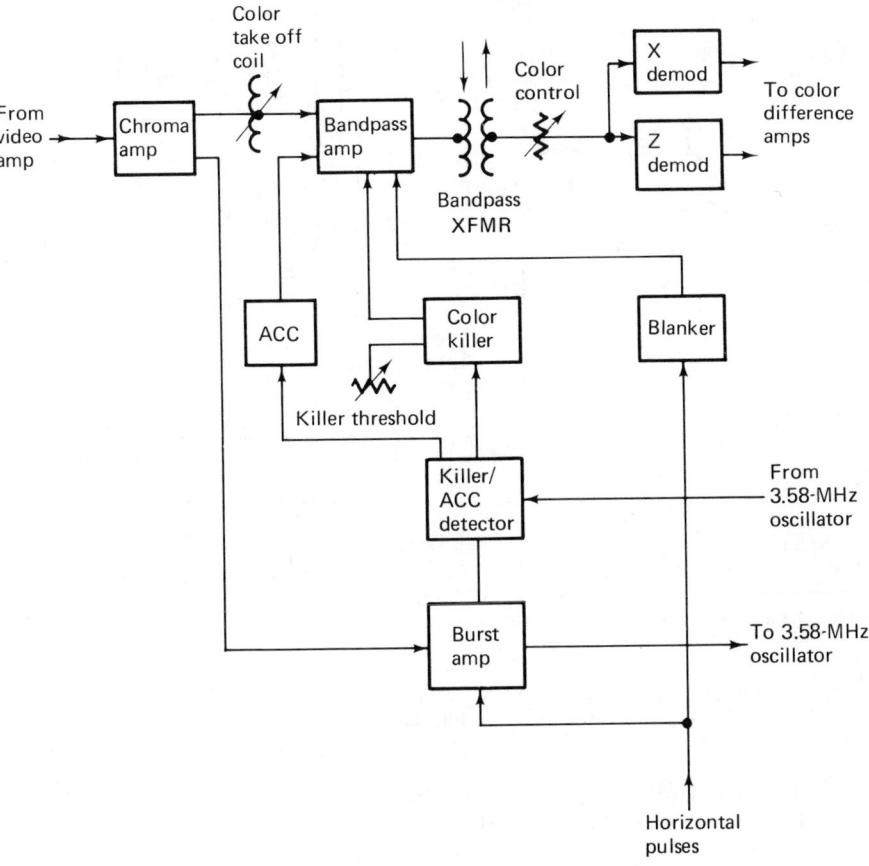

FIGURE 14-15   Circuit sections that are often involved in intermittent color reproduction.

component has been previously replaced. When a tube develops a short-
circuit, excessive current can be drawn and the cathode resistor overheated.
The damaged resistor may become unstable and cause an intermittent condition.

Cold-soldered connections are a frequent cause of intermittent
operation. A cold-soldered connection looks frosty, whereas a normal connection looks bright and shiny. Older-model color receivers occasionally utilize
neon lamps; thus, a neon lamp may be operated as a burst-gating input
clipper. If this lamp becomes defective, the burst is gated improperly. As a
neon lamp ages, it tends to become unstable and intermittent. When a
baffling intermittent problem is encountered, it is advisable to carefully study
the schematic diagram and to look for possible interactions among various

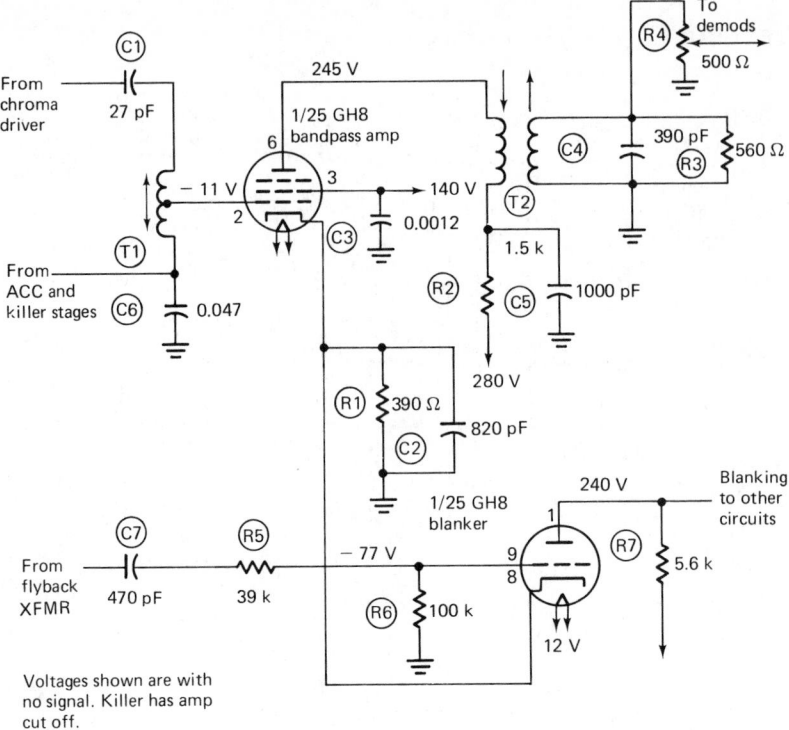

FIGURE 14-16 Bandpass-amplifier configuration showing chroma, ACC, killer and blanker inputs.

circuit sections. Minor receiver differences in this regard can be of basic importance in an intermittent-color troubleshooting situation.

EXERCISES

Questions

1. What are the functions of a television remote-control unit?

2. How does an automatic tint control section operate?

3. Explain the difference between a conventional RF tuner and a CATV tuner.

4. Describe how a remote-control unit actuates the associated control mechanism.

5. Define a *transducer*.

6. State a typical frequency utilized by a channel-change relay.

7. Briefly explain the operation of a *memory capacitor*.

8. How many control coils does a typical remote-control unit utilize?

9. Describe two methods of *tint control.*

10. Discuss the function of an ATC phase-shift capacitor.

11. What is a *preference control?*

12. Identify a possible source of X-rays in a color-TV receiver.

13. State the maximum permissible X-radiation level.

14. How many counts per minute does normal background radiation produce?

15. How many counts per minute does an X-ray field of 0.5 mR/hr produce?

True-False

1. Some present-day TV receivers are manufactured without UHF tuners.

2. A video tape recorder can be driven from the video amplifier in a TV receiver.

3. Automatic tint control involves judicious distortion of the reproduced color spectrum.

4. A CATV tuner provides 31 channels for cable reception, in addition to the UHF channels.

5. The 19 additional channels provided by a CATV tuner are allocated below channel no. 2.

6. Remote-control units are employed only with black-and-white receivers.

7. A remote-control transmitter radiates electromagnetic waves in the 12-MHz region.

8. Channel-change coils are sharply tuned.

9. Tuners are actuated by motors in remote-control mechanisms.

10. An IGFET in a memory module is biased in the range from -2 to $+20$ volts.

11. Bias variation of the IGFET in a memory module controls the associated function level, such as color intensity.

12. One method of automatic tint control entails a shift of the chroma-demodulation axes from $90°$ to $124°$.

13. ATC and AFC are equivalent terms.

14. The more elaborate ATC designs employ gates to provide expansion of orange hues without contraction of blue hues in the chroma phase spectrum.

15. X-radiation can be produced by the high-voltage regulator tube in a color receiver under conditions of malfunction.

Multiple Choice

1. ATC facilities are utilized to_____ .
   (a) avoid confetti in the picture
   (b) prevent picture jitter
   (c) stabilize reproduction of flesh tones
   (d) eliminate sound interference

2. Video tape recorders are used to _____.
   (a) monitor the line voltage
   (b) troubleshoot intermittents
   (c) test picture tubes
   (d) make a record of TV programs

3. Automatic tint control functions to_____ .
   (a) provide judicious distortion of the color spectrum
   (b) stabilize the gain of the bandpass amplifier
   (c) override the AGC control voltage
   (d) supplement AFC action

4. _____ channels were allocated when TV receivers were first introduced.
   (a) Twelve
   (b) Thirteen
   (c) Fourteen
   (d) Eighty-two

5. _____ UHF channels have been allocated by the FCC.
   (a) Twelve
   (b) Thirteen
   (c) Seventy
   (d) Eighty-two

6. A CATV tuner provides _____ channels for cable reception.
   (a) 12
   (b) 13
   (c) 31
   (d) 70

7. Typical remote-control transmitters have _____ operating controls.
   (a) 2
   (b) 4
   (c) 6
   (d) 8

8. The remote-control mechanism is actuated by _____ waves.
   (a) electromagnetic
   (b) ultrasonic
   (c) traveling
   (d) reflected

9. A frequency range of _____ is employed by a typical remote-control arrangement.
   (a) 88 to 108 MHz
   (b) 550 to 1500 kHz
   (c) 35.5 to 44.5 kHz
   (d) 20 Hz to 20 kHz

10. An example of a transducer is a _____ .
   (a) microphone
   (b) megaphone
   (c) channel switch
   (d) contrast control

11. A typical remote-control transmitter operates up to distances of _____.
   (a) 5 feet
   (b) 5 yards
   (c) 50 feet
   (d) 50 yards

12. IGFET memory circuits employ _____ .
   (a) positive and negative clamp diodes
   (b) negative feedback
   (c) positive feedback
   (d) none of the above

13. Simple ATC arrangements utilize _____.
   (a) sharply beamed antennas
   (b) $124°$ chroma demodulation axes
   (c) spark-gap protective devices
   (d) sound-converter diodes

14. X-radiation can be produced by over-voltage at the _____ .
   (a) high-voltage regulator tube
   (b) AGC rectifier
   (c) picture-tube heater
   (d) audio-output transistor

15. A Geiger-Mueller tube is employed in _____ .
   (a) black-and-white receivers
   (b) color receivers
   (c) TV-radio-phono combos
   (d) Geiger-Mueller counters

Problems

1. An ultrasonic wave travels 331.3 meters per second. Calculate the wavelength of a 37 kHz ultrasonic wave.

2. If an ultrasonic wave has a wavelength of 0.007 meter, what is its frequency?

3. A 40 kHz oscillator tank circuit has an inductance of 5 mH; what value of capacitance is in shunt to the coil?

4. What is the reactance of the inductance in problem 3 at 40 kHz?

5. What is the reactance of the capacitance in problem 3 at 40 kHz?

6. What is the reactance of the inductance in problem 3 at 1 kHz?

7. What is the reactance of the capacitance in problem 3 at 1 kHz?

8. A 1.5 μF memory capacitor has a leakage resistance of 10 megohms. How long will it take for the charge on the capacitor to decay to 37% of its original value?

9. How long will it take the charge on the capacitor in problem 8 to decay to 13.5% of its original value?

10. If a glowing neon bulb has a terminal voltage of 65 volts and draws 2 mA, how much power does the bulb dissipate?

Chapter 15

Color-Television Test Equipment Requirements

15.1 COLOR TELEVISION TEST EQUIPMENT REQUIREMENTS

All of the standard test equipment used in black-and-white TV-servicing procedures is also used in color-TV servicing. In addition, several specialized instruments are required for color-TV troubleshooting. Basic instruments include a VOM or TVM with a high-voltage DC probe, an oscilloscope with low-capacitance and demodulator probes, a signal generator, and a transistor tester. The high-voltage DC probe for color-TV service must provide a minimum range of 25 kilovolts. The oscilloscope used in color-TV testing must have a uniform vertical-amplifier frequency response to at least 4 MHz. Although a conventional AM signal generator has useful applications in color-TV service, the generator should be supplemented by a specialized sweep-and-marker generator, as explained below. Transistor testers that are satisfactory for use in black-and-white TV servicing procedures are also adequate for color TV.

    Color-TV setup procedures require the availability of a white-dot and crosshatch generator. Most of these instruments also include color-bar facilities. A color-bar generator is essential in troubleshooting procedures. Although both keyed-rainbow and NTSC-type color-bar generators are available, most technicians utilize keyed-rainbow generators. In addition to the basic instruments, a color picture-tube tester or picture-tube test jig will be convenient since a substitution test of a color picture tube is both time-consuming and tedious. Most service shops also utilize a capacitor checker. Comparatively elaborate signal-substitution instruments of the TV analyzer type are used in many shops. These are specialized multiple-generator instruments which provide test signals for any section of a color-TV receiver.

15.2   COLOR-TELEVISION OSCILLOSCOPES

Even though various types of oscilloscopes are employed in TV service, all have wide-band response in common. Chroma signals occupy the video-frequency band from 3.1 to 4.1 MHz, necessitating vertical-amplifier response up to at least 4 MHz. If full response is provided up to 4.5 MHz, the oscilloscope is also useful for checking the intercarrier-sound section of a receiver. Figure 15-1 shows the appearance of a typical color-TV oscilloscope. This is a free-running oscilloscope, which is adequate for routine color-TV servicing procedures. Note that the graticule is ruled both for conventional-waveform and vectorgram display in this example. To apply the instrument as a vectorscope, suitable test connections are made to a terminal board on the rear of the instrument (see inset). As discussed in greater detail in Chap. 16, vectorgrams provide a useful quick-check of chroma-circuit operation.

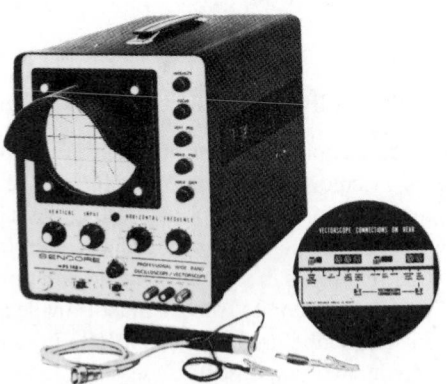

FIGURE 15-1   A service-type oscilloscope with vectorscope facilities. *(courtesy of Sencore)*

A vectorscope has maximum utility if its vertical amplifier is designed to introduce negligible phase shift into a chroma signal. This feature is provided by some vectorscopes. Although it is a practical advantage to employ a vectorscope that has a wide-band high-performance vertical amplifier and horizontal amplifier, this feature is generally restricted to lab-type oscilloscopes because of production costs. Most vectorscope tests are practical only in high-level chroma circuitry unless adequate vertical and horizontal amplifiers are available. There is a definite trend towards the use of triggered-sweep oscilloscopes in color-TV service shops. Triggered-sweep operation offers the advantage of allowing a waveform, such as a color burst, to be expanded for detailed observation. Triggered-sweep operation also enables the measurement of rise time, which is a basic factor in the operation of solid-state horizontal-sweep circuits.

15.3   COLOR-BAR, WHITE-DOT, AND CROSSHATCH GENERATORS

A typical color-bar, white-dot, and crosshatch generator is illustrated in Fig. 15-2. This instrument provides a keyed-rainbow color-bar pattern with a chroma frequency of 3.563795 MHz, as compared with the color-burst frequency of 3.579545 MHz. In other words, the generator provides a keyed-rainbow signal with a frequency that is 15,750 Hz below the color-burst or color subcarrier frequency. For this reason, it is also called a *sidelock signal*, an offset color subcarrier, or a *linear phase sweep*. Figure 15-3 depicts a keyed-rainbow waveform and the color-bar pattern that it produces. This bar pattern results from the beating of the offset subcarrier signal against the sub-carrier-oscillator signal in the receiver. In effect, a sweep of the complete color spectrum is generated during each forward-scanning interval.

FIGURE 15-2   A color-bar, white-dot, and crosshatch generator.
(courtesy of Sencore)

Note that the hues in a keyed-rainbow pattern are neither pure primary nor pure complementary colors. The mixed hues of the rainbow pattern also vary in saturation and brightness from one bar to the next. However, this situation is tolerable in servicing procedures inasmuch as it suffices to

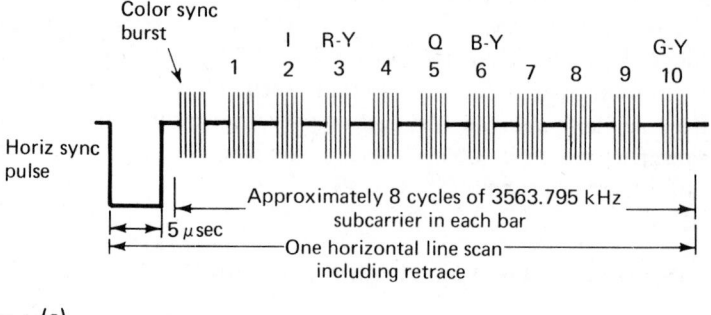

(a)

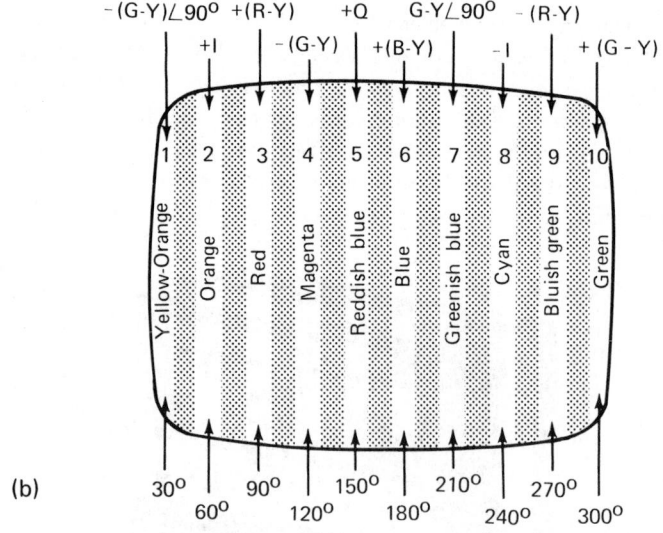

(b)

FIGURE 15-3   Signal waveform and bar pattern produced by a keyed-rainbow generator. (a) Waveform characteristics; (b) color-bar characteristics.

compare the standard pattern with the non-standard pattern that appears when there is a component defect in the receiver. As noted previously, an NTSC-type color-bar generator differs from a keyed-rainbow generator in that the NTSC instrument provides true primary and complementary colors at full saturation and brightness. However, an NTSC generator is comparatively costly and is used chiefly in laboratories and color-TV broadcast stations. Experienced technicians have often stated that they can troubleshoot a color receiver as effectively with a keyed-rainbow generator as with an NTSC color-bar generator.

White-dot and crosshatch patterns are also provided by the instrument illustrated in Fig. 15-2. Typical screen patterns are depicted in Fig. 15-4. Observe that these patterns are useful in checks of horizontal and vertical scanning linearity, as well as in convergence procedures. Note also that when a crosshatch pattern forms precise squares, the raster has a 4:3 standard aspect ratio. Many instruments, such as the one shown in Fig. 15-2, provide a choice of horizontal lines or vertical lines, in addition to the complete crosshatch pattern. The various patterns facilitate certain convergence adjustments.

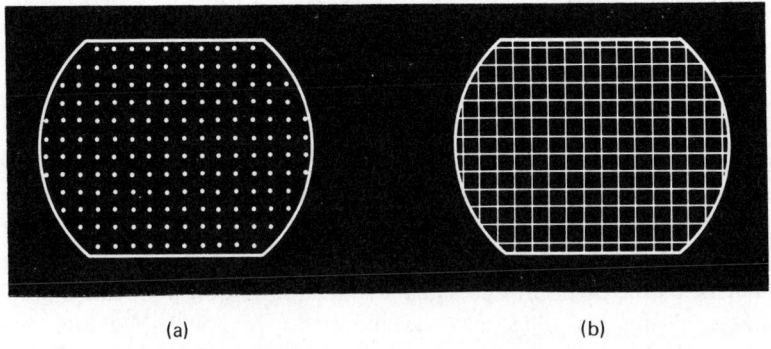

(a)   (b)

FIGURE 15-4   Patterns utilized in picture-tube convergence procedures. (a) White-dot pattern; (b) crosshatch pattern.

15.4   SWEEP-AND-MARKER GENERATORS

Sweep-and-marker generators are used in combination with the oscilloscope to align the tuned circuits in the signal channels of a receiver. A typical instrument is illustrated in Fig. 15-5. It provides VHF, IF, and video-frequency outputs. Harmonics of the VHF output can be used to sweep a UHF tuner. All markers are of the beat type in this example. However, generators that provide absorption markers for the video-frequency range are also found. To maintain uniform marker amplitude at the top of a response curve, and in traps, most modern generators employ post injection for beat markers in which the marker signal is mixed with the sweep signal after the latter has passed through the receiver circuits. The marker amplitude displayed on the scope screen thus remains independent of the signal attenuation or amplification that occurs in the tuned circuits under test.

Figure 15-6 depicts the basic principle of *sweep alignment*. An FM signal with suitable center frequency and deviation is applied to the tuned

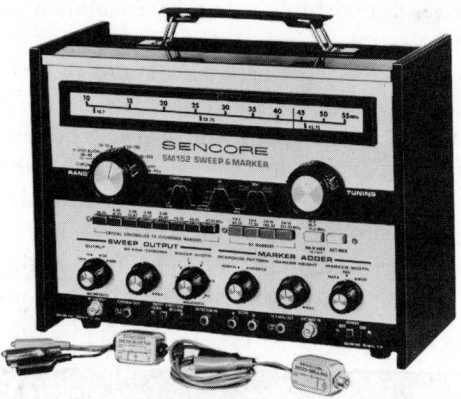

FIGURE 15-5   A sweep-and-marker generator for color-television applications. *(courtesy of Sencore)*

circuits under test. Since the tuned circuits have a certain frequency characteristic, the demodulated output appears on the scope screen as a plot or graph of output voltage versus frequency. When a CW voltage is mixed with the FM voltage, a beat marker appears on the response curve. Some sweep-and-marker generators provide single markers only; others provide multiple markers. Figure 15-7 illustrates the appearance of double and triple beat markers. If absorption markers are employed, they appear as "dips" along the response curve, as shown in Fig. 15-8.

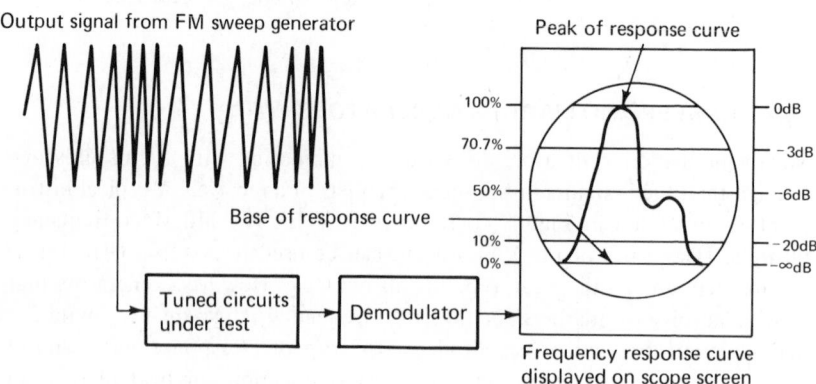

FIGURE 15-6   Basic principle of sweep alignment.

15.5   TRANSISTOR TESTERS

Transistor testers are often used in troubleshooting procedures to minimize the necessity for substitution tests. A transistor tester is also highly effective in the selection of matched pairs of replacement transistors. Figure 15-9 illustrates a typical service-type transistor tester. This instrument accommodates

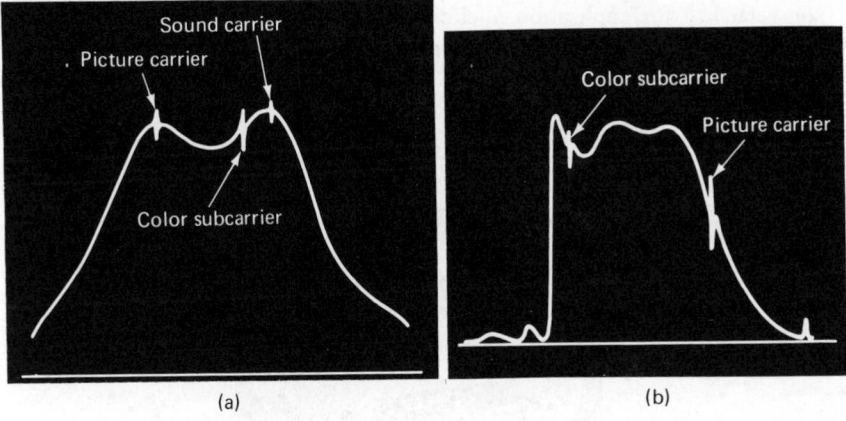

(a)   (b)

FIGURE 15-7   Appearance of beat markers. (a) Triple markers on an RF response curve; (b) double markers on an IF response curve.

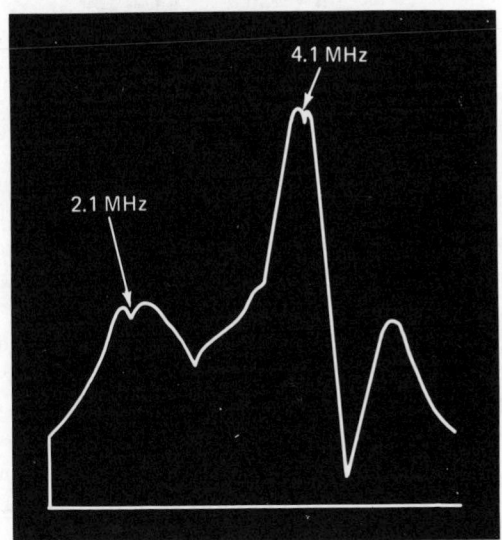

FIGURE 15-8   Appearance of absorption markers on a response curve.

both bipolar devices and unipolar field-effect transistors. It provides two basic tests for low-power and high-power bipolar transistors. An I_{CBO} leakage indication and a measurement of DC beta value are both provided. Field-effect transistors of the IGFET type are checked for leakage in an I_{GSS} configuration, and the source-drain current is measured at zero bias. Since the internal impedance of this type of tester is comparatively low, the instrument has some facility for application in-circuit. However, in-circuit tests cannot be made for leakage, and in-circuit gain measurements are likely to be misleading unless the transistor operates in a comparatively high-impedance circuit.

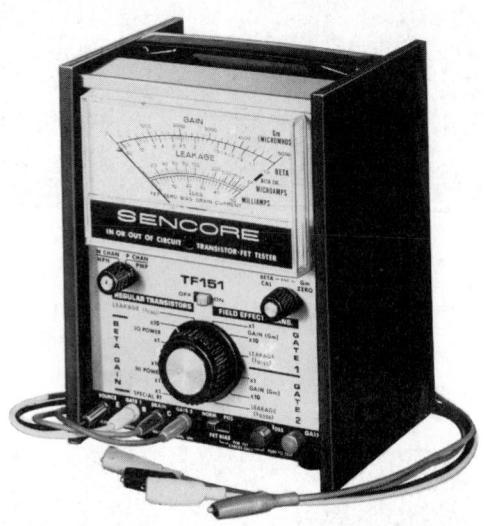

FIGURE 15-9   A transistor tester. *(courtesy of Sencore)*

An in-circuit leakage test is often made possible by employing "turn-on" and "turn-off" techniques with a conventional VOM or TVM. These techniques can also provide a practical check of the beta value. In-circuit test methods are very attractive to the television technician because most low-power transistors are soldered into printed-board circuitry. Consequently, it becomes a time-consuming and tedious procedure to unsolder transistors for out-of-circuit tests. In-circuit tests are facilitated by the availability of modern TVM's that have a *low-power ohms* function. This is an ohmmeter arrangement that applies less than 0.1 volt to the circuit under test. The advantage of this low-level test voltage is that transistors or diodes are not "turned-on" while the values of resistors that are connected to the semi-conductor terminals are being measured.

15.6   COLOR PICTURE-TUBE TESTERS

A color picture-tube tester provides considerable convenience in servicing procedures by minimizing the necessity for substitution tests. Figure 15-10 illustrates a typical service-type picture-tube tester. This instrument checks for interelectrode leakage or short-circuits and measures the emission of each electron gun. A surge-current function, which is sometimes useful in clearing interelectrode leakage or short-circuits, is also provided. A life test is made by checking emission at reduced heater voltage. In cases where the emission is subnormal under this type of operating condition, the life expectancy of the picture tube will be comparatively short. However, a picture tube with marginal emission can sometimes give appreciable additional service if the heater is operated at a voltage 15% above the rated value. This possibility of short life expectancy can also be determined with a picture-tube checker. If substantial increase in cathode emission is indicated when the heater is operated with 15% overvoltage, a picture-tube brightener can be installed to obtain extended service.

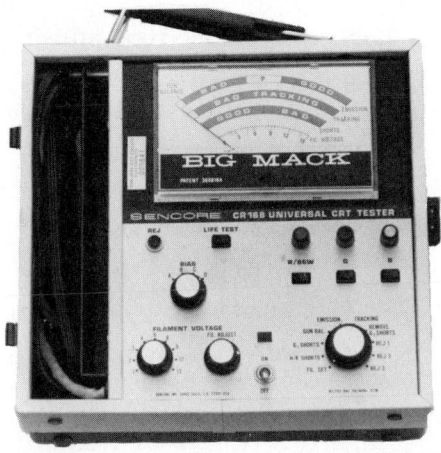

FIGURE 15-10   A color picture-tube checker. *(courtesy of Sencore)*

It is very helpful to have a color-tracking test function in a picture-tube checker. It is generally agreed that proper black-and-white tracking with normal brightness variation in a color picture tube requires that the emission of the electron guns be within a ratio of 1.5-to-1. If the emission values are out of these limits, a black-and-white picture will appear tinted in its low-lights or highlights. Basically, a color-tracking test requires an individual check of each electron gun by means of an adjustable screen-grid test voltage. With a reference control-grid bias, the screen voltage is adjusted for a reference beam-current value, and the control-grid bias is reduced to zero. In turn, the beam-current value is measured. If all three zero-bias beam-current values are within a 1.5-to-1 ratio, the picture tube will track satisfactorily.

EXERCISES

Questions

1. What are the basic color-TV troubleshooting instruments?
2. Name two test instruments that may be utilized in addition to basic instruments.
3. Why should an oscilloscope that is used in color service have vertical-amplifier response to at least 4 MHz?
4. State the chief characteristics of a keyed-rainbow color-bar generator.
5. How does an NTSC color-bar generator differ from a keyed-rainbow instrument?
6. Explain the need for white-dot and crosshatch generators.
7. Describe the general applications of sweep and marker generators.
8. What are the two basic types of transistor testers?
9. State the functions provided by a typical color picture-tube tester.
10. How do absorption markers differ from beat markers?
11. Name the types of instrument probes used in general troubleshooting procedures.
12. Can a conventional oscilloscope be employed as a vectorscope?
13. How many color bars does a keyed-rainbow generator display on the picture-tube screen?
14. Explain the term *post injection markers.*
15. Describe the display of double or triple markers on a response curve.

True-False

1. A high-voltage DC probe is used with an oscilloscope.
2. A TV analyzer is a signal-substitution type of instrument.
3. Wide-band response is required in the vertical amplifier of an oscilloscope used in chroma troubleshooting procedures.
4. A triggered-sweep oscilloscope is preferred to a free-running oscilloscope by most technicians.
5. Vectorscopes are completely different from oscilloscopes.
6. *Rainbow signal, offset color subcarrier oscillator, linear phase sweep,* and *sidelock signal* are equivalent terms.

7. Hues in a keyed-rainbow pattern are neither pure primary nor pure complementary colors.

8. An NTSC color-bar generator provides pure primary and complementary color-bar patterns.

9. Triggered-sweep oscilloscopes provide white-dot and crosshatch output signals.

10. Sweep-alignment signals are basically FM signals.

11. Marker signals are basically CW signals.

12. *Beat marker* and *absorption marker* are equivalent terms.

13. In-circuit transistor tests are less informative than out-of-circuit tests.

14. A lo-pwr ohmmeter does not "turn" on normal semiconductor junctions.

15. Color picture-tube testers measure the emission currents of electron guns in picture tubes.

Multiple Choice

1. High-voltage probes for color-TV service provide a range up to _____ kV.
   (a) 15
   (b) 25
   (c) 35
   (d) 50

2. White-dot generators are employed in _____ .
   (a) alignment procedures
   (b) transistor testing
   (c) convergence procedures
   (d) tube testing

3. TV analyzers are utilized for _____ .
   (a) alignment procedures
   (b) signal tracing
   (c) field-strength measurements
   (d) signal injection

4. An oscilloscope must have vertical-amplifier frequency response to _____ if intercarrier-sound checks are to be made.
   (a) 455 kHz
   (b) 10.7 MHz
   (c) 4.5 MHz
   (d) 41.25 MHz

5. Triggered-sweep oscilloscopes are provided with_____ .
   (a) vectorscope screens
   (b) calibrated time bases
   (c) AGC networks
   (d) AFC networks

6. A keyed-rainbow color-bar display has _____ bars.
   (a) 8
   (b) 10
   (c) 11
   (d) 12

7. Keyed-rainbow color-bar signals contain a_____ .
   (a) window
   (b) horizontal sync pulse
   (c) suppressed sideband
   (d) suppressed carrier

8. Bursts in a keyed-rainbow signal differ progressively by _____ in phase.
   (a) 90°
   (b) 180°
   (c) 30°
   (d) 45°

9. Sweep and marker generators are used in combination with an oscilloscope to _____ .
   (a) check rise time
   (b) align tuned circuits
   (c) display vectorgrams
   (d) display color-bar patterns

10. A –3 dB level on a response curve corresponds to_____ amplitude.
    (a) 10%
    (b) 50%
    (c) 70.7%
    (d) 100%

11. Lo-pwr ohmmeters are useful in solid-state troubleshooting because this type of instrument does not _____ .
    (a) radiate interference
    (b) turn on normal semiconductor junctions
    (c) load the circuit excessively
    (d) use excessive current

12. Proper picture-tube tracking requires that the emission of the three electron guns be within a ratio of _____ .
    (a) 1 to 1

(b) 1.5 to 1
(c) 2 to 1
(d) none of the above

13. An in-circuit transistor tester cannot check for _____ .
    (a) collector-junction leakage
    (b) short-circuited junctions
    (c) open-circuited junctions
    (d) burned-out transistors

14. A sweep-signal generator has a/an _____ output.
    (a) CW
    (b) AM
    (c) FM
    (d) PM

15. A marker generator has a/an _____ output.
    (a) CW
    (b) AM
    (c) FM
    (d) PM

Problems

1. A keyed-rainbow generator can operate either on the high side or the low side of the color subcarrier. Calculate the chroma-signal output frequency of a high-side keyed-rainbow generator.

2. A white-dot generator provides a horizontal row of 16 dots. Calculate the repetition rate of the dot signal.

3. The master oscillator in a white-dot and color-bar generator has a frequency of 189 kHz. How many times is this frequency divided down to generate the horizontal sync pulse?

4. How many times is the master-oscillator frequency in problem 3 divided down to generate the vertical sync pulse?

5. If color bars are produced in a keyed-rainbow generator at the rate determined by a 189-kHz master oscillator, how many bars will be displayed on the picture-tube screen?

6. The third harmonic of the channel-13 output from a sweep generator is used to sweep certain UHF channels. What are these channel numbers?

7. When the fourth harmonic of the channel-13 output from a sweep generator is used to sweep a UHF tuner, what channel numbers are covered?

8. If the deviation of the sweep signal on channel 5 is 6 MHz, what is the deviation of the third harmonic of this VHF signal?

9. If the deviation of the sweep signal on channel 10 is 6 MHz, what is the deviation of the third harmonic of this VHF signal?

10. If the deviation of the sweep signal on channel 13 is 6 MHz, what is the deviation of the fourth harmonic of this VHF signal?

Chapter 16

Color-Television Receiver Installation And Setup Procedures

16.1   INSTALLATION AND SETUP REQUIREMENTS

Color-TV receivers are designed to operate from all-channel antennas, such as the unit depicted in Fig. 16-1. Good color reproduction requires an adequate antenna installation. Indoor antennas and some outdoor types that are satisfactory for black-and-white reception may not have sufficient signal pickup or requisite bandwidth for good color response. Most antennas have a 300-ohm terminal impedance, although some have a 75-ohm impedance. An antenna is usually connected to a receiver via 300-ohm twin lead, although sometimes a 75-ohm coaxial cable is used. Figure 16-2 depicts how the three chief types of lead-in are connected to receivers. The antenna often needs to be oriented in various directions for optimum color reception from various stations. Accordingly, a rotor is frequently desirable or essential. Note that excessive signal strength can be as objectionable as insufficient signal strength. In cases where the incoming signal overloads the receiver, the signal can be attenuated as required by means of resistive pads, shown in Fig. 16-3.

When the incoming signal is too weak to reproduce a satisfactory

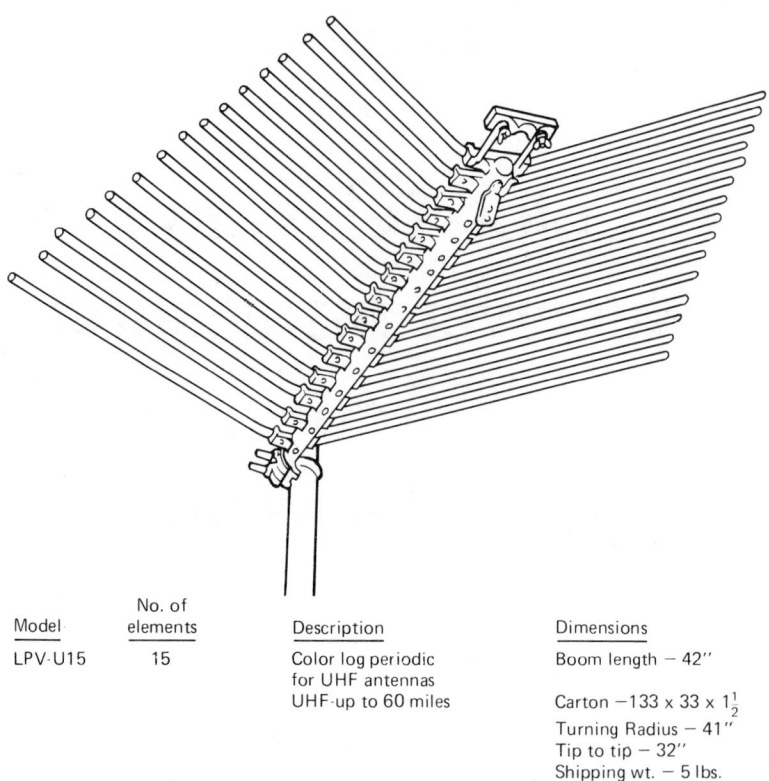

Model	No. of elements	Description	Dimensions
LPV-U15	15	Color log periodic for UHF antennas UHF-up to 60 miles	Boom length − 42″ Carton −133 x 33 x 1½ Turning Radius − 41″ Tip to tip − 32″ Shipping wt. − 5 lbs.

FIGURE 16-1   Typical antenna for color-television reception. *(courtesy of JFD Manufacturing Company)*

color image, it is often possible to improve reception substantially by installing a preamplifier or booster, such as the one illustrated in Fig. 16-4. A preamplifier is most effective when it is mounted on the antenna mast so that the available signal-to-noise ratio does not deteriorate due to the additional noise pickup of the lead-in between the antenna and receiver. Thus, it is impossible to improve the signal-to-noise ratio by the installation of a preamplifier. Both the noise voltage and the signal voltage are amplified by the same factor. Therefore, preamplification should take place at or near the antenna terminals. The signal-to-noise ratio can be improved, when necessary, by installation of a more sharply directional antenna. However, practical limits in this regard are imposed by the high cost of elaborate antenna arrays.

Commercial color-TV antennas are designed to provide uniform response across each channel, as exemplified in Fig. 16-5(a). If the lead-in is

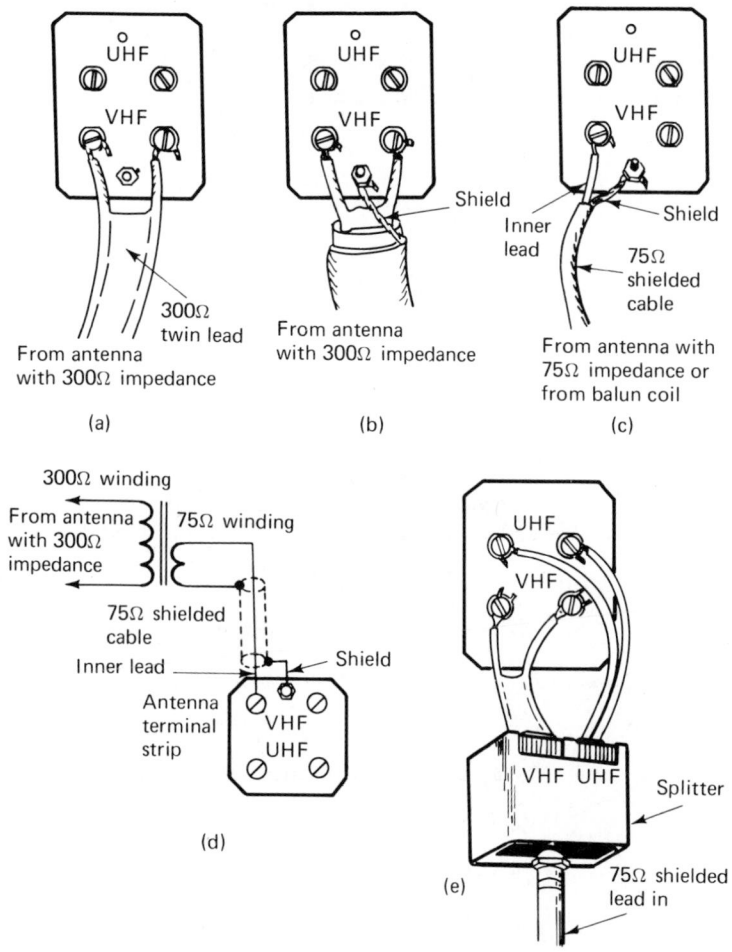

FIGURE 16-2   Connection of lead-in to receiver-input terminal block. (a) Flat twin lead; (b) shielded twin lead; (c) coaxial cable; (d) connection of balun transformer; (e) connection of balun with signal-splitter.

properly installed, the same uniform frequency response will be provided at the input terminals of the receiver. On the other hand, a home-built antenna, or a poorly installed lead-in with a commercial antenna, will sometimes have a very nonuniform frequency response, such as shown in Figure 16-5(b). In turn, the color signal and the sound signal are considerably attenuated. Reproduction of the color signal will be impaired more than that of the sound signal. If the technician suspects that poor color reproduction may be

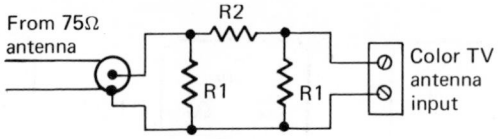

(resistors are ½W composition type)

For a small amount of attenuation (6dB)
R1 = 220Ω
R2 = 56Ω

For a medium amount of attenuation (12dB)
R1 = 120Ω
R2 = 150Ω

For a large amount of attenuation (18dB)
R1 = 100Ω
R2 = 300Ω

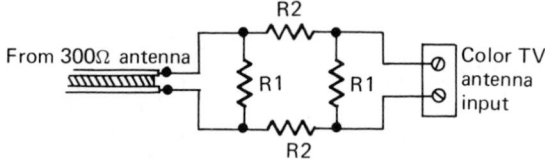

(resistors are ½W composition type)

For a small amount of attenuation (6dB)
R1 = 910Ω
R2 = 100Ω

For a medium amount of attenuation (12dB)
R1 = 510Ω
R2 = 270Ω

For a large amount of attenuation (18dB)
R1 = 390Ω
R2 = 560Ω

FIGURE 16-3   Resistive attenuator pad arrangements. *(courtesy of Heath Company)*

caused by a poor antenna system, it is advisable to try another antenna, or to check the receiver operation from a color-bar generator signal.

After the receiver is connected to a suitable antenna, all of the usual black-and-white receiver adjustments are made. That is, the receiver is tuned to a TV station, and adjustments are made of the AGC, horizontal-hold, vertical-hold, centering, height, noise, and width controls. These basic adjustments may need touch-up even if the receiver has been set up at the factory. Next, it is good practice to measure the accelerating voltage for the color picture tube with a DC voltmeter and high-voltage probe. If the measured

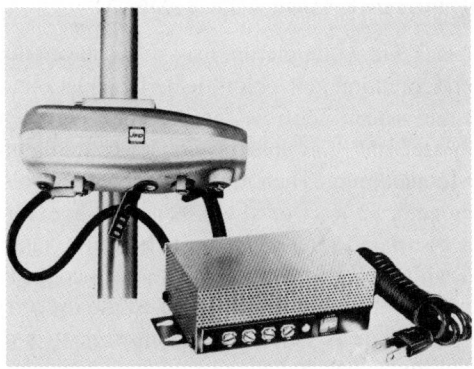

FIGURE 16-4   A mast-mounted preamplifier and remote power supply. *(courtesy of JFD Manufacturing Company)*

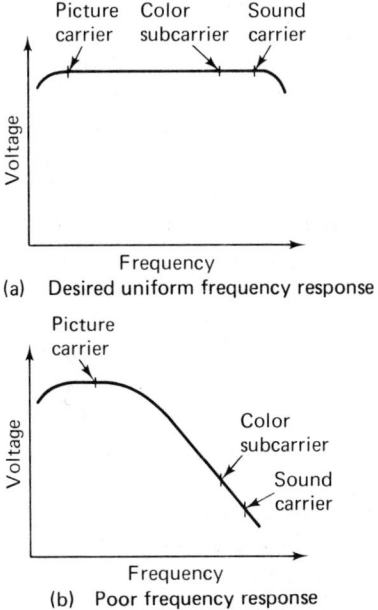

(a)   Desired uniform frequency response

(b)   Poor frequency response

FIGURE 16-5   Comparative antenna frequency-response characteristics. (a) Desired uniform frequency response; (b) poor frequency response.

value does not check closely with the value specified in the receiver service manual, the high-voltage control should be adjusted accordingly.

16.2   COLOR PICTURE-TUBE ADJUSTMENTS

At this time, a black-and-white picture may be displayed normally, or, it may be displayed on a predominant color field. If a black-and-white picture is tinted, the bias and screen controls for the color-picture tube must be adjusted. At the outset, the bias controls are set to minimum, and the screen controls are set to maximum. Then, one of the bias controls is advanced until its corresponding color appears on the screen. The bias control is left in this position, and its associated screen control is turned to minimum. This procedure is repeated with the remaining bias and screen controls. Then the screen controls are adjusted for "black level." That is, the red screen control is now advanced until a red hue is visible. Final adjustment is determined at the point that the red hue is just below visibility. This procedure is repeated for the green and blue screen controls.

Although a black-and-white picture may now be displayed without background tinting, *misconvergence* (color-bleed) may be observed surrounding edges in the image. This is an indication that the color picture tube is in need of convergence. Misconvergence should not be confused with poor color purity. To check the purity adjustments, observe the raster with no input signal to the receiver. If the raster exhibits tinted splotches, it indicates that purity adjustments are needed. Since purity and convergence adjustments interact, it is essential to start with correct purity. The purity adjustments are generally made with a red field (green and blue electron guns disabled). The basic procedure is as follows:

Slide the deflection yoke back in its mount as far as possible. Then adjust the purity ring magnets (see Fig. 16-6) to obtain an uncontaminated red area in the center of the screen. If the yoke is now moved forward, it may be possible to obtain an entirely uncontaminated color field. However, since there is some interaction between the purity-magnet and yoke-position adjustments, it may be necessary to work back-and-forth to obtain a uniform red field. Inability to do so indicates that the static convergence adjustments are considerably in error. Accordingly, leave the purity adjustments at optimum settings and proceed to check the static convergence. After the static convergence adjustments have been corrected, the purity adjustments can be rechecked and finalized. Of course, in this situation, the static convergence adjustments will require touch-up also.

Note in passing that poor color purity should also not be confused with magnetization of the color picture tube or its supporting structures. Modern color receivers have built-in degaussing arrangements. However, if an old-model receiver is being serviced, it may be necessary to employ an external degaussing coil to demagnetize the picture-tube assembly before good color purity can be obtained. A degaussing coil is merely a large ring wound with many turns of insulated wire. The coil is energized from a 117-volt, 60-Hz power outlet, and its AC field effectively removes residual magnetism from the color picture-tube structure.

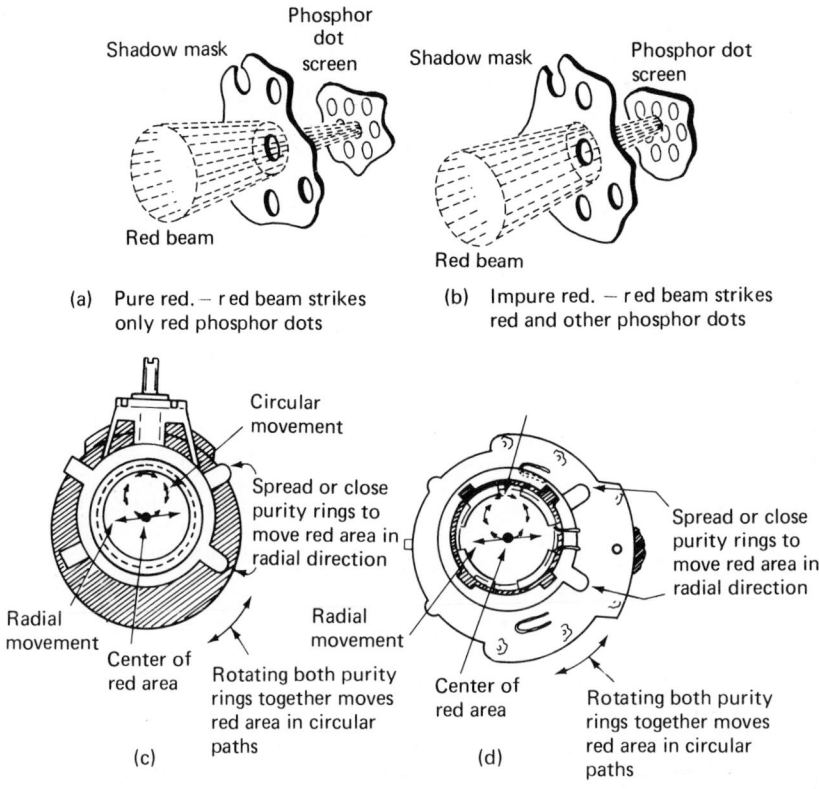

FIGURE 16-6   Principles of purity-magnet adjustments. *(courtesy of Heath Company)*

As noted previously in Chap. 10, convergence adjustments are required, and are routinely made with the aid of a white-dot or crosshatch generator. With a dot pattern displayed on the picture-tube screen, the static convergence magnets are set to positions that provide good convergence in the central screen area, denoted by the dotted circle in Fig. 16-7. Disregard the convergence conditions on the screen toward the top and bottom, and the left and right sides; consider only the center. Figure 16-8 shows how the three electron beam directions change in response to static convergence-magnet adjustments. There is some interaction between these adjustments, especially in cases where one or more happen to be far out of normal setting. It is often helpful to disable the blue gun at the outset and to merge the green and red dots to form yellow dots. Then the blue gun can be enabled, and the two blue convergence magnets are easily adjusted to form white dots. If only a slight touch-up is required, technicians usually disregard the blue gun operating while converging the red and green beams.

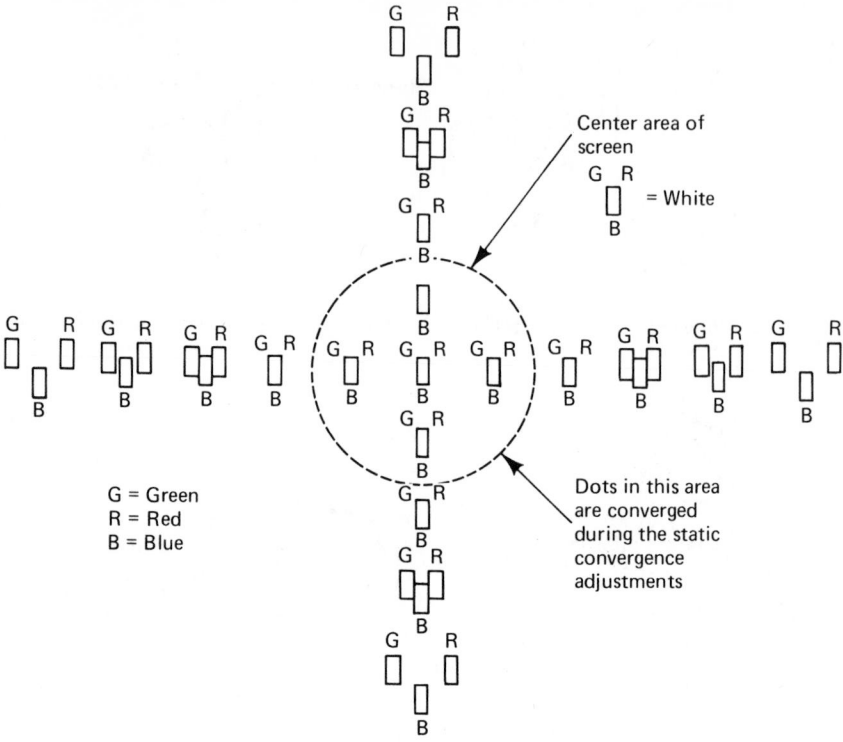

FIGURE 16-7   Static convergence adjustments are made with reference to the central screen area.

After static convergence procedures are completed it is usually necessary to proceed with dynamic convergence adjustments. Although the central area of the screen displays white dots, the outer region is likely to have objectionable "color bleed." Static and dynamic convergence adjustments tend to interact. Accordingly, it is often necessary to discontinue the dynamic procedure temporarily and to touch up the static adjustments. Thereupon the dynamic procedure may be resumed. Experience is required to become proficient in this technique, even when the instructions provided in field-service manuals are followed explicitly. Therefore, merely an overview of the dynamic-convergence procedure will be presented in this chapter.

Figure 16-9 shows a dynamic control-board layout. Twelve controls, which can be roughly divided into four groups, are provided. As indicated in the diagram, controls A, B, and C affect edge convergence chiefly in the lower screen area; conversely, controls D, E, and F affect edge-convergence chiefly in the upper screen area. Left-edge convergence is dominated by controls G, H, and J; whereas right-screen convergence is dominated by controls K, L

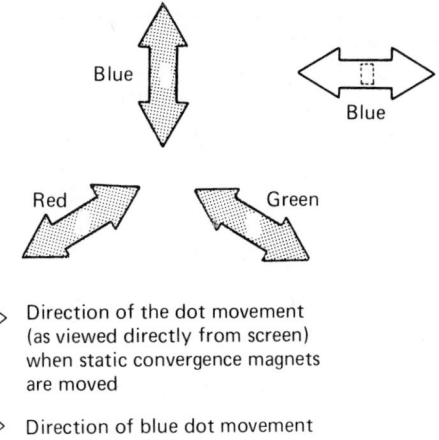

Direction of the dot movement
(as viewed directly from screen)
when static convergence magnets
are moved

Direction of blue dot movement
when shaft of blue lateral
magnet is turned

FIGURE 16-8   Electron-beam response to static convergence
adjustments. *(courtesy of Heath Company)*

and M. As in the static-convergence procedure, the blue gun is disabled at the
outset of the dynamic convergence procedure. After working back-and-forth
between the four basic groups of dynamic controls (and the static controls,
if required), a clean display of yellow dots is obtained by precise convergence
of the red and green dots.

At this time, the blue gun is enabled, and if a uniform vertical array
of white dots is obtained referring to Fig. 16-10, note that two basic steps are
involved in this procedure, supplemented by touch-up of the static adjustments, if necessary. Finally, the adjustments shown in Fig. 16-11 are made to
obtain a uniform horizontal array of white dots. Often "bringing in the blue
gun" causes more or less misconvergence of the red and green beams, particularly when considerable adjustment is required of the blue dynamic-convergence controls. Therefore, it becomes necessary to make touch-up adjustments of the red and green dynamic controls before the adjustment of the
blue dynamic controls can be finalized.

16.3   MISCELLANEOUS SETUP ADJUSTMENTS

In addition to the basic setup adjustments that have been described, some
additional controls may require attention in particular receivers. For example,
large-screen receivers provide pincushion-correction facilities, as noted previously. Pincushion-control adjustments are generally made with a white-dot or
crosshatch pattern displayed on the picture-tube screen. The amplitude and
phase controls are set to positions that remove the curvature from normally

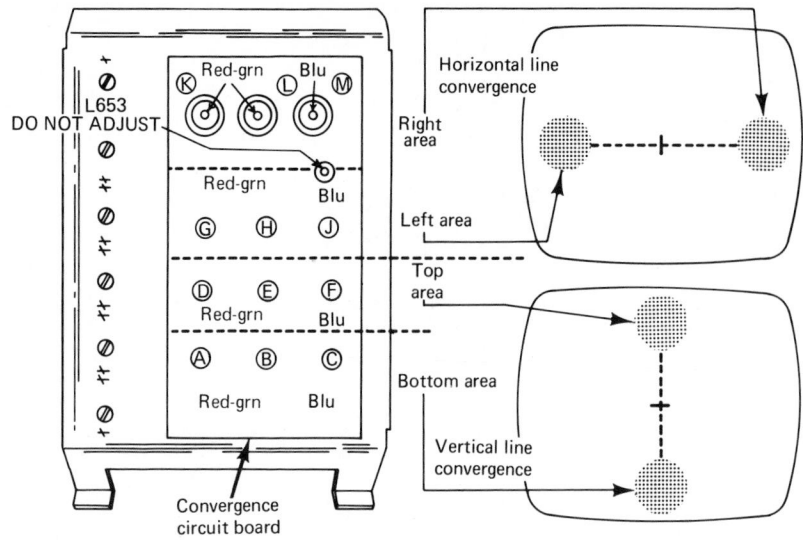

FIGURE 16-9   Plan of the dynamic-convergence procedure. *(courtesy of Heath Company)*

straight lines. This condition is evaluated to best advantage by viewing the screen considerably off-center so that the eye can "sight" along the line of interest.

It is sometimes observed that when the brightness control is turned to a high position, the raster tends to become tinted. Conversely, the raster might show a tint when the brightness control is turned to a low position. In this latter situation, it is the gray scale that is not tracking properly. Poor tracking is corrected by the alternate adjusting of the screen and drive controls. This procedure is generally carried out while the receiver is operated at a normal brightness setting and tuned to a strong black-and-white TV signal. This provides an image with bright highlights and dim lowlights. The drive control for a given hue affects its appearance in the highlights. For example, if green is evident in lowlights, the green tint is eliminated by adjusting the green drive control.

EXERCISES

Questions

1. Why is it sometimes desirable or necessary to install a rotor for an antenna?

2. Explain why a preamplifier should be mounted on the antenna mast.

CONVERGING YELLOW AND BLUE TO MAKE WHITE
(BLUE GUN SWITCH TO NORMAL)

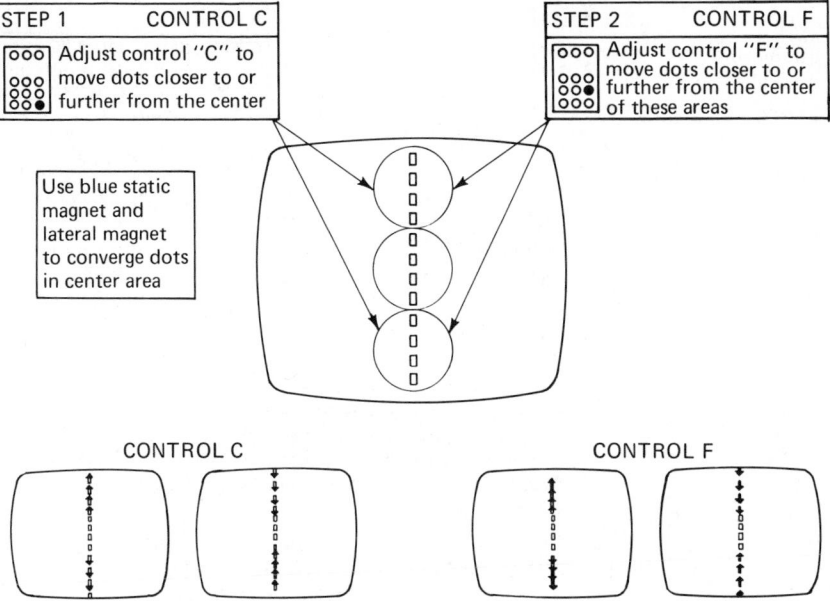

FIGURE 16-10   Vertical dynamic convergence of the blue electron beam. *(courtesy of Heath Company)*

3. What receiver controls does the technician adjust when a receiver is installed?

4. Define a *balun*.

5. How is a color picture tube set up?

6. State the number of dipole elements provided in a typical log periodic antenna.

7. What can the technician do to improve the signal-to-noise ratio of an antenna?

8. Why might an attenuation pad be installed in an antenna lead-in?

9. Do all color receivers provide the same picture-tube adjustments?

10. Describe a *degaussing coil*.

True-False

1. Color receivers cannot be operated from all-channel antennas.

2. Most antennas have a 300-ohm terminal impedance.

3. An antenna preamplifier can improve the incoming signal-to-noise ratio.

CONVERGING YELLOW AND BLUE TO MAKE WHITE
(BLUE GUN SWiTCH TO NORMAL)

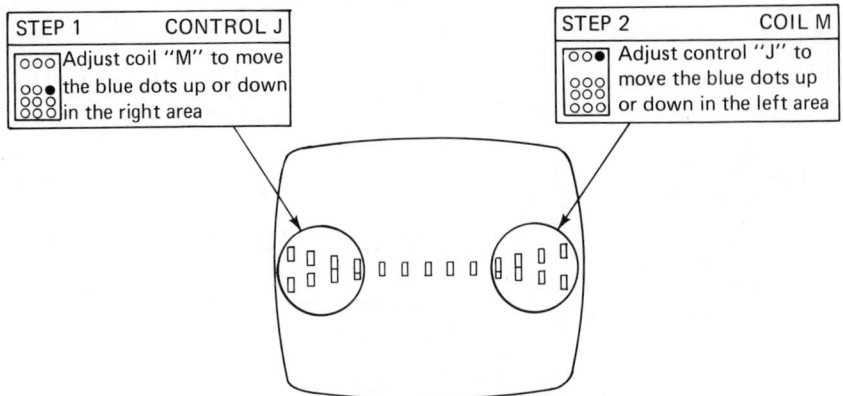

Note: Blue dots can be moved horizontally
only by turning the shaft of the blue
lateral magnet. Slight misconvergence
at the edge of the screen may be
considered normal.

FIGURE 16-11   Horizontal dynamic convergence of the blue electron beam. *(courtesy of Heath Company)*

4. The signal-to-noise ratio can be improved by installation of a more
   sharply directional antenna.

5. Standard coaxial cable has a 75-ohm characteristic impedance.

6. A balun converts a 75-ohm single-ended source into a 300-ohm
   double-ended source.

7. An attenuation of 90% reduces a signal to a -20 dB level.

8. Background tinting is corrected by adjustment of the convergence
   controls.

9. *Purity* and *convergence* are equivalent terms.

10. Pincushion-correction controls are generally adjusted with the aid
    of a white-dot or crosshatch pattern.

Multiple Choice

1. A typical color log periodic antenna has _____ elements.
   (a) 2
   (b) 4
   (c) 5
   (d) 15

2. A large all-channel antenna is typically rated for reception up to
   _____ miles.
   (a) 2
   (b) 10
   (c) 60
   (d) 500

3. Signal splitters operate to _____ .
   (a) improve the signal-to-noise ratio
   (b) separate VHF and UHF signals
   (c) eliminate adjacent-channel interference
   (d) suppress ignition noise

4. Resistive attenuator pads are employed to _____ .
   (a) avoid receiver overloading
   (b) minimize cochannel interference
   (c) separate VHF and UHF signals
   (d) eliminate FM interference

5. A balun is utilized to _____ .
   (a) improve the signal-to-noise ratio
   (b) eliminate standing waves
   (c) reduce airplane flutter
   (d) match a 75-ohm single-ended source to a 300-ohm double-
   ended load

6. Misconvergence is caused by _____ .
   (a) sound interference in the picture
   (b) misadjustment of the fine-tuning control
   (c) poor signal-to-noise ratio
   (d) incorrect adjustment of the convergence controls

7. A degaussing coil may be utilized to _____ .
   (a) align the tuned circuits
   (b) demagnetize the picture-tube assembly
   (c) trap out FM interference
   (d) prevent IF oscillation

8. Impedance is a/an _____ ratio.
   (a) front-to-back
   (b) imbalance
   (c) AC voltage/current
   (d) resistance

Problems

1. A resistive attenuator pad reduces the incoming signal level by 18 dB. Calculate the corresponding percentage reduction in signal level.

2. What is the impedance of a series arrangement consisting of 300 ohms of resistance and 300 ohms of reactance?

3. Calculate the impedance of 300 ohms of resistance connected in parallel with 300 ohms of reactance.

4. If a 0.00025-μF capacitor is connected in series with a 300-ohm resistor, and a current of 0.1 ampere rms flows through the circuit, how much power does the capacitor dissipate?

5. How much power does the resistor dissipate in problem 4?

6. If a 75-ohm resistance is connected in parallel with a 300-ohm resistance, what is the resistance of the combination?

7. Calculate the impedance of a series circuit comprising 300 ohms of capacitive reactance, 300 ohms of inductive reactance, and 75 ohms of resistance.

8. Calculate the impedance of a parallel circuit consisting of 300 ohms of capacitive reactance, 300 ohms of inductive reactance, and 75 ohms of resistance.

Chapter 17

Color-Television Troubleshooting Procedures

17.1 PRELIMINARY CONSIDERATIONS

Troubleshooting procedures start with analysis of picture and sound symptoms. Three general classifications of symptoms are recognized in this preliminary analysis; picture reproduction may be unsatisfactory, sound reproduction may be unacceptable, or, both picture and sound reproduction may be objectionable. For example, if the picture has poor contrast, the intercarrier sound section of the receiver is not at fault. On the other hand, if the picture is reproduced normally, but the sound output is weak, the video section is not suspect. However, if both the picture and sound are weak, it is logical to conclude that the trouble will be found in the signal channel prior to the sound-takeoff point. Note that in the latter situation, the antenna is included in the signal-channel system.

When picture reproduction is unsatisfactory, the problem lies in one of three basic subclassifications of trouble symptoms. Black-and-white pictures may be displayed normally, whereas color reproduction may be absent or distorted. On the other hand, sometimes color reproduction may be normal, with the black-and-white images absent or distorted. These trouble symptoms are illustrated in Fig. 17-1. Of course, the trouble symptoms may include both black-and-white and color reproduction. When black-and-white reproduction is normal, but color reproduction is distorted, the black-and-white section of the receiver is not checked at the outset of the troubleshooting procedure. On the other hand, if color reproduction is normal, but black-and-white images are distorted or absent, the chroma section of the receiver is disregarded. Again, in cases where both black-and-white and color reproduction are unsatisfactory, the trouble will be found in the signal channel prior to the chroma-takeoff point.

FIGURE 17-1   Subclassifications of picture trouble symptoms. (a) Normal color-bar pattern; (b) color distortion, with black-and-white reproduction normal, (hues are shifted in phase); (c) black-and-white distortion with color reproduction normal, (Y component absent); (d) both color and black-and-white distortion, (hues are shifted in phase and Y component is absent). (See inside back cover for figure.)

A no-raster symptom should not be confused with a no-picture symptom. In other words, if the raster is absent, proceed to check the horizontal-deflection section, the high-voltage section, and the picture tube. On the other hand, if a raster is displayed, but a picture display is absent, proceed to check the picture-signal channel in the receiver. Picture trouble symptoms include low or excessive brightness, low or excessive contrast, poor focus, excessive snow, off-center raster, smearing, ghosts, negative picture, ringing, poor sync action, incorrect hues, color impurity, misconvergence, interference, and various intermittent conditions. Sound trouble symptoms include distortion, weak sound output, sync buzz, noise, hum, and intermittent operation. Fluctuation or instability of either picture or sound reproduction is a trouble symptom that falls somewhere between normal and intermittent operation.

17.2   SECTIONALIZATION OF TROUBLE SYMPTOMS

Since a particular trouble symptom does not necessarily correspond to an individual component defect in a specific section of the receiver, sectionalization is often required in preliminary analysis. For example, a weak-color symptom can be caused by a fault in the chroma bandpass amplifier, in the automatic chroma-control section, or in the subcarrier-oscillator section. Therefore, suitable tests are required to localize the defective section. There are two fundamental types of tests that are employed for this purpose. One technique involves signal-tracing procedures; the other entails signal-substitution procedures. The basic signal-tracing instrument is the wide-band oscilloscope, and the basic signal-injection instrument is a specialized color-signal generator.

Receiver service manuals generally specify normal waveforms and their peak-to-peak voltage values at key test points in the receiver system. Signal-tracing procedures involve waveform checks at these points, using an oscilloscope with a suitable probe and comparison of the displayed waveform with the specifications in the service manual. As explained in greater detail below, a distortion in waveshape and/or display at incorrect peak-to-peak voltage indicates circuit trouble in the vicinity of the test point. An experienced technician can often analyze a distorted waveform to determine the particular circuit that is operating incorrectly. Sometimes, waveform distortion can be interpreted in terms of individual component defects.

Signal-injection procedures employ the picture tube in the receiver as an indicator. Suitable test signals are injected at progressive points through the receiver system, and the resulting displays on the picture-tube screen are observed. The specialized color-signal generators utilized in troubleshooting procedures are called analyzer-type instruments. They provide color-bar or black-and-white test patterns, as depicted in Fig. 17-2. As the injection point is moved step-by-step back from the picture tube toward the antenna-input terminals of the receiver, the screen pattern may suddenly disappear, or, it may become weak or distorted. This change from a normal display indicates that there is circuit trouble in the vicinity of the signal-injection point.

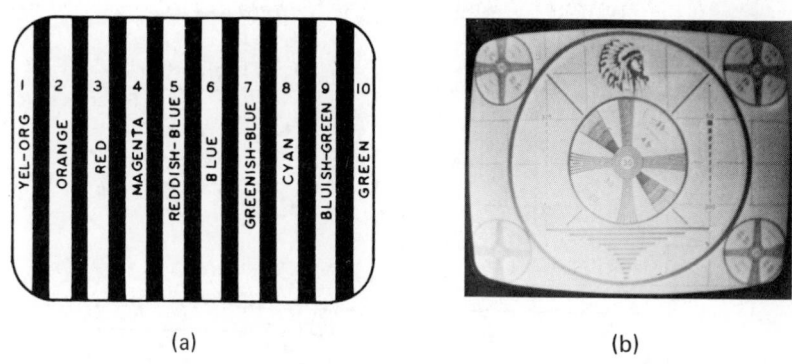

(a)   (b)

FIGURE 17-2   Screen patterns produced by an analyzer-type instrument. (a) Keyed-rainbow pattern; (b) black-and-white test pattern.

17.3   COLOR-WAVEFORM ANALYSIS

Consider the basic waveform characteristics which are involved in troubleshooting procedures. All waveforms can be classified into either sinusoidal types or complex types. Figure 17-3 shows examples of a sine wave and several complex waves. A sine wave has rms, peak, and peak-to-peak values, as depicted in Fig. 17-4. Its rms value is numerically equivalent to a DC value that produces the same amount of power. The positive- and negative-peak values of a sine wave are equal, and the wave's peak-to-peak value is double its peak value. Occasionally an rms voltage must be converted into its corresponding peak-to-peak sine wave voltage when an oscilloscope is being calibrated, as shown in Fig. 17-5. This must be done because a VOM indicates rms values of sine waves, whereas an oscilloscope is calibrated in terms of peak-to-peak values.

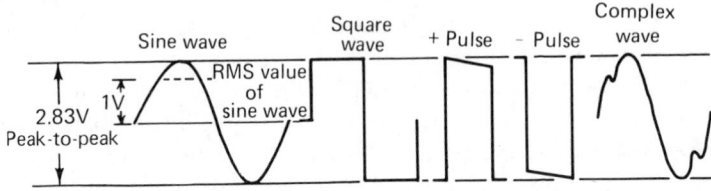

FIGURE 17-3   A sine waveform and four complex waveforms.

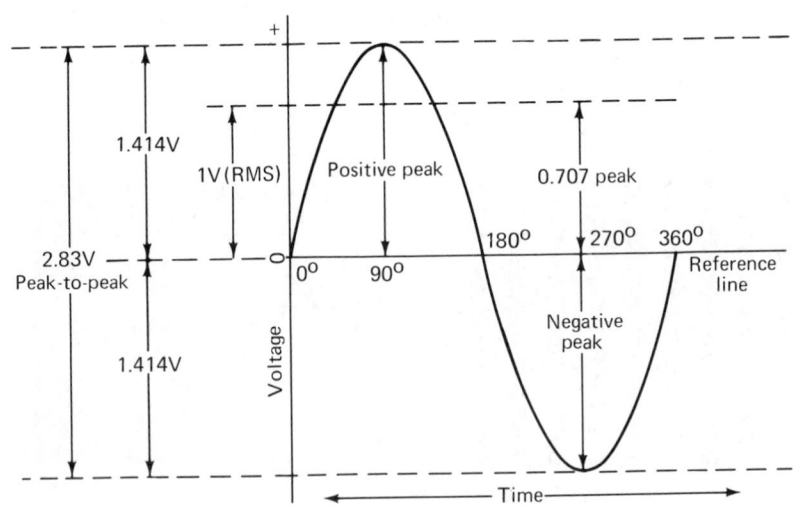

FIGURE 17-4   Basic values and their relations in a sine waveform.

Sine waves are characterized by frequency as well as by amplitude. Figure 17-6 shows how the variation of amplitude and frequency of a sine wave appears on a scope screen. Sine waves are also characterized by phase, as recalled from the previous discussion of chroma demodulation. The meaning of lead and lag are depicted in Fig. 17-7. Lagging waveforms start later in time than reference waveforms, and leading waveforms thus start earlier in time than the reference waveforms. Note that the waveforms depicted in Fig.

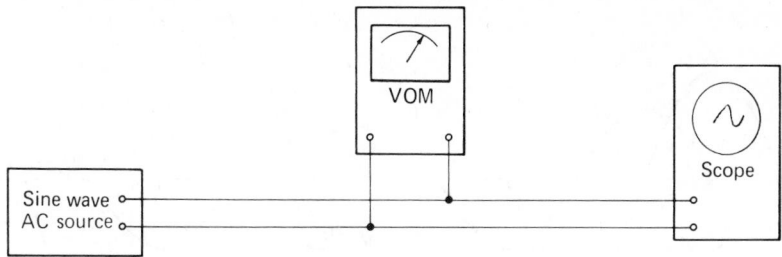

FIGURE 17-5   Connection arrangement for oscilloscope calibration.

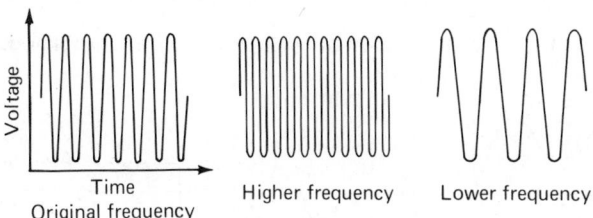

(a) Sine waveforms with same amplitude

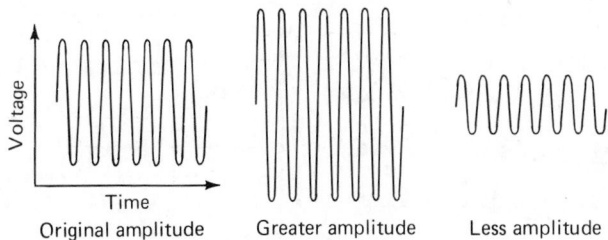

(b) Sine waveforms with same frequency

FIGURE 17-6   Variation in amplitude and frequency of a sine waveform.

17-7 have the same frequencies, but different amplitudes. Waveforms that differ in frequency will have a fixed phase relation, provided that their frequencies are integrally related.

A square wave is the most basic of the complex waves. From the viewpoint of waveform analysis, it is often helpful to regard a square wave as the sum of a large number of sine waves, as illustrated in Fig. 17-8. Note that a square wave contains odd harmonics only, and that the harmonics are in phase with the fundamental and with one another in the ideal square waveform. Video amplifiers or Y amplifiers are sometimes tested for square-wave response. It is standard practice to employ a 100-kHz square wave in this test.

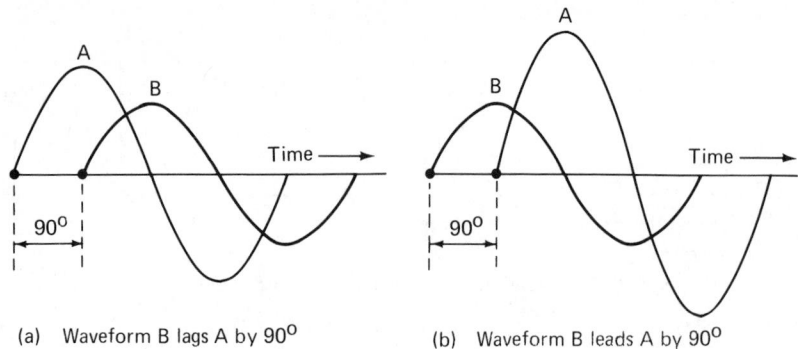

(a) Waveform B lags A by 90°   (b) Waveform B leads A by 90°

FIGURE 17-7 Phase relations in leading and lagging voltages.

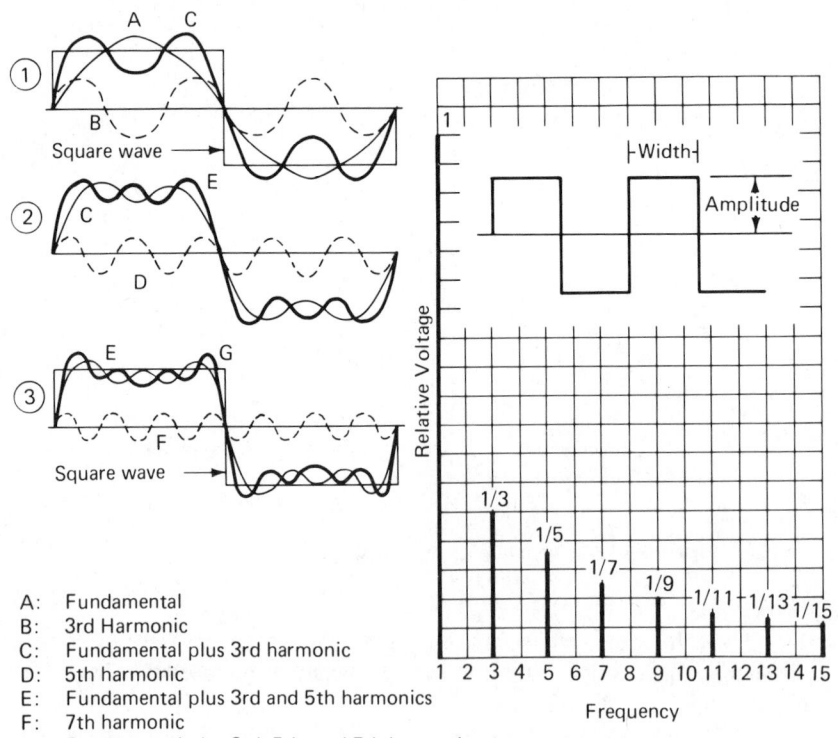

A: Fundamental
B: 3rd Harmonic
C: Fundamental plus 3rd harmonic
D: 5th harmonic
E: Fundamental plus 3rd and 5th harmonics
F: 7th harmonic
G: Fundamental plus 3rd, 5th, and 7th harmonics

FIGURE 17-8 Synthesis of a square wave from a fundamental sine wave and its odd harmonics.

Harmonics up to the 39th will be passed by a video amplifier with 4-MHz bandwidth. Distortion in square-wave reproduction is interpreted in Fig. 17-9.

For example, if the fundamental frequency of a square wave is attenuated during its passage through a circuit, the reproduced square wave will appear as it does in Fig. 17-9(b). If the harmonics are attenuated and the fundamental is unaffected, the reproduced square wave appears as in Fig. 17-9(c).

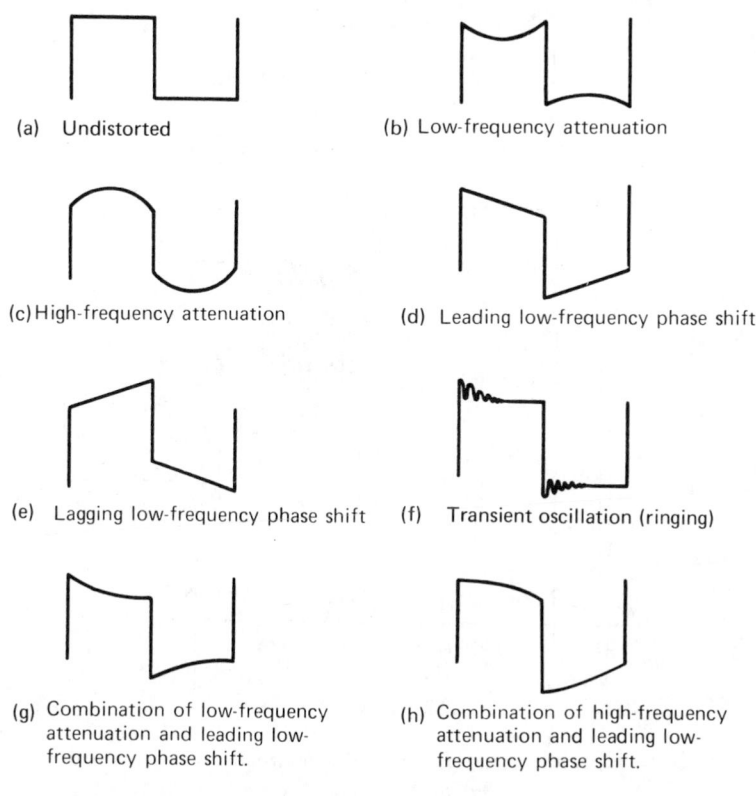

(a)   Undistorted

(b) Low-frequency attenuation

(c) High-frequency attenuation

(d)   Leading low-frequency phase shift

(e)   Lagging low-frequency phase shift

(f)   Transient oscillation (ringing)

(g) Combination of low-frequency attenuation and leading low-frequency phase shift.

(h) Combination of high-frequency attenuation and leading low-frequency phase shift.

FIGURE 17-9   Typical square-wave distortions and their interpretations.

Nonsymmetrical square waves are called *rectangular waves*. A highly nonsymmetrical square wave is called a *pulse*. Figure 17-10 indicates the positive-peak voltage, negative-peak voltage, and peak-to-peak voltage of such a pulse. Observe in Fig. 17-11 that a horizontal sync pulse with color burst has two pulse components and a sine-wave component. The sync tip is formed from a narrow pulse, and the pedestal is formed from a wider pulse. In turn, the sync pulse has a front porch and a back porch, and the color burst occupies most of the back porch. Note also that a practical sync pulse departs somewhat from an ideal pulse waveform, the practical sync pulse having rounded corners instead of sharp right-angled corners. Moreover, a practical sync pulse has sloping sides instead of perfectly vertical sides.

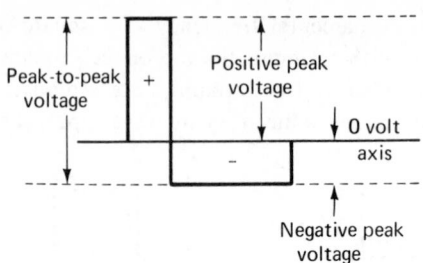

FIGURE 17-10   Characteristics of a pulse waveform.

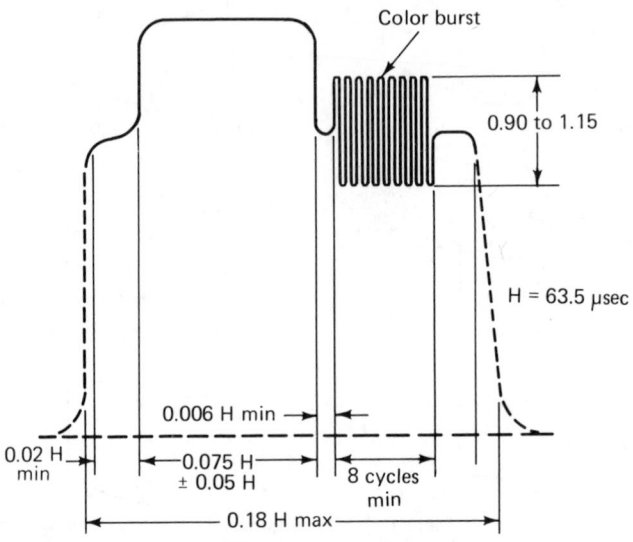

FIGURE 17-11   Characteristics of a horizontal sync pulse with color burst.

17.4   VECTORSCOPE PATTERNS

A vectorscope is basically an oscilloscope that is operated to display Lissajous patterns of chroma waveforms. The particular type of Lissajous patterns used are called *vectorgrams*. It is instructive to consider the development of vectorgrams, most of which are produced by keyed-rainbow signals. Referring to Fig. 17-12, note that a keyed-rainbow signal produces a train of color bursts from the output of the bandpass amplifier. These bursts produce the color-bar pattern depicted in Fig. 17-13(a), with the corresponding R–Y, B–Y, and G–Y chroma-demodulator output waveforms seen in Fig. 17-13(b). A vectorgram is displayed by connecting an oscilloscope to the out-

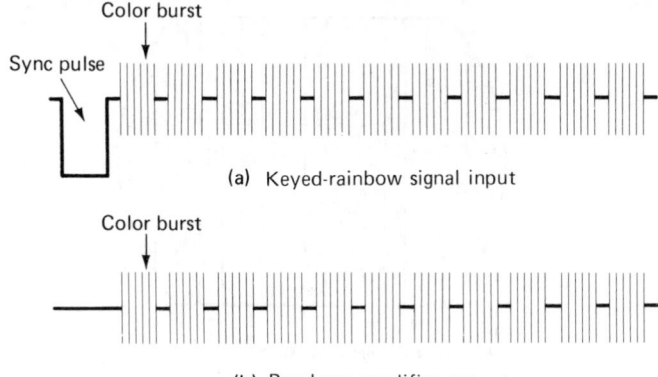

(a) Keyed-rainbow signal input

(b) Bandpass-amplifier output

FIGURE 17-12   Waveforms at input and output of bandpass amplifier.

puts of the R–Y and B–Y demodulators, as shown in Fig. 17-14. The result is a display of a pattern as illustrated in Fig. 17-15.

Suppose that the R–Y and B–Y signal channels have subnormal bandwidth. The practical result is distortion of the waveforms, and one of these distorting actions results in the introduction of baseline *curvature*. Figure 17-16 shows examples of baseline curvature and its effect on the displayed vectorgram. Note that the "spokes" in the vectorgram pattern do not extend down to the center of the pattern. Instead, there is an open space in the center portion of the vectorgram. This space has a circular boundary when the R–Y and B–Y waveforms have equal baseline curvatures. If the R–Y and B–Y waveforms have unequal baseline curvatures, the central portion of the vectorgram will have an elliptical boundary, as shown in Fig. 17-17.

Practical vectorgrams differ from ideal vectorgrams in many respects. For example, let us briefly consider the pattern shown in Fig. 17-18. The "petals" in the pattern are not entirely straight (being rounded at their tops), and no two "petals" have exactly the same shape. The central portion of the pattern is bounded by an irregular ellipse. There are residual phase errors indicated by the lack of angular uniformity from one "petal" to the next. As might be anticipated from Chap. 9, the pattern comprises 10 "petals," inasmuch as the 11th and 12th positions are blanked in the receiver. Numerous factors are accounted for in the "petal" shape details. For example, when corresponding R–Y and B–Y pulses have slightly different shapes, the resulting "petal" shape is affected as Fig. 17-19 shows. Vectorgram analysis is an extensive topic, and interested students are referred to specialized texts on this subject.

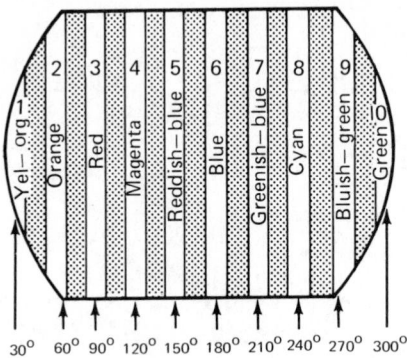

(a)   Picture tube display of
color-bar signal

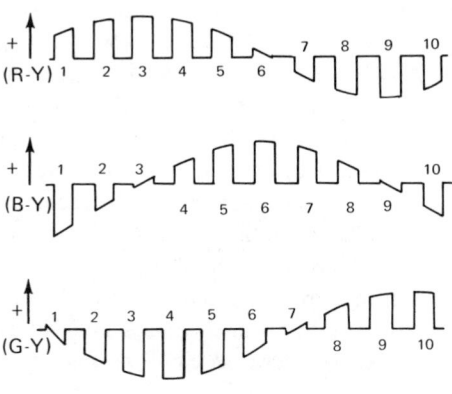

(b)   Ideal signals at red, blue and
green guns with bar pattern
input to receiver

FIGURE 17-13   Keyed-rainbow screen pattern and corresponding
R—Y, B—Y, and G—Y chroma-demodulator out-
puts.

17.5   FLOW-CHART ANALYSIS

Sectionalization and localization procedures are often facilitated by consider-
ing the signal flow chart for the receiver under test. Figure 17-20 shows a
flow chart for a widely used design of color-TV receiver. As noted previously,
the basic troubleshooting problem is to determine the source of a trouble
symptom by analysis of receiver response and test data. In a flow chart, four

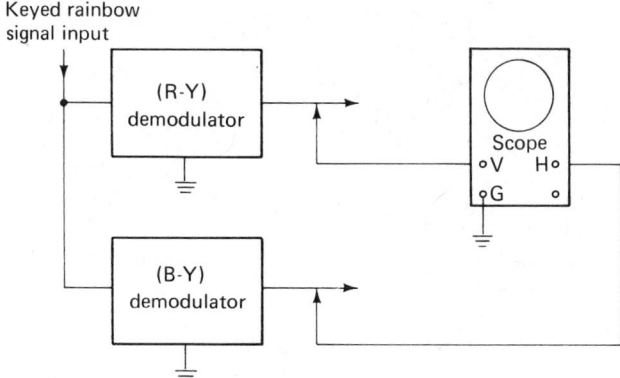

FIGURE 17-14   Basic vectorscope test arrangement.

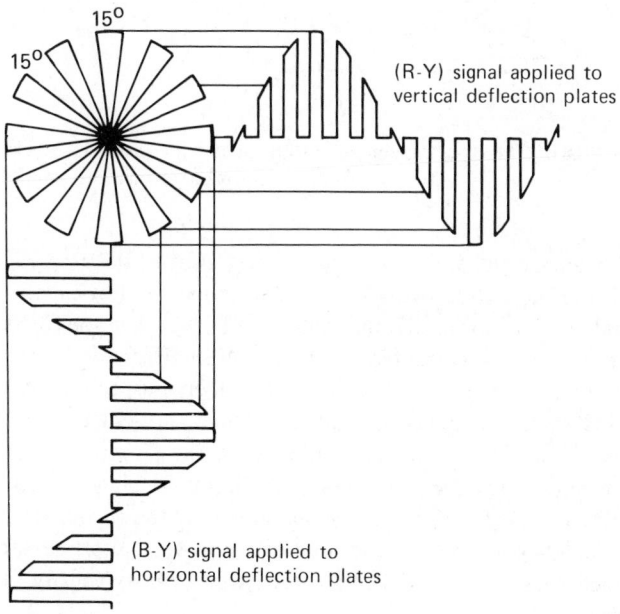

FIGURE 17-15   Development of an ideal vectorgram pattern.

signals can be recognized: the chroma signal, the Y signal, the sync signal, and the sound signal. Observe that all four signals normally pass through the tuners and the IF amplifiers, thereupon branching into the individual receiver sections.

Suppose that picture reproduction is normal, but sound reproduction is absent. In such a case, do not expect to find the trouble in the tuners

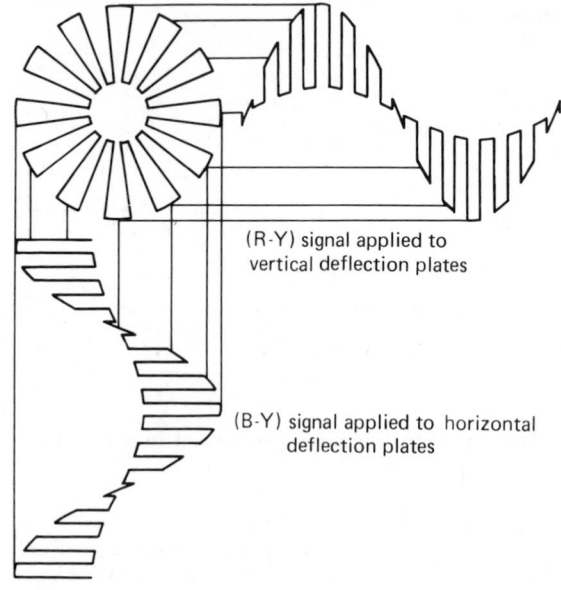

(R-Y) signal applied to
vertical deflection plates

(B-Y) signal applied to horizontal
deflection plates

FIGURE 17-16   Baseline curvature in the R−Y and B−Y waveforms produces distortion in the corresponding
vectorgram.

or the IF amplifiers, and in conclusion test IC201, Q801, or Q201 of Fig.
17-20. However, if tests with a wide-band scope and low-capacitance probe
show that there is no signal input voltage to IC201, then the first supposition
was incorrect, and the trouble will be found in either the IF section or the
tuner section. In this situation, the most likely suspect is the intercarrier
sound-detector diode, which is located near the picture-detector diode on the
IF circuit board. Associated components on the IF circuit board can also
"kill" the sound signal. For example, the output coupling capacitor from the
sound-detector diode might be open-circuited, or the IF bypass capacitor at
the output of the sound-detector diode might be short-circuited. Either
signal-tracing tests or signal-injection tests will serve to identify the defective
component.

Although the foregoing tests usually serve to localize the fault when
a picture-but-no-sound symptom occurs, there are also situations in which the
sound signal is "killed" in the IF amplifier prior to the sound detector. This
is called a *separated-sound-and-picture condition.* To check for this possibility, turn the fine-tuning control and observe the resulting picture and
sound reproduction. Suppose that the picture reception with no sound is
found to occur at one setting of the fine-tuning control, and that sound
reception with no picture occurs at another setting of the fine-tuning control.

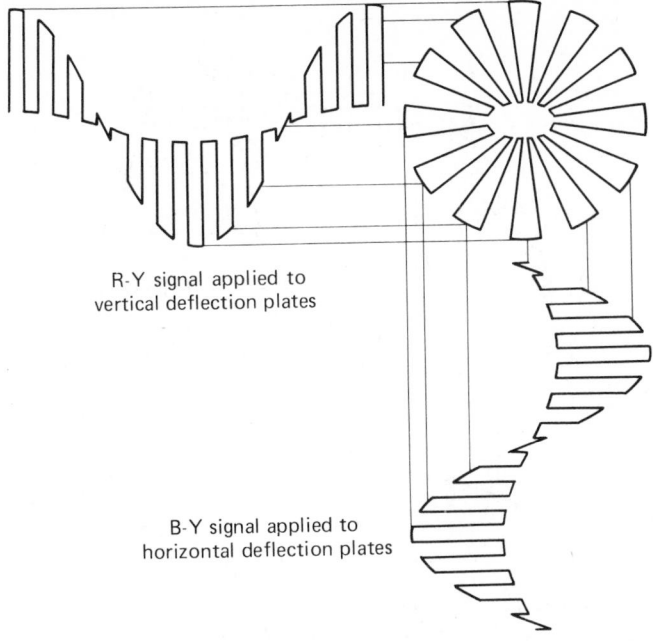

R-Y signal applied to
vertical deflection plates

B-Y signal applied to
horizontal deflection plates

FIGURE 17-17   Unequal baseline curvatures in the R–Y and
B–Y waveforms and the corresponding vector-
gram distortion.

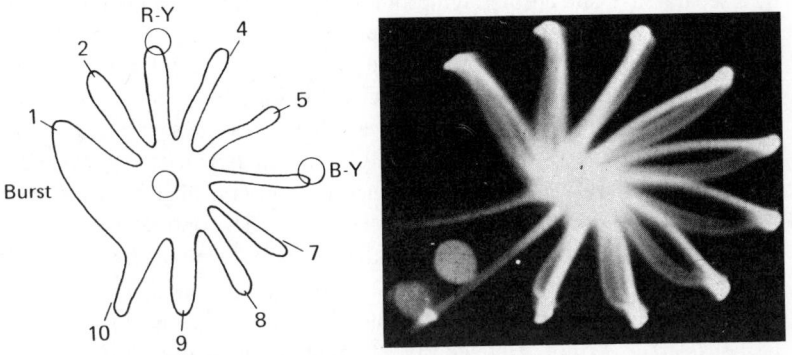

FIGURE 17-18   Normal vectorgram pattern for a Sony receiver.

It is then possible to conclude that the sound signal is being trapped out completely in the IF amplifier when the receiver is tuned so that the picture signal
is passed by the IF amplifier. The cause of this abnormal IF-amplifier action

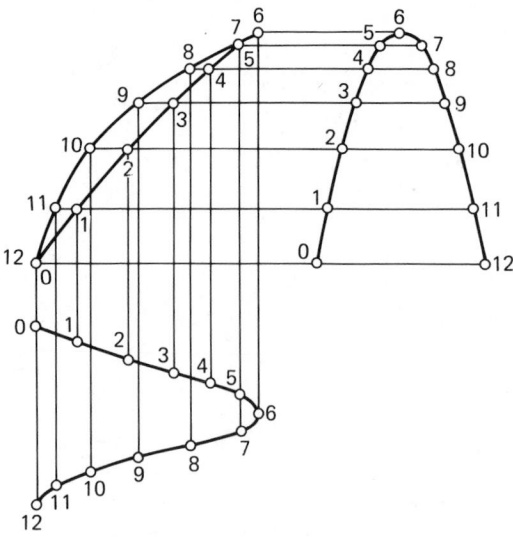

FIGURE 17-19 Vectorgram petal formed by R−Y and B−Y pulses with different shapes.

is a badly distorted frequency-response curve. Unless the alignment adjustments had been tampered with, a defective bypass or decoupling capacitor would generally be found in the IF system.

Note that a separated-sound-and-picture symptom is accompanied by a separated Y-and-chroma symptom. In the separated Y-and-chroma symptom, when the fine-tuning control is set to obtain a sharp picture, the color disappears from the image. When the fine-tuning control is set to obtain normal color in the image, the picture definition becomes very poor. Although this trouble symptom is not as obvious as the separation of sound and picture signals, it serves as a useful confirmation of the first analysis. Thus, a general rule at the service bench is to reserve alignment procedures until all component defects have been corrected. The only exception to this rule will occur when it is known that the viewer has tampered with the alignment adjustments.

The more detailed area trouble chart shown in Fig. 17-21 will now be considered. As noted previously, if there is sound output but either no picture or an unsatisfactory picture, next observe whether or not a raster is present. If a raster is displayed, possible trouble areas include the vertical-deflection section; and the pincushioning section is suspect in cases where there is distorted, little, or no deflection. On the other hand, if the raster geometry is normal with no picture visible, it may be concluded that there is

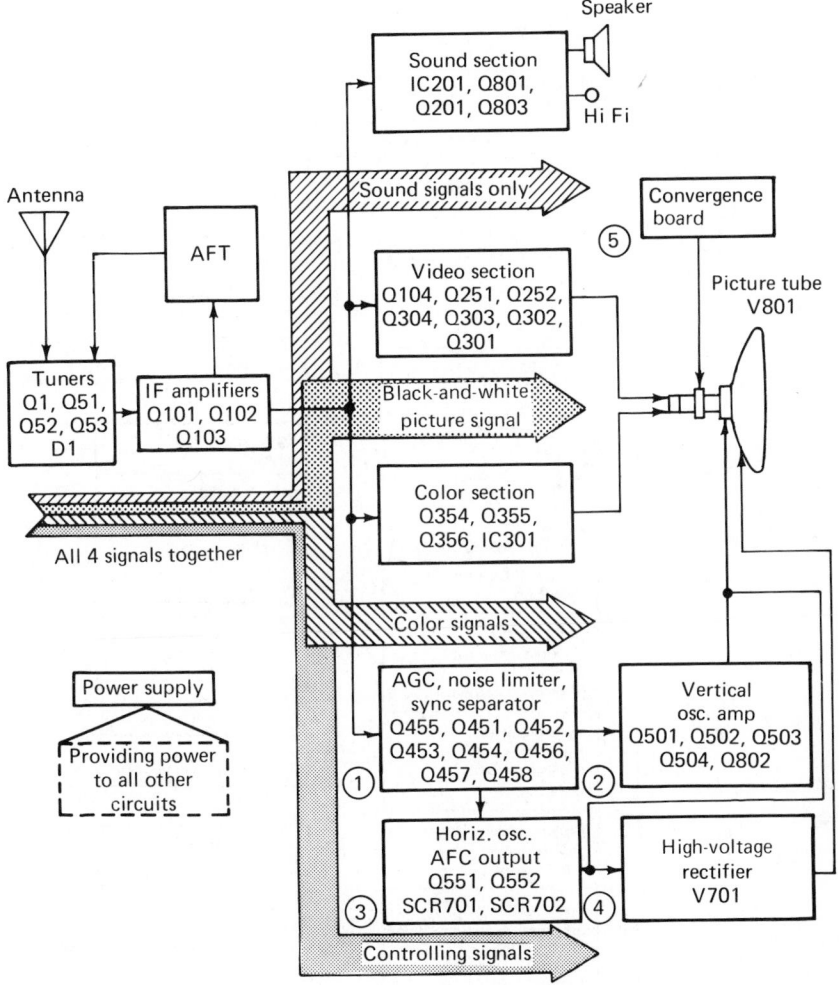

FIGURE 17-20 A signal flow chart. *(courtesy of Heath Company)*

a fault in the picture-signal channel. Unsatisfactory picture reproduction can be localized to the Y section, to the chroma section, or to both the Y and chroma sections. Picture trouble symptoms include weak or incorrect colors, misconvergence, smeary image, "snow," loss of color sync, loss of black-and-white horizontal or vertical sync, and poor focus. Note that if there is a dark screen, a picture signal is usually present, as shown by a waveform check at the output of the Y amplifier.

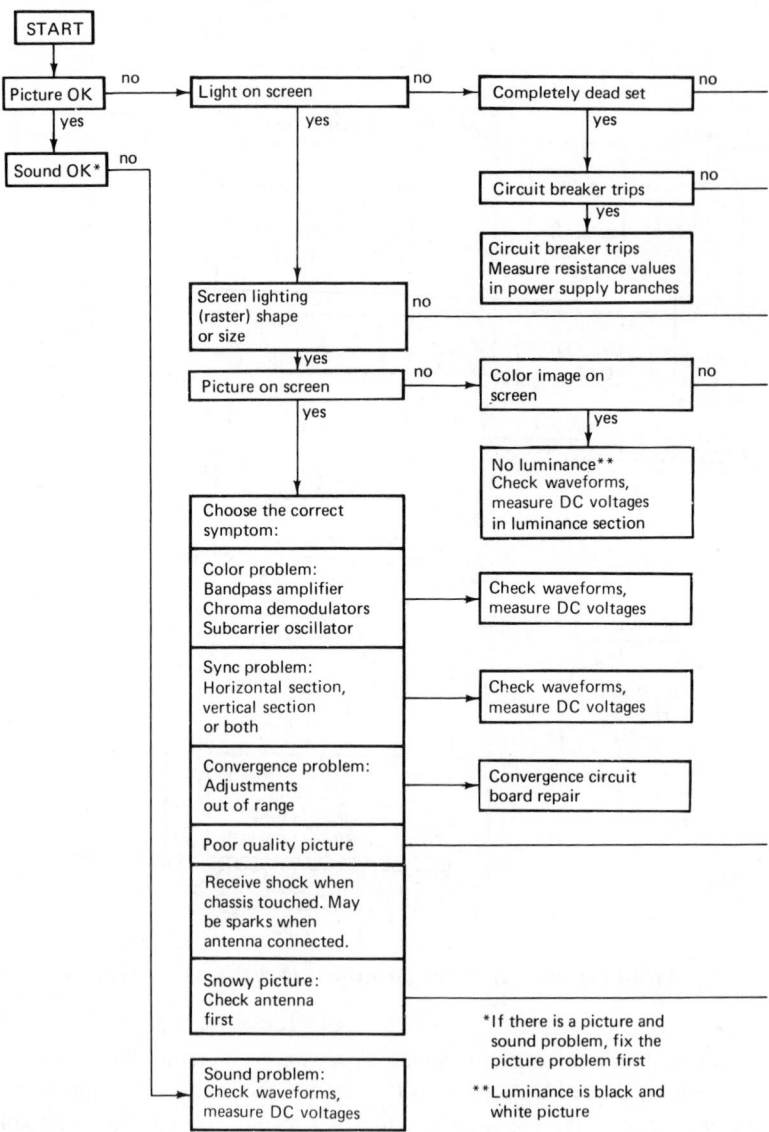

FIGURE 17-21   A trouble area chart.

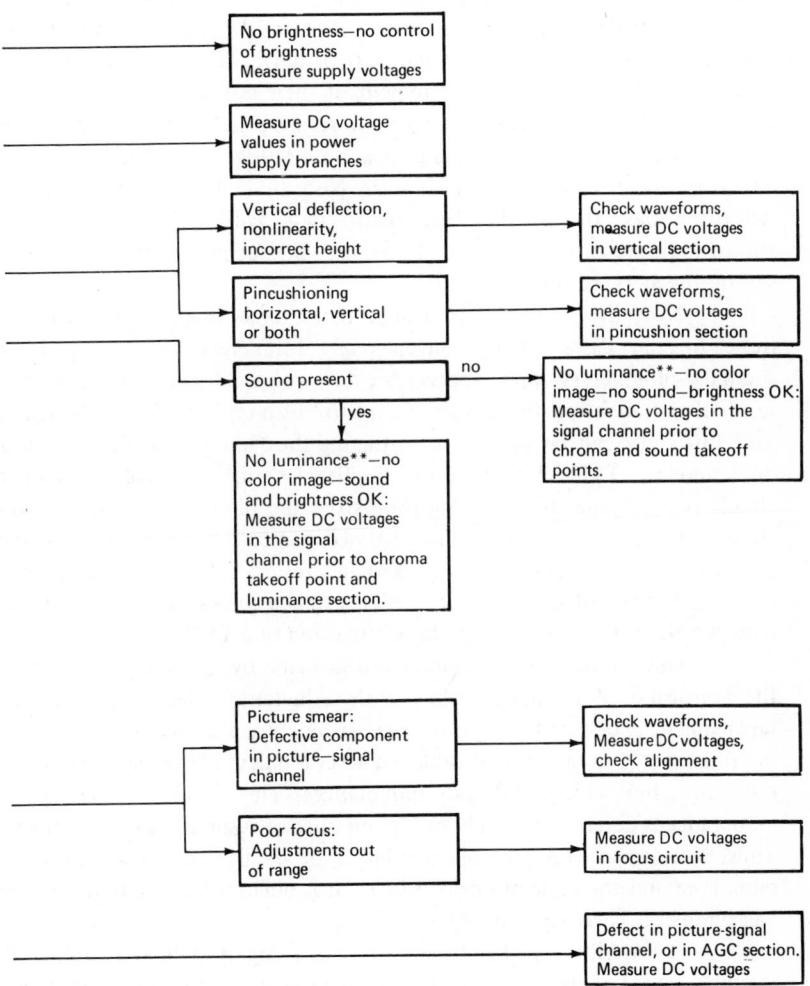

FIGURE 17-21 (Continued.)

17.6   PINPOINTING DEFECTIVE COMPONENTS

Defective components are generally pinpointed by means of DC-voltage measurements. Resistance measurements are often helpful in supplementing voltage data, or in checking components that do not carry DC current. Sometimes, a final determination cannot be made without substituting a known good component for the suspected part or device. Defective transistors can usually be pinpointed by measurement of their terminal voltages, and comparison of the measured values with those specified in the receiver service manual. DC-voltage measurements can be usefully supplemented in many situations by turn-on and turn-off tests. Note that a leaky coupling capacitor will often make the associated transistor "look bad" by bleeding DC voltage through the defective capacitor. Therefore, the technician should keep himself on the alert for this possibility.

Note that in-circuit resistance measurements are most likely to be misleading in cases where conventional ohmmeters are employed since ordinary ohmmeters apply appreciable test voltage to the circuit, with the result that transistors and diodes are apt to "turn on" and falsify the reading. This difficulty can be avoided by utilizing the "low-power ohms" function on a modern TVM. This function applies less than 0.1 volt between the circuit points under test, causing transistors and diodes to "look like" open circuits during the test procedure. Service manuals often provide resistance charts for quick checks of resistance values to ground from all key points in the chassis networks. These specified resistance values can be checked only with the aid of the "low-power ohms" function of a TVM.

One of the most common errors made by beginning technicians is the confusion of resistance values with inductance values. For example, a replacement choke or transformer cannot be chosen on the basis of its winding resistance because it is possible for several units to have the same winding resistance, but widely different inductances. This is due to the fact that inductance is determined solely by the number of turns and the core material. Thus, if a manufacturer employs a larger or smaller size of wire, using the same core and the same number of turns, it is obvious that winding-resistance measurements become meaningless.

Color-television troubleshooting is a highly-skilled discipline that requires considerable study and experience. Most of the difficulties encountered in practice are caused by insufficient knowledge of circuit action. It is not necessary for a competent technician to know higher mathematics. On the other hand, it is essential for him to be familiar with arithmetic, algebra, trigonometry, and electrical and electronic theory. Electrical theory necessarily includes both low- and high-frequency circuit action because factors that can be neglected at low frequencies often become dominant considerations in high-frequency operations. Solid-state theory is indispensable in the electronics area. In summary, the more technical education we acquire, the more successful we become in the field of color television.

EXERCISES

Questions

1. How do color-TV troubleshooting procedures start?
2. What is the difference between a no-picture symptom and a no-raster symptom?
3. Describe the basic *signal-tracing* procedures.
4. Briefly discuss the basic *signal-injection* procedures.
5. Explain the relations of rms, peak, and peak-to-peak values for a sine wave.
6. How does a square wave differ from a sine wave?
7. Discuss the development of a *vectorgram.*
8. Can a distorted IF response curve cause separated picture and sound?
9. List some typical picture trouble symptoms.
10. How are defective components generally pinpointed?
11. What is a *turn-off test*?
12. What is a *turn-on test*?
13. Explain the advantage of a lo-pwr ohmmeter in transistor circuit testing.
14. Why is the resistance of a coil unrelated to its inductance value?
15. Can a color-TV receiver be serviced without the aid of servicing data?

True-False

1. Troubleshooting procedures start with DC voltage measurements.
2. Tubes are checked first in tube-type receivers when malfunction occurs.
3. A picture trouble symptom can occur with sound reproduction unimpaired.
4. The antenna is included in the signal-channel system of a TV receiver.
5. A *no-raster* symptom is the same as a *no-picture* symptom.
6. Particular trouble symptoms do not necessarily correspond to individual component defects.
7. Faulty components can always be pinpointed by oscilloscope tests.
8. *Signal injection* and *signal tracing* are equivalent terms.

9. The peak-to-peak voltage of any waveform is equal to 2.83 times its rms voltage.

10. A service-type VOM indicates the rms values of sine waves.

11. Video amplifiers are often checked for 100-kHz square-wave response.

12. Oscilloscopes are often used with high-voltage DC probes.

13. A *TVM* indicates the rms value of any voltage waveform.

14. Vectorgrams can be displayed only by triggered-sweep oscilloscopes.

15. Color-bar generators are used in chroma alignment procedures.

Multiple Choice

1. If a color receiver reproduces color programs in black-and-white, the _____ should be checked first.
   - (a) antenna
   - (b) picture tube
   - (c) line voltage
   - (d) color-killer setting

2. When sync buzz occurs in the sound output, a technician suspects malfunction in the _____ .
   - (a) audio amplifier
   - (b) intercarrier-IF amplifier
   - (c) ratio detector
   - (d) RF tuner

3. Display of an overloaded or muddy and filled-up picture is likely to be caused by a fault in the _____.
   - (a) AGC section
   - (b) AFC section
   - (c) ATC section
   - (d) none of the above

4. Receiver service manuals generally specify_____ .
   - (a) wave propagation conditions
   - (b) normal waveforms with p-p voltages
   - (c) standing-wave ratios
   - (d) local-oscillator frequencies

5. All waveforms have _____ .
   - (a) rms, peak, and peak-to-peak values
   - (b) the same ratio of rms/peak values
   - (c) equal average values
   - (d) the same ratio of DC/AC values

6. Amplitude and frequency values for any waveform are _____ .
   (a) directly related
   (b) partially related
   (c) independent parameters
   (d) none of the above

7. The voltage across a capacitor _____ the capacitor current.
   (a) leads
   (b) lags
   (c) is simultaneous with
   (d) has no relation to

8. Video amplifiers and Y amplifiers are tested for _____ square-wave response.
   (a) 100 MHz
   (b) 10 MHz
   (c) 4.5 MHz
   (d) 100 kHz

9. A vectorgram is a particular type of _____ .
   (a) Lissajous figure
   (b) time base
   (c) color-bar pattern
   (d) none of the above

10. A vectorscope is connected at the outputs of the _____ .
    (a) RF and IF amplifiers
    (b) video and sound detectors
    (c) R–Y and B–Y demodulators
    (d) horizontal and vertical oscillators

Problems

1. Color bars in a keyed-rainbow pattern are formed by a 189-kHz square wave. If the bandwidth of a chroma demodulator circuit is 0.5 MHz, how many harmonics of the keyed-rainbow signal are passed?

2. If a keyed-rainbow signal were to be passed through a chroma-demodulator circuit with negligible distortion, the 15th harmonic would have to be reproduced. What bandwidth would be required?

3. When a video amplifier with 4-MHz bandwidth is tested for 100-kHz square-wave reproduction, how many harmonics of the square wave pass?

4. The bandwidth of a circuit is equal to $1/(3T)$, where T is the rise time of a reproduced square wave. If a chroma-demodulator circuit has a rise time of 0.66 μs, what is its bandwidth?

5. If a chroma demodulator has a rise time of 0.66 μs, what is its fall time?

6. When a square wave passes through two successive circuit sections, its rise time becomes equal to the square root of the sum of the squares of the individual rise times. If a square wave passes through a bandpass-amplifier circuit with a rise time of 0.66 μs, and then through a chroma-demodulator circuit with a rise time of 0.66 μs, what is the rise time of the reproduced square wave?

7. If a peaking coil has an inductance of 1 mH, and resonates at 0.5 MHz, what value of capacitance is in shunt to the coil?

8. What is the reactance of the coil in problem 7 at 0.5 MHz?

9. What is the reactance of the capacitance in problem 7 at 0.5 MHz?

10. What is the reactance of the coil in problem 7 at 50 kHz?

Chapter 18

Community Antenna Television Systems

18.1 PRINCIPLES OF CATV

Formerly, community antenna television (CATV) systems were utilized only in fringe or far-fringe areas where reception was made difficult or impossible because of low-level signal conditions. However, CATV systems are now also being used in high-level signal areas where multipath propagation is a serious problem. In either case, a CATV system often serves an entire town or city. A single antenna site, which may be on top of a hill, mountain, or skyscraper, is employed. Separate high-gain and properly oriented antennas are commonly utilized for each active channel. Thereby, compromises are avoided and reception is optimized across the entire TV spectrum. The plan of a typical CATV system is shown in Fig. 18-1. Signals from the antenna site are conducted to the TV-receiver locations via coaxial cable.

A basic CATV system is composed of a preamplifier or booster for each antenna, and a mixer or combiner for the amplifier output signals. When UHF reception is provided in addition to VHF reception, the signal from each UHF channel is processed by a *translator*. A translator is a frequency converter that heterodynes the UHF signal frequency down to a VHF

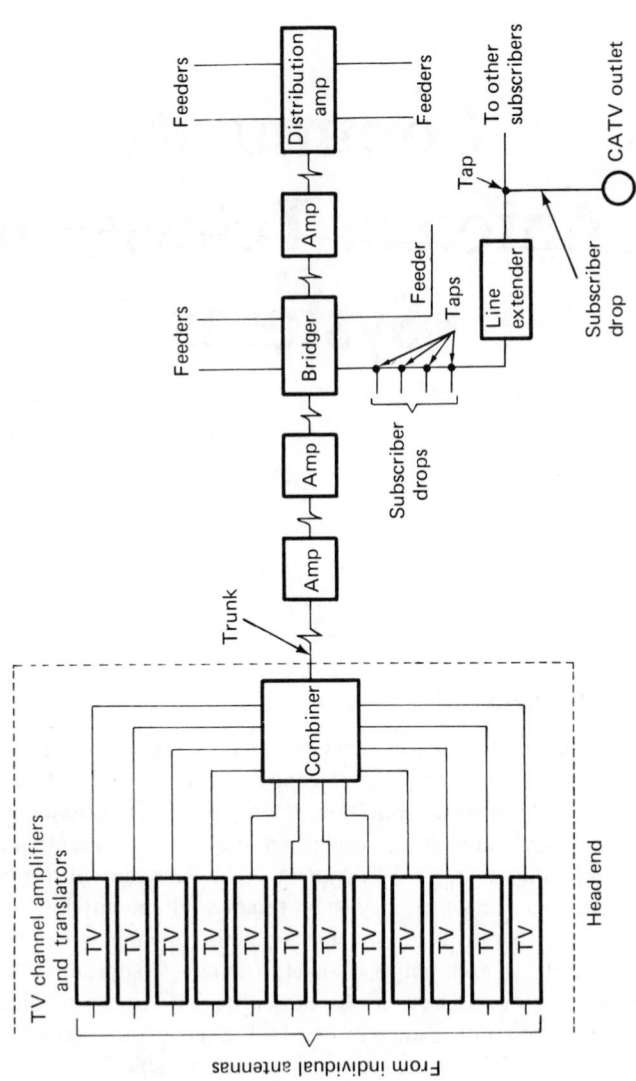

FIGURE 18-1 Plan of a typical CATV system.

frequency. Translation is advantageous inasmuch as a CATV system necessarily operates with lengthy coaxial cables, and the transmission loss through the cables is much greater at UHF than at VHF. For example, if an 800-MHz signal is heterodyned down to 100 MHz, the cable transmission loss is reduced to a fraction of the amount that would occur at 800 MHz. Figure 18-1 shows that FM radio transmissions are also processed by a typical CATV system. Numerous other services may be provided by more elaborate and complex systems, as explained in greater detail below.

The CATV output signal voltages from the combiner network are fed into one or more trunk cables. A trunk cable conducts the signal voltages from the antenna site to the utilization site(s), which may be many miles distant. As noted previously, a progressive signal attenuation occurs and is due to cable losses. Therefore, amplifiers must be inserted at suitable points along the cable run. At a frequency of 150 MHz, a coaxial cable imposes a signal loss of approximately 1 dB per 100 feet of cable. Thus, an amplification of about 50 dB per mile of cable is required. Voltage and power ratios with their corresponding dB values are listed in Table 18-1.

TABLE 18-1

ELECTROMAGNETIC WAVE SPECTRUM

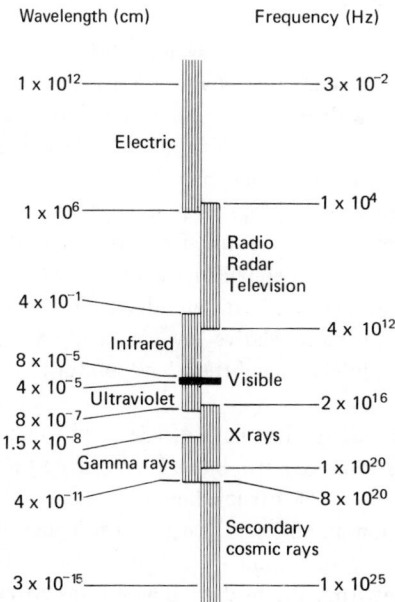

In most CATV installations, trunk lines are routed from the head end to the beginning of the distribution system, which may be mounted on poles or buried beneath the surface of the ground. A *head end* is the generic term for the preamplifiers or boosters, translators, and combiners at the antenna site. In cases where the cable is installed in a duct or trench, it is brought above ground and into a pedestal at the location of a cable amplifier. From the pedestal, the amplified signals are fed into the next length of buried cable. It is standard practice to utilize cables with 75 ohms of characteristic impedance. Therefore, it is essential to maintain 75-ohm input and output impedances at each amplifier or other unit in the cable system. Unless carefully matched impedances are employed throughout, reflections will occur, and line ghosts will appear on the picture-tube screens.

Note that the distribution system starts with a bridging amplifier, or *bridger*. The basic bridger comprises a passive network consisting of a VHF transformer with associated capacitors and resistors. This bridger splits or divides the incoming signal into several portions for supplying the feeder lines. Signal attenuation occurs during this process, and a bridging amplifier is generally included to bring up the output signal levels to approximately those of the input signal level. This technique causes the insertion loss of the splitter network to be effectively zero. A cable amplifier or a bridging amplifier must be inserted at each point in the system where the signal-to-noise ratio falls to a comparatively low level. In other words, if the signal-to-noise ratio is permitted to deteriorate excessively at any point in the system, subsequent amplification becomes useless since the noise (snow) is amplified to the same extent as the signal.

In an extensive distribution system, feeder amplifiers may also be installed to maintain an adequate signal-to-noise ratio. Like a trunk amplifier, a feeder amplifier is a wide-band VHF amplifier. A typical amplifier provides 25-dB gain over a frequency range from 50 to 220 MHz. Inasmuch as cable losses are greater at high frequencies than at low frequencies, it is evident that high-band attenuation will be greater than low-band attenuation. Therefore, a rising frequency characteristic is required, and the amplifiers or signal splitters are often supplemented by an equalizer, or *tilt control*. A tilt control consists of a bandpass-filter arrangement with an adjustable frequency characteristic. It operates by introducing a relative low-frequency loss so that the output signals from the amplifiers or splitters have uniform amplitude across the entire VHF band.

Bridging amplifiers must have at least one trunk output in addition to its feeder outputs; that is, a trunk line is continued through a bridger. The final amplifier at the end of a trunk line is termed a *distribution amplifier*. Although a distribution amplifier is basically a bridging amplifier, it serves as a termination point for the trunk line. A feeder line can be operated with runs up to 1,000 feet from the bridger. Beyond this distance, a line extender is required to maintain an adequate signal-to-noise ratio.

Feeder cables are sometimes called *distribution cables*. A cable installation inside of a building may also be termed a distribution cable. Technically, a distribution cable is a tap-off from a feeder cable. The tap point is called a *subscriber tap*. A tap at a point along a feeder line is called a *subscriber drop*. As noted previously, this branch connection is necessarily made in a manner that maintains a 75-ohm impedance in the system. In most installations, the signal level at the subscriber drop is sufficiently high so that substantial attenuation can be tolerated. In turn, elaborate impedance-matching arrangements are not required. Instead, a high-value series resistance called an *isolating resistor*, may be connected between the tap point and the cable run to the TV receiver. This high resistance introduces negligible change of cable impedance at the tap point. Another method employs a high value capacitive reactance, called an *isolating capacitor*, connected between the tap point and the cable to the TV receiver. This method has the advantage of providing a compensating tilt for the subscriber drop cable. In any case, the signal level provided to a TV receiver by the CATV system is on the order of 1,500 microvolts. This level provides good-quality reception without causing accompanying radiation problems from the CATV system, which could cause interference to other installations and services.

18.2   METROPOLITAN CATV

There is a marked trend toward CATV installation in densely populated areas where the signal level is high, but ghosts are a vexing problem. In addition to provision of high-quality VHF, UHF, and FM reception, various other communication services are involved in this trend. For example, a subscriber may be offered two-way communication between his location and a central processor. In turn, the coaxial cable can be required to carry as many as 40 broad-band signals in both directions simultaneously. The proposed services for a complete "wired city" include two-way education, two-way security alarms, and one-way entertainment programs supplementing TV broadcast programs. Automatic utility meter reading, shopping by TV, special on-demand information, and computer-aided instruction are in the process of field-testing at this time.

   CATV offers two kinds of service—one to individual residences, and the other to the city government and its functions. A burglar alarm system can be provided in combination with a fire alarm system. Metropolitan areas now have police and fire call-box installations, and are developing computer-controlled traffic-light systems, automatic vehicle-locating systems, and specialized communication facilities, which could be integrated by an elaborate cable system linked to each street intersection. Commercial institutions, such as banks, can benefit by better communication between branches. Similarly, within the medical community, hospitals have a need for better video and digital facilities. At present, high-resolution (1,000-line) color television is already in wide use for surgical classroom instruction by closed-circuit installations. It is only a small step from closed-circuit operation to CATV utilization.

Figure 18-2 depicts a comparatively simple two-way CATV system that has been field-tested in Reston, Virginia. This installation employs telephone-line links in lieu of bidirectional CATV facilities for effective two-way operation. The head-end computer utilizes a polling technique to integrate the entire network and process information from each terminal of the system. Thus, the subscriber might choose to order merchandise from some retail outlet, to vote for a certain TV program, or to play a game of chess with another subscriber. Again, the subscriber might choose to avail himself of the computer to solve mathematical problems or to obtain computer-aided instruction in a particular field. Information retrieval from any encyclopedia or dictionary can be provided if desired.

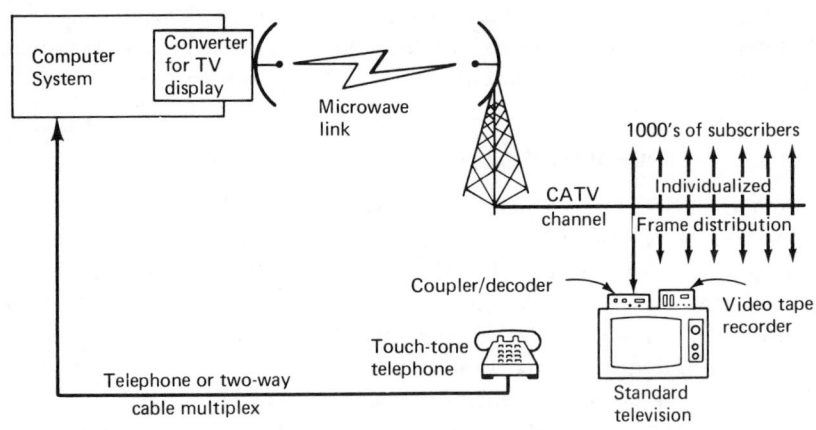

FIGURE 18-2   A simple two-way CATV system. *(courtesy of Radio Electronics)*

To summarize briefly, metropolitan CATV seeks to supplement conventional CATV facilities with household, business, educational, and governmental services. Household services include security, alarm, surveillance, banking, shopping, news, weather, time, mail, special programs, and opinion polls. Business services include commercial data retrieval, document reproduction, meter reading, credit-card validation, and market testing. Educational services include computer-aided instruction, research and data retrieval, centralized library services, and personalized instruction from a central location. Governmental services include burglar and fire protection, criminal identification by photographic and fingerprint record retrieval, televised line-ups of criminal suspects, agency interconnections, and automated traffic-control systems.

18.3   TROUBLESHOOTING CATV SYSTEMS

Customer relations are more of a problem in CATV servicing than in other areas of television service. For example, if a subscriber reports trouble symptoms, the technician might need to check an amplifier or cables in a neighbor's backyard. Since the neighbor regards a technician as a nuisance in this situation, tact and diplomacy are prime requisites. It is essential for the technician to wear a uniform with his name in plain sight, and to carry an identification card. This ID card should be presented immediately when the doorbell is answered. After permission is obtained to enter the property, the technician must beware of dogs that might attack him. If possible, it is desirable to have the property owner standing by while the technician makes his tests and/or repairs. The following troubleshooting procedures are typical.

Identification of Trouble.   When a complaint of poor or no reception is called in, it is necessary to determine whether the trouble is being caused by a cable fault or by a receiver defect. A quick check can be made by connecting the cable to another receiver that is in good operating condition. If the trouble is in the cable, the technician should first inspect the cable run from the receiver to the wall outlet. In many cases, dogs chew coaxial cables, causing short-circuits and open-circuits. Exposed cable runs in a backyard may also have been chewed by dogs. Underground cable runs may become defective owing to leakage of water into the cable, or to mechanical damage. Common trouble symptoms caused by moisture are weak and snowy pictures, no high-band station reception, and weak or no color reception.

Trouble Along the Right-of-Way.   A common cause of outage in portions of a CATV system is damage owing to car accidents, particularly at night to trunk lines running into sparsely populated towns. This type of damage is customarily repaired by the public-utility company. Sometimes the utility workmen may switch the circuits and erroneously leave the power supply for the cable system "dead." In these situations, it is difficult for the subscriber to understand why his television receiver does not work after the power line has been repaired. Again, this is an example of the requirement for good public relations.

Tampering by Subscribers.   When a house is remodeled by a do-it-yourselfer, he may try to modify the cable installation. Sometimes the tyro will attempt to extend a cable run by means of ordinary twin lead, with the result that ignition noise and other forms of interference are displayed in the picture. If coaxial cable is utilized, a complaint of weak reception is made occasionally, owing to the requirement for increased amplification. "Pirates," also called "bootleggers," sometimes attempt to evade subscriber fees by tapping into a cable run. In some installations, the cable also carries the supply voltage for the line extenders, and grounding of the cable by a "pirate" will blow the fuse and stop reception by all subscribers along one or more streets.

Sweep-Frequency Testing.   Both troubleshooting and preventive maintenance are facilitated by sweep-frequency testing. A wide-band sweep generator is applied at the head end of the system, and access points along the cable run are checked progressively with an oscilloscope. A scope technician may need to work on amplifiers up in the air, with the aid of a "bucket" truck. If the high-frequency end of the response pattern shows excessive attenuation, or vice versa, the equalizers or tilt controls may require readjustment. When poor frequency response is caused by line reflections (standing waves), there is a cable fault somewhere that must be located. Most line reflections are caused by water leakage into cables or boxes.

Amplifier Overload.   An excessively high signal level is as undesirable as an excessively low level. When amplifier overload occurs, cross-modulation results. Typical overload trouble symptoms are grain in the picture, herringbone interference, and the "windshield-wiper" effect (a weak interfering picture may appear with the desired picture, with a horizontal blanking bar drifting back and forth across the image). Sometimes cross-modulation causes a weak interfering picture to be displayed in which the vertical blanking bar drifts up and down through the image. Therefore, it is advisable for the technician to carry a field-strength meter along with an oscilloscope. If an abnormally high signal level is measured at an access point, the gain control of the associated amplifier should be adjusted accordingly.

Ghosts and Fuzzy Images.   Ghosts and fuzzy images in receivers connected to a cable system are usually caused by a cable break or similar defect. However, some types of amplifier overload can simulate ghosts. Cable damage can be caused by construction crews who are unaware of buried cables, by vandalism, or by sabotage. Sometimes irresponsible people with rifles shoot into the air at CATV cables. Disgruntled subscribers have been known to chop cables with axes. Less serious image deterioration can result from "bootlegger" taps along a cable run; these are sometimes hard to find, unless a thorough and systematic checkout is made.

Hum in the Picture.   In older types of CATV installations that employ tube-type amplifiers, 60-Hz hum bars occasionally appear in the picture. This trouble symptom is caused in most instances by heater-cathode leakage in an amplifier tube. Occasionally, an open filter capacitor in the power supply for a solid-state system will cause hum bars to appear in the image.

Amplifier Troubleshooting.   Troubleshooting CATV amplifiers involves the same basic principles that have been noted previously for conventional high-frequency amplifiers. However, more stringent requirements are placed on amplitude linearity in a CATV amplifier. Linearity can be accurately checked with a lab-type signal generator that has a calibrated output meter. As the signal level is varied over its normal range, the amplifier output should remain precisely proportional to the input level. Nonlinearity is generally caused by incorrect bias on a transistor, or by a defective transistor. As noted previously, an incorrect bias voltage will often be tracked down to a leaky coupling capacitor.

EXERCISES

Questions

1. Why are CATV systems utilized?
2. What is the difference between a preamplifier and a combiner?
3. How does a *translator* function?
4. Describe a *trunk cable*.
5. State the approximate signal loss per 100 feet of coaxial cable at 150 MHz.
6. Define a *head end* in a CATV system.
7. Explain the function of a *bridger*.
8. How much gain does a typical feeder amplifier provide?
9. Discuss the function of a *tilt control*.
10. What is a *distribution amplifier*?
11. Describe a *subscriber tap*.
12. Why is the maximum signal level in a CATV system limited to approximately 2000 μV?
13. How far can a feeder line be run without a line extender?
14. What is the approximate signal level supplied to a TV receiver by a CATV system?
15. Define a *subscriber drop*.

True-False

1. CATV systems are used only in fringe or far-fringe areas.
2. An elaborate antenna system is employed in a CATV installation.
3. *Preamplifier* and *booster* are equivalent terms.
4. A translator heterodynes a low-frequency signal to a higher frequency.
5. *Combiner* and *mixer* are equivalent terms.
6. At 150 MHz, a coaxial cable imposes a signal loss of approximately 1 dB per 100 feet.
7. Trunk lines run from the combiner to the beginning of the distribution system.
8. A head end comprises preamplifiers, translators, and combiners.
9. A pedestal is a base mounting for an antenna.
10. CATV systems maintain a 300-ohm impedance throughout.

11. A distribution system starts with a bridging amplifier.

12. Bridgers are basically signal splitters.

13. Noise filters are used in a cable system to improve the signal-to-noise ratio.

14. Trunk and feeder amplifiers typically provide 100 dB gain.

15. A tilt control introduces a relative low-frequency loss.

Multiple Choice

1. CATV systems are used to overcome _____ problems.
   - (a) multipath propagation
   - (b) stormy weather
   - (c) airplane traffic
   - (d) none of the above

2. Signal lines in CATV systems employ_____ .
   - (a) television lead-in
   - (b) coaxial cable
   - (c) high-voltage cable
   - (d) public-utility cable

3. Translators are used to _____ .
   - (a) heterodyne foreign-language programs
   - (b) convert low frequencies to high frequencies
   - (c) decode foreign color programs
   - (d) convert high frequencies to low frequencies

4. Coaxial cable losses are _____ .
   - (a) the same at all frequencies
   - (b) greater at high frequencies
   - (c) greater at low frequencies
   - (d) inversely proportional to the square of the distance

5. CATV output signal voltages are fed from the combiner into _____.
   - (a) trunk cables
   - (b) utilization sites
   - (c) line extenders
   - (d) subscriber drops

6. A CATV system requires an amplification of approximately_____ dB per mile of cable.
   - (a) zero
   - (b) 5
   - (c) 50
   - (d) 500

7. The head end comprises _____ .
   (a) boosters, translators, and combiners
   (b) antennas, preamplifiers, and bridgers
   (c) trunk amplifiers, feeder amplifiers, and line extenders
   (d) none of the above

8. Cable amplifiers are located in _____ .
   (a) satellites
   (b) pedestals
   (c) blanking pulses
   (d) color receivers

9. An impedance of _____ohms is maintained throughout a CATV system.
   (a) 300
   (b) 600
   (c) 75
   (d) 10

10. A distribution system starts with a _____ .
    (a) bridger
    (b) line extender
    (c) preamplifier
    (d) integrator

11. Feeder amplifiers are _____ .
    (a) wide-band VHF amplifiers
    (b) single-channel VHF amplifiers
    (c) wide-band UHF amplifiers
    (d) narrow-band UHF amplifiers

12. A tilt control operates by introducing a _____ .
    (a) weighted signal voltage
    (b) low-frequency loss
    (c) high-frequency loss
    (d) progressive phase shift

13. Feeder cables are also called _____ .
    (a) low-noise lines
    (b) distribution cables
    (c) regenerative loops
    (d) degenerative loops

14. A CATV system provides a level of about _____ at each receiver.
    (a) 1500 volts
    (b) 1500 millivolts
    (c) 1500 microvolts
    (d) none of the above

15. Bridgers are essentially _____ .
    (a) signal splitters
    (b) impedance measuring devices
    (c) capacitor checkers
    (d) logic field analyzers

Problems

1. A television wave travels 328 yards/μs. How long does it take the wave to travel one mile?

2. If a ghost signal is delayed 22 μs with respect to the direct signal, how far is the ghost image displaced on a picture-tube screen 15 inches wide?

3. An 800-MHz signal is heterodyned down to 100 MHz in a translator. If a low-side oscillator is utilized, what is the oscillator frequency?

4. A cable has an attenuation of 1 dB/100 ft at 150 MHz. What is the ratio of output voltage to input voltage for one mile of cable?

5. What is the ratio of output power to input power for the cable in problem 4?

6. If a CATV amplifier provides a gain of 50 dB, what is its ratio of output signal voltage to input signal voltage?

7. What is the ratio of output power to input power for the amplifier in problem 6?

8. A distribution amplifier provides a gain of 20 dB. What is its ratio of output signal voltage to input signal voltage?

9. What is the ratio of output power to input power for the amplifier in problem 8?

10. A passive device introduces an insertion loss of 6 dB. What is its ratio of output signal voltage to input signal voltage?

Appendix 1

Digital
Color-TV Receiver

A digital color-TV receiver (Fig. A1-1) features on-screen channel readout with the time in hours, minutes, and seconds. The chassis employs a total of 33 integrated circuits and 71 transistors, with 20 modules. To display a number on the picture-tube screen, the scanning beam must be turned on during the appropriate intervals. Seven-segment digits are used, driven by a character generator. The number of time intervals that are required for turning the electron beam on and off to form the digits determines the complexity of the character generator. For this reason, the seven-segment digits are formed as shown in Fig. A1-2. Each segment in a digit is a straight line that lies on either the horizontal or vertical axis. In turn, the four vertical segments occupy only two time periods during the horizontal scan, which occur at the same time on each line. The three horizontal segments also occupy the same time periods on the horizontal scan.

Also shown in Fig. A1-2 is the relative positioning of adjoining segments, so that their ends overlap to provide a continuous character. Both the horizontal and vertical scanning times for each digit are divided into eight time slots. Each vertical time slot consists of an even number of horizontal

FIGURE A1-1   View of a digital color-TV receiver. *(courtesy of Heath Company)*

scan lines. The first two horizontal time slots and the first vertical time slot of each digit are blank. The top, center, and bottom horizontal segments occupy the third through the eighth horizontal time slots during the second, fifth, and eighth vertical time slots. The left and right vertical segments occupy the third and eighth horizontal time slots. Also, the left and right vertical segments occupy the second through the fifth, and the fifth through the eighth vertical time slots, with overlap during the fifth time slot.

The display circuit is designed to indicate the time of day, as well as the channel number that is in use. These time data are supplied from an external clock source, and fill eight digit spaces below the channel number. A block diagram of the display integrated circuit that switches the electron beam on and off as required is shown in Fig. A1-3. Horizontal and vertical sync signals for display-circuit operation are derived from the horizontal and vertical retrace pulses in the TV receiver. The position of the display on the screen can be changed by adjusting the horizontal and vertical positioning controls for the monostable multivibrators. Basic timing pulses for the electronic switching circuits are provided by a pulse generator, called the clock generator. Figure A1-4 shows a block diagram for the entire display system, including the clock generator and connections to the picture-tube input circuit.

Sync vertical pulses are gated into the display circuit for an adjustable period of time which can be varied from several seconds up to half a minute or more: each time that the channel is changed, or when the recall switch (part of the volume-control circuit) is activated. In turn, the digital

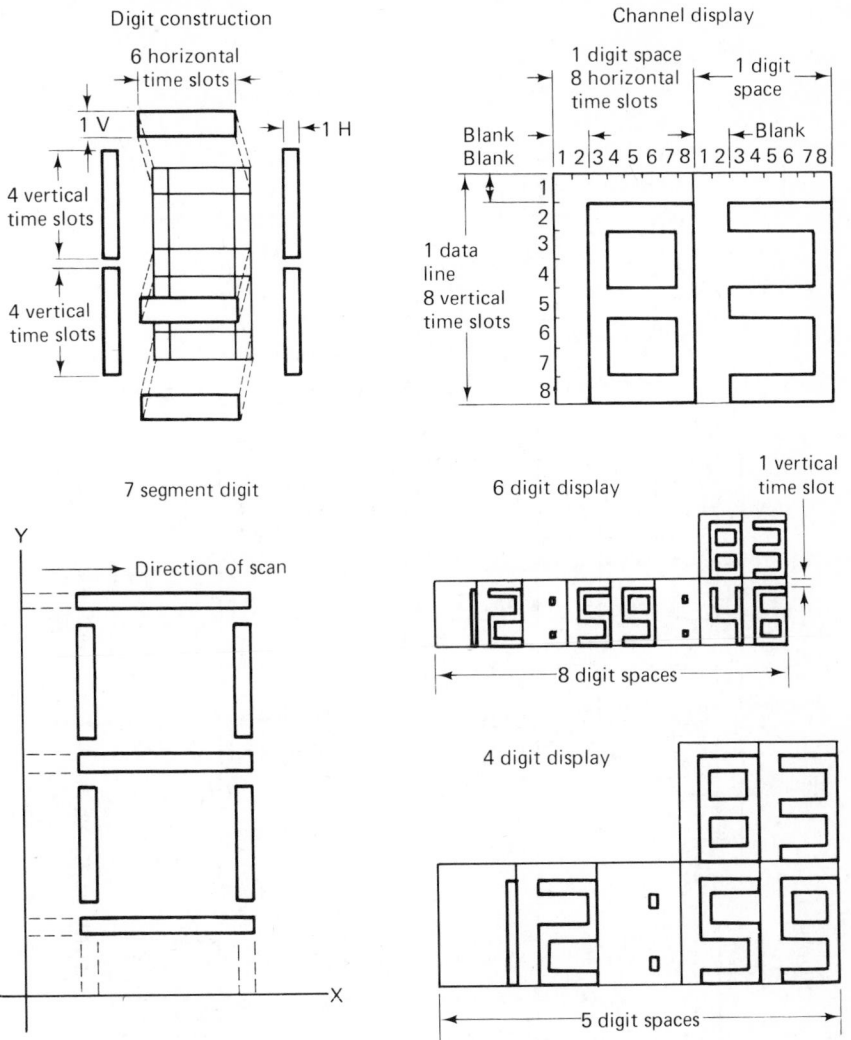

FIGURE A1-2 How digits are formed on the picture-tube screen.

output signal from the display circuit is coupled to a display-driver transistor which is connected in parallel with the luminance-driver stage. Thus, the display signal enters the video stages as a Y (black-and-white) signal which forms a white display on the picture-tube screen. The display driver circuit is adjustable, thus controlling the display brightness without affecting the reproduction of the program; the digital display is superimposed over the video signal.

The circuit for the digital clock is depicted in Fig. A1-5. Electronic-switching circuitry in the IC provides 6-digit, 12- or 24-hour time data to the

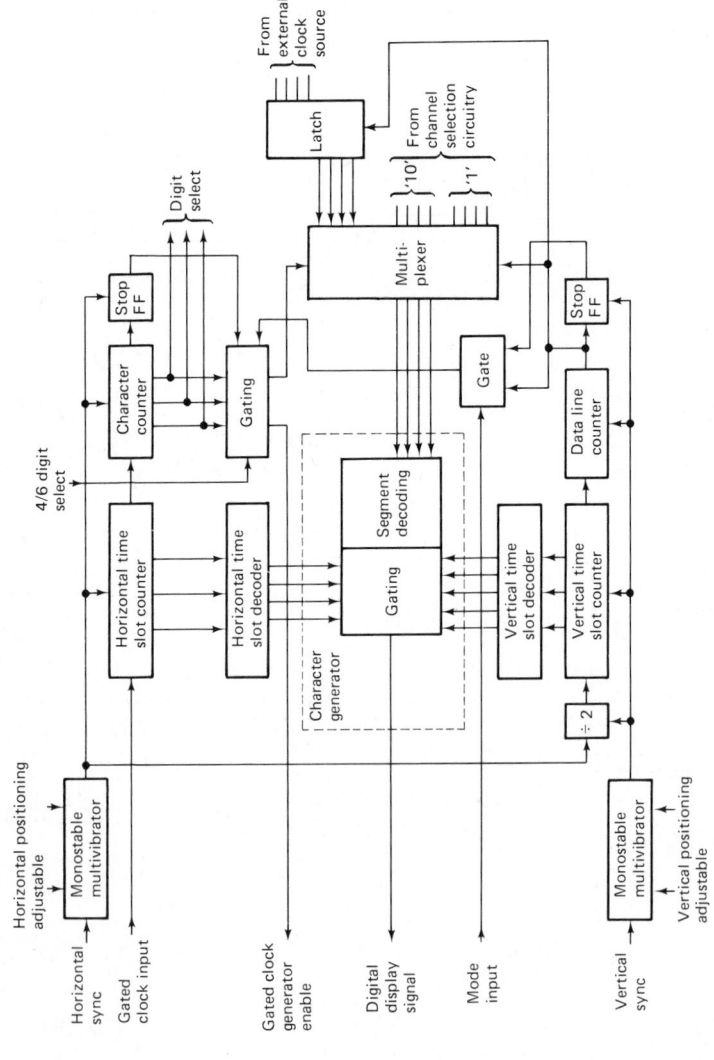

FIGURE A1-3 Block diagram of the display IC that forms the characters which appear on the picture-tube screen.

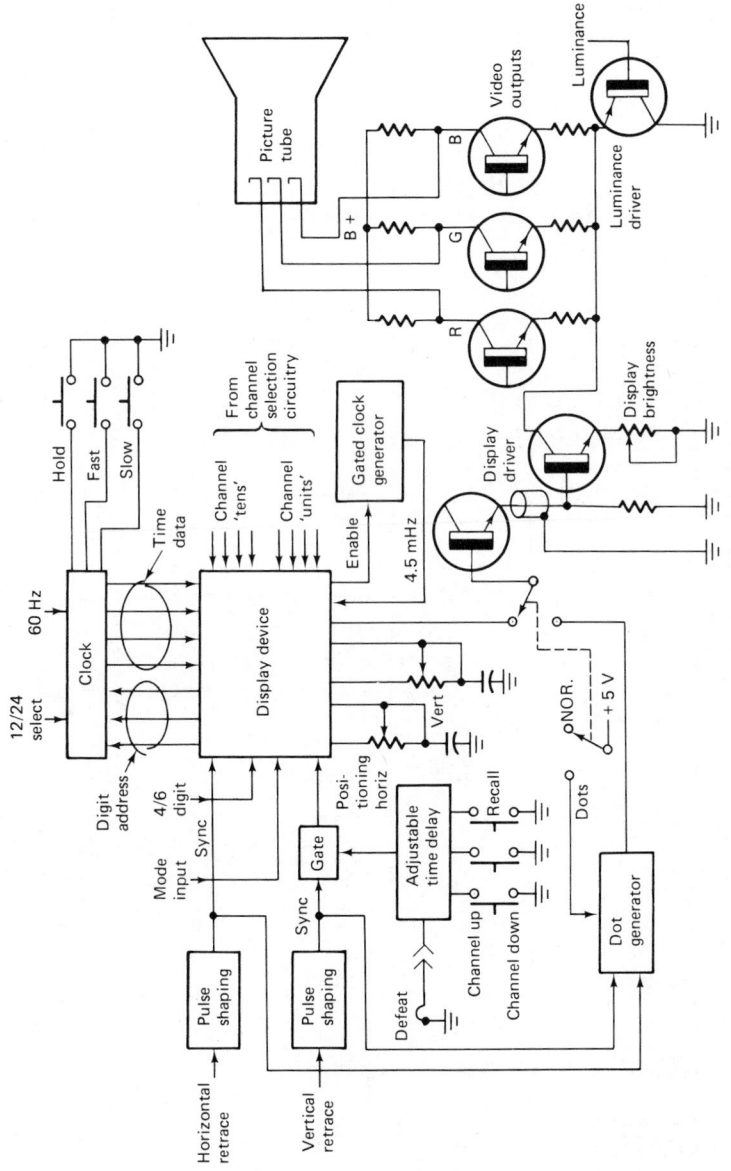

FIGURE A1-4 Block diagram of the complete digital display system.

373

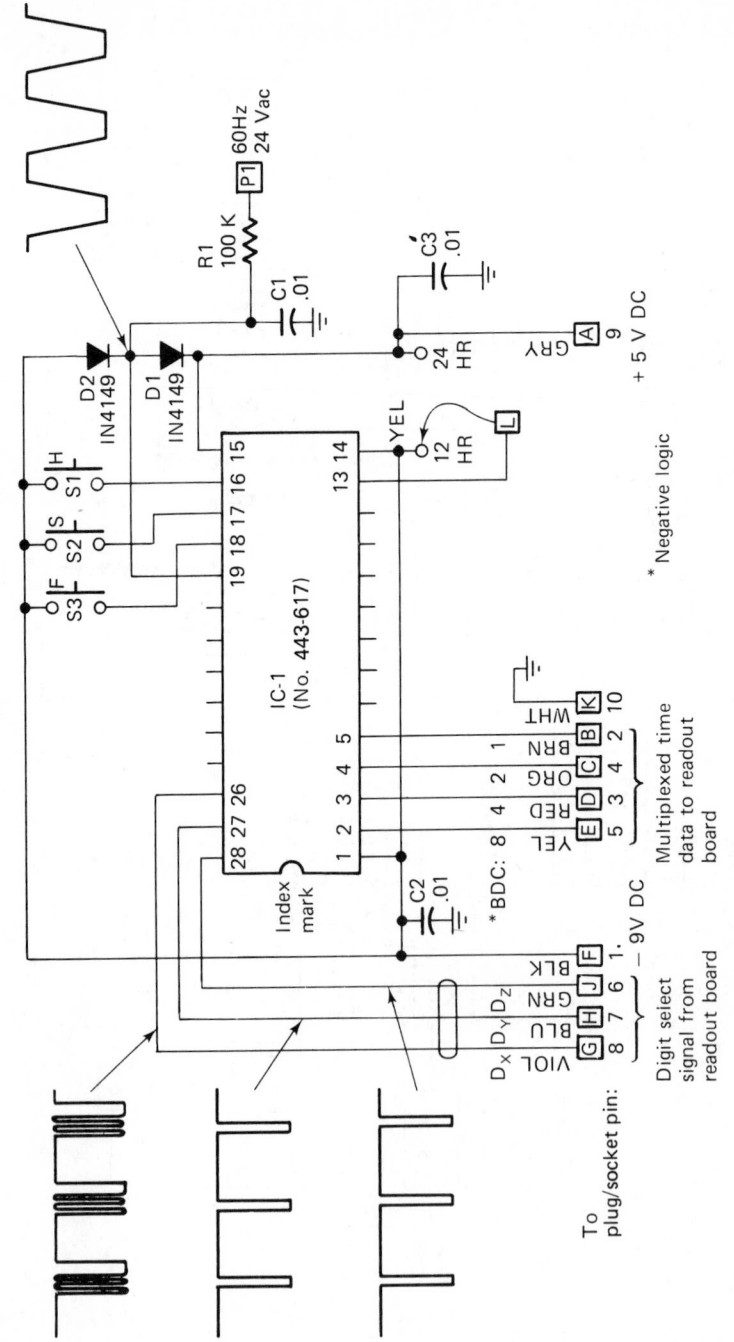

FIGURE A1-5 Digital clock circuit for the digital display system.

readout circuit. The clock operates from a 24-volt AC, 60-Hz source that connects to pin P1. It is coupled by resistor R1 to pin 19 of IC1. Diode D1 clamps the positive half-cycles to +5 volts DC, and diode D2 clamps the negative half-cycles to +9 volts DC. Capacitors C1, C2 and C3 are bypasses. Three momentary switches are utilized to set the time. Pushbutton H (switch S1) stops the clock to allow actual time to catch up to the display. Pushbutton S (switch S2) advances the minutes, and pushbutton F (switch S3) advances the hours. Either a 12- or 24-hour display mode can be selected by connecting pin 13 to pin 14 for a 12-hour display, or pin 13 to pin 15 for a 24-hour display. The voltages that cause the digits to be selected and displayed by the readout circuit appear on wires G, H, and J. Multiplexed (time-shared) data are fed back to the readout board on lines B, C and D. Views of the various circuit boards are shown in Figs. A1-6 through A1-10.

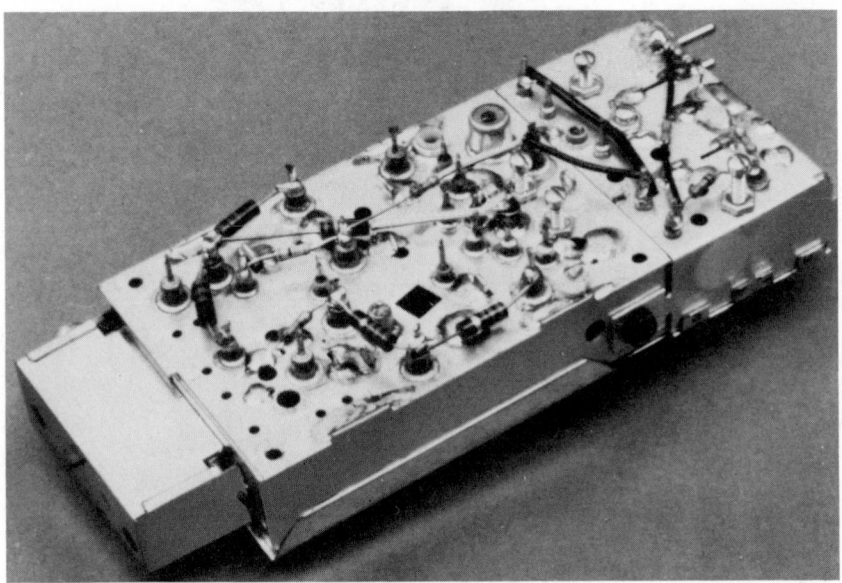

FIGURE A1-6   Appearance of the varactor tuner. *(courtesy of Heath Company)*

FIGURE A1-7   View of the channel-selection board. *(courtesy of Heath Company)*

FIGURE A1-8   Fixed-tuned IF filter and frequency-response curve. *(courtesy of Heath Company)*

FIGURE A1-9   Appearance of clock readout board. *(courtesy of Heath Company)*

FIGURE A1-10   View of the digital channel readout board. *(courtesy of Heath Company)*

Appendix 2

SI Units

Quantity	Unit	Symbol
Time	second	s
Current	ampere	A
Frequency	hertz	Hz
Energy	joule	J
Power	watt	W
Charge	coulomb	C
Voltage	volt	V
Resistance	ohm	Ω
Capacitance	farad	F
Inductance	henry	H

Answers to
Review Questions

CHAPTER 1

Questions

1. The ratio of the visible electromagnetic spectrum interval to the total electromagnetic spectrum interval is approximately 1 to 3×10^{22}.

2. Spectrum limits of visible electromagnetic radiation are 400 and 700 millimicrons.

3. Red, green, and blue are the additive primary colors; yellow, cyan, and magenta are the additive complementary colors.

4. From the physical standpoint, white is a brightness characteristic; black is the absence of light.

5. *Saturation* denotes the extent to which a hue is or is not blended with white light.

6. Orange hues require more definition than blue hues in the reproduction of a color image.

7. Brightness, hue, and saturation are the three basic characteristics of a color.

8. Four primary colors may be employed in the reproduction of a color image.

9. $\pm(R-Y)$ and $\pm(B-Y)$ are the basic transmission primaries.

10. A *color-difference* signal is the difference between a hue and its brightness.

11. Red has a comparative brightness (with respect to white) of 0.30; green, 0.59; and blue, 0.11.

12. White has a brightness of unity; black has a brightness of zero.

13. $(R-Y) = 0.70R - 0.59G - 0.11B$

14. *Compatibility* denotes that a color receiver will reproduce a color broadcast in color and a black-and-white broadcast in black-and-white, and that a black-and-white receiver will reproduce either a color broadcast or a black-and-white broadcast in black-and-white.

15. In general, an $R-Y$ signal corresponds to a red hue; and a $-(R-Y)$ signal to a bluish-green hue.

16. A $B-Y$ signal corresponds in a general way to a blue hue; a $-(B-Y)$ signal corresponds in general to a greenish-yellow hue.

17. In principle, a color scanning disc comprises three transparent segments that transmit red, green, and blue hues; the scanning disk rotates in front of the screen of a black-and-white picture tube.

18. A color scanning disk is large, clumsy, and subject to mechanical wear; rapidly moving objects are reproduced with more or less blur.

19. Objectives realized by the NTSC system include compatibility in an all-electronic system that provides high-quality color reproduction without perceptible jerkiness or blurring of rapidly-moving objects, and compact construction with reliable operation.

20. Color scanning disk systems are utilized in considerably simplified form in outer-space programs wherein high-quality color-TV reproduction is not a dominant consideration.

True-False

1. False	5. False
2. True	6. False
3. True	7. True
4. False	8. True

9. False

10. True

11. True

12. False

13. False

14. True

15. False

16. True

17. True

18. True

19. True

20. True

21. True

22. True

23. False

24. True

25. True

Multiple Choice

1. lower

2. 550 millimicrons

3. less

4. red, green, and blue

5. red and green

6. color map

7. red and cyan

8. orange

9. independent

10. $\pm(R-Y)$ and $\pm(B-Y)$

11. purplish-red

12. purplish-blue

13. bluish-green

14. greenish-yellow

15. $-(R-Y)$ and $-(B-Y)$

16. $+(B-Y)$ and $-(R-Y)$

17. brightness

18. an absence of light

19. a blend of R, G, and B signal voltages

20. R, $-G$, and $-B$

Problems

1. 28 millionths of an inch

2. 10^{12} to 1, approximately

3. 4.3×10^8 MHz, approximately

4. 1 to 10^{12}, approximately

5. 6.5×10^8 MHz, approximately

6. 575 millimicrons, approximately

7. unity

8. zero

9. zero

10. 0.30

CHAPTER 2

Questions

1. A color-TV transmitter utilizes three color-TV camera tubes.

2. A trimming filter is an optical color filter.

3. The basic plan of the NTSC system is to add hue and saturation information to the brightness information contained in a standard black-and-white TV signal.

4. Clusters are groups of energy that are concentrated at particular video frequencies.

5. Frequency interleaving is a staggered distribution of Y-signal clusters and chroma-signal clusters.

6. The color-subcarrier frequency is 3.579545 MHz.

7. To obtain frequency interleaving, the subcarrier frequency is chosen as an odd multiple of one-half the horizontal-scanning frequency.

8. A Y signal contains most of its energy at low frequencies.

9. A chroma signal contains most of its energy in the vicinity of the color-subcarrier frequency.

10. The video-frequency channel has an approximate bandwidth of 4.2 MHz.

11. Basically, the chroma signal has a double-sideband form.

12. A difference of 920 kHz separates the color subcarrier from the intercarrier-sound center frequency.

13. Although a black-and-white TV system employs a 60-Hz vertical-scanning frequency, a color-TV system utilizes a 59.94-Hz vertical-scanning frequency.

14. Whereas a black-and-white TV system employs a 15,750-Hz horizontal-scanning frequency, a color-TV system utilizes a 15,734.264-Hz horizontal-scanning frequency.

15. To *encode* means to translate an electrical signal into another form.

16. While scanning a red strip, the red-camera output is 100%; while scanning a green strip, its output is zero; while scanning a blue strip, its output is zero.

17. A Y matrix receives outputs from the red, green, and blue cameras, and blends these outputs to form an output voltage that is a Y, or black-and-white, signal.

18. A red chroma signal has an angle of 67° with respect to burst.

19. Unadjusted chroma values are produced by an encoder, prior to readjustment of these original values for modulation on the picture carrier.

20. It is impractical to transmit unadjusted chroma values because the transmitter would be repeatedly overmodulated during program transmission.

21. Readjusted chroma values are obtained by reducing R–Y to 0.877 of its original value, and reducing B–Y to 0.493 of its original value.

22. Readjustment of chroma values results in change of all phase angles with the exception of the R–Y and B–Y signals themselves.

23. Both the chroma amplitude and the Y amplitude are changed by desaturation of a hue.

24. Readjusted chroma values are converted to unadjusted chroma values at the color receiver.

25. A color burst consists of at least 8 cycles of 3.58-MHz voltage on the back porch of the horizontal sync pulse. This burst has the same phase as the color subcarrier and its amplitude is the same as the amplitude of the tip of the horizontal-sync pulse.

True-False

1. False	11. False
2. True	12. False
3. False	13. False
4. False	14. True
5. False	15. True
6. True	16. True
7. True	17. True
8. False	18. False
9. True	19. False
10. True	20. False

Multiple Choice

1. simpler than

2. image-orthicon or vidicon

3. red, green, and blue

4. staggering the clusters of Y and chroma signals

5. reduction of interference between the Y and chroma signals

6. color-subcarrier frequency
7. 920 kHz
8. 15.734264 kHz
9. 59.94 Hz
10. intercarrier-sound
11. three color cameras
12. RGB; ±(B−Y) and ±(B−Y)
13. 100%
14. changing with each change in color
15. 0.59G, 0.30R, 0.11B
16. 90° out of phase
17. 0.877 to 0.493
18. 67° to 76.5°
19. on the back porch of the horizontal sync pulse
20. 3.579545 MHz

Problems

1. 3.571678 MHz and 3.587413 MHz, approximately
2. 3.579545 MHz, approximately
3. zero
4. 76%
5. 0.83 to 0.59
6. 45° to 61.1°
7. 0.81 to 0.20, approximately
8. 0.89
9. 0.60, approximately
10. 0.74

CHAPTER 3

Questions

1. A complete color signal has a Y component and a chroma component.
2. Components of the complete color signal are chiefly separated by the Y amplifier and the chroma section in the color receiver.
3. A chroma channel has a narrower bandwidth than the IF amplifier.
4. Suppression of the color subcarrier at the transmitter assists in reducing interference between the Y and chroma signals.
5. If a black-and-white image is being transmitted by a color-TV station, the value of the chroma signal component is zero.
6. No change of amplitude occurs in the color burst when the color-TV camera shifts from a color scene to a black-and-white scene.

7. A delay line is employed in the Y amplifier to delay the passage of the Y signal.

8. Approximately 1 microsecond of delay is provided by a Y-amplifier delay line.

9. Quadrature signals have a $90°$ phase relation to each other.

10. A burst amplifier develops a train of 3.58-MHz output bursts at a repetition rate of 15,734.264 Hz.

11. A chroma signal is reconstituted in a color receiver by inserting a locally-generated 3.58-MHz subcarrier into the chroma-sideband signal.

12. During reception of a black-and-white TV broadcast, the chroma channel in a color receiver is automatically disabled to prevent confetti interference.

13. Circuit sections and stages used only in color receivers include the delay line, burst amplifier, chroma sync and phase detector, chroma bandpass amplifier, R—Y demodulator, B—Y demodulator, G—Y amplifier, R—Y amplifier, B—Y amplifier, color killer, subcarrier oscillator, subcarrier-oscillator AFC circuit, and convergence circuit.

14. A chroma signal is decoded in the chroma section by insertion of the color subcarrier and synchronous demodulation.

15. If chroma signals are applied to the grids, and the Y signal is applied to the cathodes in a color picture tube, the tube operates as a final decoding device.

16. Six classes of chroma-demodulator-and-matrix arrangements employed in color receivers are: R—Y and B—Y demodulated with G—Y matrixed; R—Y and G—Y demodulated and B—Y matrixed; twin-device demodulation and matrixing; X and Z demodulation with R—Y, B—Y, and G—Y matrixing; triple demodulation; and demodulator/RGB matrix arrangement.

17. R—Y and B—Y signals are separated in a pair of chroma demodulators by phase-detector action.

18. A G—Y signal can be developed either by demodulation or by matrixing.

19. X and Z demodulation is a non-quadrature demodulation process, whereas R—Y and B—Y demodulation is a quadrature demodulation process.

20. The earliest demodulator-and-matrix arrangement utilized an RGB matrix to drive the color picture tube; a modernized version of this design has been developed and is in use by various color receiver manufacturers.

True-False

1. True	14. False
2. True	15. True
3. True	16. True
4. False	17. True
5. True	18. False
6. False	19. True
7. True	20. True
8. True	21. True
9. True	22. True
10. False	23. True
11. True	24. True
12. True	25. True
13. True	

Multiple Choice

1. $-(B-Y)$	11. during demodulation
2. greater	12. the video amplifier
3. less	13. disabling the bandpass amplifier
4. yellow and cyan	14. after the chroma demodulators
5. red and blue	15. 15,734.264 Hz
6. 3.1 to 4.1 MHz	16. greater
7. delaying the Y signal	17. 4
8. poor color fit	18. 920 kHz
9. rise time	19. 4
10. G–Y signal	20. less

Problems

1. 0.16 μs, approximately	6. 920 kHz
2. 0.32 μs, approximately	7. 100 pF, approximately
3. 0.64 μs, approximately	8. 120 pF, approximately
4. 3.3 MHz, approximately	9. 56 kHz
5. −25 volts	10. 100

CHAPTER 4

Questions

1. Basic constructional features of a shadow mask color picture tube include an electron gun assembly, a shadow mask, and a phosphor-dot screen.

2. An aperture-grille color picture tube contains an electron-gun assembly, an aperture grille, and a phosphor-stripe screen.

3. Shadow-mask and aperture-grille color picture tubes operate from the same type of color signal.

4. Convergence is the process of bringing the three electron beams in a color picture tube into focus at any spot on an aperture grille or shadow mask during the scanning process.

5. Static convergence provides beam focus by permanent-magnet fields, whereas dynamic convergence provides beam focus by electromagnetic fields.

6. A degaussing coil is a large coil of wire that produces a strong AC field; a 60-Hz field is typical.

7. An aperture-grille color picture tube is not used as a final matrix.

8. Aperture-grille color picture tubes typically employ a 19-kV second-anode accelerating voltage.

9. A typical shadow mask has 400,000 perforations, approximately.

10. Approximately 12,000,000 phosphor dots are deposited on the screen.

11. Converging plates in an aperture-grille color picture tube assist in bringing the three electron beams into focus at the same spot on the aperture grille.

12. Color picture tubes employ magnetic shields to avoid impairment of color purity by the earth's magnetic field, or other external fields.

13. Screen purity denotes a display of a primary hue without contamination by other primary hues.

14. A blue lateral corrector is a permanent magnet that is utilized to shift the horizontal positions of the blue dots on the picture-tube screen.

15. A beam magnet is a permanent magnet that functions to shift the diagonal position of a red dot, or of a green dot, and to shift the vertical position of a blue dot on the picture-tube screen.

16. Internal pole pieces are provided inside a color picture tube to shield each electron gun from adjacent guns.
17. Rim magnets are movable permanent magnets mounted around the screen edge of some color picture tubes; a rim magnet is adjusted for optimum screen purity in the vicinity of the screen edges.
18. White-dot and crosshatch patterns are generally utilized in convergence procedures.
19. The location of the deflection yoke on the picture-tube neck is critical because purity is greatly affected by the magnetic field of the yoke.
20. A neck twist coil on an aperture-grille picture tube serves the same general function as the purity magnet assembly on a shadow-mask picture tube.

True-False

1. False
2. True
3. False
4. False
5. True
6. True
7. True
8. True
9. False
10. True
11. True
12. True
13. True
14. False
15. True
16. False
17. True
18. False
19. True
20. True

Multiple Choice

1. cannot
2. three
3. never
4. a slightly different point
5. blue gun and green
6. the same
7. red, green, and blue guns
8. a shorter
9. physically complete saturation of a hue
10. white-dot or crosshatch
11. automatic degaussing action
12. horizontal adjustment of the blue dots
13. a shadow mask
14. parabolic
15. the same waveform at different peak-to-peak voltages

Problems

1. 20 megohms
2. 20 watts
3. 5%
4. 22%, approximately
5. 0.73 inch, approximately
6. 396 megohms
7. 990 megohms
8. 50 μA
9. 20 μA
10. 1 watt, approximately

CHAPTER 5

Questions

1. The VHF frequency range extends from 54 MHz to 216 MHz. The UHF frequency range extends from 470 to 890 MHz.
2. When a UHF tuner is switched into operation, the VHF tuner is switched to operate as an IF amplifier.
3. A UHF tuner provides no amplification.
4. Automatic fine-tuning action in a tuner is provided by a varactor diode in the local-oscillator circuit; the varactor diode is biased from a discriminator in the IF section.
5. VHF tuners usually provide one stage of RF amplification.
6. A tuner has the basic function of preselection, with provision of some gain to avoid reduction of the signal-to-noise ratio.
7. Sound and picture carriers are usually located on the peaks of the RF-tuner frequency-response curve.
8. An adjacent-channel sound trap eliminates the sound signal that might gain entry from the neighboring lower-frequency channel.
9. A tube-type tuner operates with considerably higher DC voltages and has substantially higher circuit impedances than a solid-state tuner.
10. An IF amplifier has the basic functions of providing the major portion of the gain and selectivity for a color receiver.
11. Harmonics from the picture detector, if permitted to radiate to the lead-in, can cause picture interference.

12. All but the last stage in an IF amplifier are usually AGC-controlled.

13. 6-dB bandwidth denotes the number of Hz between the 50% of maximum response points on a frequency response curve.

14. The sound signal proceeds through the IF amplifier at much lower level than the picture signal.

15. An RF tuner has more than twice the bandwidth of an IF amplifier.

16. In most color receivers, the color subcarrier is located half-way up the side of the IF frequency response curve.

17. High-side local-oscillator operation avoids the possibility of the oscillator frequency falling within any VHF channel when the receiver is tuned through channels 2 to 13.

18. A high-side oscillator produces a lower sound-signal difference frequency than the picture-signal difference frequency.

19. Noise cannot be trapped out of an IF signal.

20. An RF amplifier and mixer section has a typical gain of 40 dB. An IF amplifier section has a typical gain of 58 dB.

True-False

1. False	14. False
2. True	15. True
3. True	16. True
4. True	17. False
5. False	18. True
6. True	19. False
7. True	20. False
8. True	21. False
9. False	22. True
10. False	23. False
11. True	24. True
12. True	25. True
13. True	

Multiple Choice

1. cascode configuration

2. generation of sum and difference frequencies

3. IGFET

4. common-base

5. appreciable loss

6. tuned circuits

7. variable capacitor

8. 9 MHz

9. most important

10. picture-detector diode

11. 58 dB

12. add

13. all but the last of the

14. neutralization

15. the same number of

16. rectifier followed by incomplete filtering

17. DC component in the video signal

18. tubes are checked first

19. about the same as in solid-state IF amplifiers

20. reversal of the IF frequency progression

Problems

1. 7500 times

2. 68 dB

3. 100

4. 200 ohms

5. 5 bars displayed horizontally

6. 5 bars displayed vertically

7. 60 Hz; origin is in the vertical-sync pulse

8. 10 times

9. 7.7 μA

10. 0.19 mA

CHAPTER 6

Questions

1. A video amplifier in a color receiver functions to step up the amplitude of the complete color signal.

2. Video amplifiers have greater bandwidth than Y amplifiers.

3. A color-subcarrier trap is included in the Y-amplifier configuration to minimize the amount of chroma signal at the picture-tube cathodes.

4. The video amplifier processes the complete color signal, whereas the Y amplifier processes the Y signal only.

5. A delay line functions to slow up the passage of the Y signal for approximately 1 μsec.

6. DC coupling is employed in video-amplifier and Y-amplifier circuitry to retain the DC component of the video signal.

7. Brightness-control limiter circuitry functions to prevent excessive beam-current flow in the color picture tube.

8. Retrace-blanking circuits operate to cut off the picture-tube beam current during flyback time.

9. A Darlington pair consists of two series-connected transistors operating in the CC mode; a beta value of over 500 is typical.

10. Sync clippers function to slice the sync tips off the tops of the blanking pedestals.

11. Signal-developed bias provides an advantage in sync-clipper operation by permitting the cut-off bias level to "follow" the average signal-amplitude level.

12. Horizontal sync pulses are separated from vertical sync pulses by means of low-pass and high-pass filters called integrating and differentiating circuits.

13. Serrations are provided in the vertical sync pulse to enable the horizontal locking action to be maintained during passage of the vertical sync pulse.

14. Equalizing pulses function to ensure that the stored charge in the vertical integrator is practically zero at the start of the vertical sync pulse.

15. To minimize the disturbing action of high-level noise pulses, a "hole"-punching or noise-cancellation circuit may be employed.

16. A typical video-and-Y-amplifier arrangement provides a voltage gain of 100.

17. Video-amplifier stage gain can be easily measured by comparing input-output signal amplitudes with an oscilloscope.

18. Tubes are checked first when malfunction occurs in a tube-type video amplifier.

19. The DC-voltage distribution in a tube-type video amplifier is considerably higher than in solid-state video amplifiers.

20. Both tube-type and solid-state video-and-Y-amplifier arrangements provide essentially the same gain.

True-False

1. False	3. False
2. True	4. True

5. False

6. False

7. False

8. True

9. False

10. True

11. False

12. True

13. True

14. True

15. False

16. False

17. True

18. False

19. True

20. False

Multiple Choice

1. 2 MHz

2. 2 MHz

3. 3 MHz

4. inherent nonlinearities in the output system

5. 1 microsecond

6. 70%, approximately

7. two series-connected transistors in the CC mode

8. emitter-follower mode

9. bias stabilization

10. protect the color picture tube

11. 5.1 μs

12. 190.5 μs

13. diode or transistor

14. end of the vertical-sync pulse

15. maintain horizontal locking action during passage of the vertical-sync pulse

16. reversing the polarity of noise pulses

17. punches "holes" in the clipper-output pulse train

18. operates only when enabled

19. flyback pulses

20. its collector voltage is very low and its base current is very high

Problems

1. 40 dB

2. 0.83 μs

3. 3.3 MHz

4. 0.1

5. 1500 kHz

6. 0.1 μs

7. 0.05 μH, approximately

8. 10 kilohms, approximately

9. 10 kilohms, approximately

10. 45°

CHAPTER 7

Questions

1. A chroma amplifier has the basic function of separating the chroma signal from the Y signal and stepping up the chroma signal.

2. Typical bandpass amplifiers have a frequency response from 3.1 to 4.1 MHz.

3. An I/Q signal can be processed as an (R−Y)/(B−Y) signal.

4. An I axis is $33°$ from the R−Y axis, and the Q axis is $90°$ from the I axis.

5. Compensation for the slope of the IF response curve is accomplished by means of an opposite slope in the bandpass-amplifier response curve.

6. An ACC circuit functions by varying the gain of the bandpass amplifier in accordance with the amplitude of the color burst.

7. The color-killer circuit enables the bandpass amplifier when a color burst is present, but disables the bandpass amplifier when a color burst is absent.

8. Enabling and disabling of the color killer action is effected by the color burst.

9. An I signal is a vestigial-sideband signal; a Q signal is a double-sideband signal.

10. To process the entire I signal requires a bandpass-amplifier frequency response from 2.1 to 4.1 MHz, or a bandwidth of 2 MHz.

11. Limitation of the bandpass amplifier bandwidth to 1 MHz results in some loss of color detail, but with an advantage of simplified circuitry and lower production cost.

12. A G−Y signal can be recovered from I and Q signals.

13. Tube-type bandpass amplifiers operate at higher supply voltage than their solid-state counterparts. Tubes cause the majority of malfunctions in tube-type amplifiers, whereas transistors are comparatively reliable devices.

14. An I signal is not centered on the color-subcarrier frequency; the Q signal is centered on the color-subcarrier frequency.

15. Adjustable tuned circuits are provided in a bandpass amplifier.

16. Blanking pulses are applied to the bandpass-amplifier transistor to prevent passage of the color burst.

17. Most bandpass amplifiers have more than one stage.

18. A technician can determine whether signal stoppage is being caused by color-killer trouble or by bandpass-amplifier trouble by measuring the color-killer control voltage.

19. A color-killer circuit has an adjustable threshold.

20. An ACC section controls the bandpass-amplifier gain, whereas an AGC section controls the RF and IF gain.

True-False

1. False	11. True
2. False	12. True
3. True	13. True
4. True	14. False
5. True	15. False
6. True	16. True
7. True	17. True
8. True	18. False
9. True	19. True
10. False	20. True

Multiple Choice

1. steps up the chroma signal
2. I/Q chroma system
3. 3.1 to 4.1 MHz
4. rejected
5. vestigial-sideband
6. double-sideband
7. double-sideband
8. gated
9. uniform overall response
10. varying the gain
11. disabling the bandpass amplifier during black-and-white reception
12. ACC section
13. voltage rises to the supply-voltage value
14. significant sidebands
15. significant sidebands

Problems

1. 10 k; 100 k
2. 50 μH, approximately
3. 1100 ohms, approximately
4. 1100 ohms, approximately

5. 6 μH, approximately

6. 10 ohms

7. 28%, approximately

8. 300 ohms, approximately

9. 5 ohms, approximately

10. 20 dB

CHAPTER 8

Questions

1. A color sync system maintains the 3.58-MHz output from the sub-carrier oscillator on-frequency and in-phase with the color burst.

2. If the subcarrier oscillator operates off-frequency, rainbows are displayed on the picture-tube screen; if the oscillator is off-phase, incorrect hues are displayed in the image.

3. A burst amplifier is enabled during the color burst, thereby separating the burst from the color signal.

4. Typical burst amplifiers have a bandwidth of 0.5 MHz, approximately.

5. Gating of the burst amplifier effectively eliminates passage of any camera signal into the color-sync system.

6. If the burst-gating pulse becomes partially mistimed, color sync action is impaired.

7. As the horizontal-hold control is adjusted, the timing of the burst-gating pulse changes with respect to the color burst.

8. Two basic types of subcarrier-oscillator configurations are the ringing-crystal design and the free-running/APC design.

9. A ringing crystal is followed by a limiter to clip the decaying waveform from the subcarrier oscillator to a uniform amplitude.

10. The chief subsections in an APC network are the burst amplifier, APC section, varactor control circuit, and subcarrier oscillator.

11. A green-red-blue downward sequence of colors in an out-of-sync color area denotes that the subcarrier oscillator is running too slow. On the other hand, a green-blue-red sequence indicates that the oscillator is running too fast.

12. A two-phase 3.58-MHz output is developed from a single-phase source by means of LCR circuitry.

13. Tube-type color-sync systems operate at comparatively high DC voltages, and various of the circuits have higher internal resistance than in their solid-state counterparts.

14. Either the video amplifier or the bandpass amplifier may be employed as a source for the burst-amplifier input signal.

15. An oscilloscope is the most useful instrument for preliminary troubleshooting of the color-sync system.

16. A color burst contains from 8 to 11 cycles of 3.58-MHz sine-wave voltage.

17. If the burst amplifier has subnormal bandwidth, the output waveform becomes attenuated and colors are likely to drift in the reproduced scene.

18. Excessive burst-amplifier bandwidth results in sudden color shifts, owing to contamination of the burst signal with the camera signal.

19. An ACC detector employs a reference 3.58-MHz signal that is 90° out of phase with the reference signal utilized by an APC detector.

20. A triggered-sweep oscilloscope with a calibrated time base can be used to measure the subcarrier-oscillator frequency.

True-False

1. False	11. True
2. True	12. False
3. False	13. True
4. True	14. True
5. True	15. True
6. False	16. False
7. False	17. True
8. True	18. True
9. True	19. False
10. True	20. False

Multiple Choice

1. complete color signal	6. gated
2. less	7. wider
3. 8 to 11	8. horizontal-flyback pulses
4. video amplifier and the burst amplifier	9. horizontal-hold control
5. 3.35 to 3.90 MHz	10. significant sideband frequencies

11. either positive or negative	16. high DC voltages
12. either positive or negative	17. reverse-biased
13. 90°	18. always
14. quartz crystals	19. fundamental
15. small-signal diodes	20. slug-tuned

Problems

1. 2550 cycles, approximately
2. 700 μs, approximately
3. 3.579485 MHz
4. 3.579605 MHz
5. 3.579425 MHz
6. 40 volts
7. 40 volts
8. 1100 ohms, approximately
9. 1100 ohms, approximately
10. 3.58 MHz, approximately

CHAPTER 9

Questions

1. Basically, the chroma-demodulator section functions to decode the chroma signal into R–Y, B–Y, and G–Y components, or equivalents.
2. Separation of the R–Y, B–Y, and G–Y signals is accomplished by an amplitude/phase demodulation process with associated subcarrier reinsertion.
3. A G–Y signal can be formed by combining suitable proportions of negative R–Y and B–Y signals.
4. R, G, and B signals can be matrixed from I and Q signals.
5. X/Z demodulation is utilized extensively because it is economical to manufacture and because a symmetrical design is employed that contributes to long-term stability.
6. A color picture tube is operated as an RGB matrix by applying the Y signal to its cathodes, and applying R–Y, B–Y, and G–Y signals to its grids.

7. RGB demodulation and matrixing is accomplished by demodulating on the R, G, and B axes, while the Y signal is applied to the demodulator diodes.

8. Tubes are checked first when malfunction occurs in a tube-type chroma-demodulator and matrix system.

9. Integrated circuits are utilized to some extent as chroma-demodulator devices.

10. It is feasible to demodulate R−Y and G−Y signals, and to matrix the B−Y signal.

11. Low-pass filters are included in chroma-demodulator output circuits to reject feedthrough subcarrier.

12. Demodulation of the G−Y signal takes place at 3.58 MHz, whereas matrixing of the G−Y signal takes place at a comparatively low video frequency.

13. G−Y and G demodulation axes are not identical; they are separated 4.4°.

14. There is no technical distinction between a synchronous detector and a product detector.

15. Hue is a function of chroma phase; saturation is a function of chroma amplitude.

16. The output from a chroma demodulator may have either positive or negative polarity.

17. A chroma matrix may have either positive or negative output polarity.

18. An R demodulator-matrix output may be either positive or negative.

19. X phase corresponds approximately to magenta-red and cyan-green hues; Z phase corresponds approximately to reddish blue and green hues.

20. Low-level chroma demodulators operate at a low signal level and are followed by active devices; high-level chroma demodulators operate at a high signal level and are followed by a color picture tube.

True-False

1. True	4. True
2. True	5. False
3. True	6. False

7. False

8. False

9. True

10. True

11. False

12. True

13. True

14. False

15. True

16. True

17. False

18. False

19. True

20. False

Multiple Choice

1. decoding the chroma signal

2. insertion of the color sub-carrier

3. B−Y signal

4. phase

5. amplitude

6. a phase error in the sub-carrier reference voltage

7. G−Y matrix

8. X and Z signals

9. active devices

10. a color picture tube

11. −R

12. 20

13. semiconductor diodes

14. ratio detector

15. unadjusted chroma values

Problems

1. zero to 0.5 MHz

2. 3.1 to 4.1 MHz

3. zero to 0.5 MHz

4. 1.14 and 2.23, respectively

5. 0.74 μs, approximately

6. 0.125 MHz

7. 75%

8. 3.58

9. 358

10. 35.8

CHAPTER 10

Questions

1. Convergence denotes the focusing of the three electron beams in a color picture tube at the same point on the shadow mask.

2. Unless the deflection yoke is precisely positioned on the picture-tube neck, the screen purity will be poor.

3. Static convergence controls provide convergence at center screen, whereas dynamic convergence controls provide convergence around the outer regions of the screen.

4. A color picture tube requires dynamic convergence because the edge areas of the screen cannot be converged by adjustment of the static convergence controls.

5. Static-convergence assemblies employ permanent magnets.

6. Dynamic-convergence assemblies utilize electromagnets.

7. Purity magnets are located behind the convergence assembly.

8. Parabolic convergence waveforms are obtained by integrating saw-tooth deflection waveforms.

9. Diodes are included in dynamic-convergence circuitry to develop a DC component in the convergence waveform.

10. Second-harmonic convergence waveforms are produced by second-harmonic ringing coils.

11. Dynamic-convergence controls tend to interact because each pole piece in a color picture tube affects its adjacent electron beams to some extent.

12. The convergence assembly is mounted behind the deflection yoke.

13. A blue static-convergence control produces vertical motion of the blue beam; a blue lateral corrector produces horizontal motion of the blue beam.

14. It is possible to have good horizontal convergence with poor vertical convergence.

15. A critical value of DC component in the convergence waveforms is essential to ensure that the convergence current is zero as the scanning beam passes through the center of the screen.

True-False

1. True	9. True
2. True	10. False
3. False	11. False
4. False	12. True
5. False	13. True
6. False	14. False
7. True	15. False
8. True	

Multiple Choice

1. at the shadow mask
2. produce dynamic beam convergence
3. provide center-screen convergence
4. shadow mask
5. display of a pure color, such as an uncontaminated red field
6. scanning system
7. triplet
8. horizontal scanning lines
9. develop a DC component
10. second-harmonic ringing circuits
11. parabolic
12. twelve
13. potentiometers and slug-tuned coils
14. maintenance controls
15. it provides red-beam adjustment

Problems

1. $7°\ 38'$
2. 1150 pF, approximately
3. 1100 ohms, approximately
4. 8000 ohms, approximately
5. 3000 ohms, approximately
6. 3000 ohms, approximately
7. 3.8 ohms, approximately
8. 2.2 megohms, approximately
9. 2200 ohms, approximately
10. 4000 ohms, approximately

CHAPTER 11

Questions

1. Basically, an AFC section prevents horizontal jitter by rejecting noise pulses.
2. An AFC section operates by comparing the repetition rate of the sync pulses with the repetition rate of the flyback pulses.
3. Integrating action is employed in an AFC section to average out noise pulses.
4. Most receivers employ blocking oscillators in the horizontal section; the oscillating frequency is controlled by the AFC section.
5. An SCR is a semiconductor device that functions comparably to a thyratron.
6. Tube-type horizontal-sweep and high-voltage systems operate at higher DC potentials and generate higher voltage AC waveforms than their solid-state counterparts.

7. High-voltage vacuum diodes and semiconductor rectifier stacks are employed in various receivers to rectify high-voltage AC pulses.

8. A high-voltage regulating arrangement employs a Zener-diode reference voltage and a regulator transistor to maintain the high-voltage value constant.

9. A phase splitter changes a single-polarity sync-pulse source into a double-polarity source.

10. A horizontal-hold control establishes the free-running frequency of the horizontal oscillator by adjusting the base bias of the oscillator transistor.

11. The comparison waveform in an AFC system derives from the flyback pulse.

12. Jitter is minimized by ringing-coil action in a horizontal-oscillator circuit.

13. Temperature stability is provided by thermistor action.

14. Negative bias on the SCR during the forward-scan interval eliminates the possibility of false triggering.

15. Triggering a horizontal oscillator directly from the sync-separator output results in picture jitter, tearing, and occasional total loss of sync lock.

True-False

1.	False	9.	True
2.	False	10.	True
3.	True	11.	True
4.	False	12.	False
5.	False	13.	False
6.	True	14.	True
7.	True	15.	False
8.	True		

Multiple Choice

1.	two-phase circuit	5.	higher
2.	a DC pulse	6.	semiconductor diodes
3.	stabilize the AFC operation	7.	opposite directions
4.	minimize picture jitter	8.	opposite directions

9. retrace switch/diode and the trace switch/diode

10. an adjustable resonant circuit

11. tubes and semiconductors

12. shorter

13. equivalent terms

14. exponential

15. automatic frequency control

Problems

1. 14 μs, approximately

2. 1000 VARS, or volt-amperes reactive

3. 1000 watts

4. 1000 volt-amperes

5. 0.007 μF, approximately

6. 325 ohms, approximately

7. 65 ohms, approximately

8. 325 ohms, approximately

9. 1400 ohms, approximately

10. 4 watts

CHAPTER 12

Questions

1. Basically, the vertical-sweep system functions to scan the picture-tube screen in a vertical direction.

2. Vertical deflection requires a peaked-sawtooth waveform because the deflection coils have significant resistance in addition to inductance.

3. A vertical integrator develops a trigger pulse by building up a charge on its shunt capacitor during passage of the vertical sync pulse.

4. Maximum output amplitude from the vertical integrator occurs as the vertical sync pulse ends.

5. Equalizing pulses are located prior to and subsequent to the vertical sync pulse. Serrations are contained within the vertical sync pulse.

6. A sawtooth waveshaper configuration is energized by a pulse voltage and generates a sawtooth output voltage.

7. Peaked-sawtooth waveforms are commonly generated by pulsing a transistor arrangement that includes a series RC branch in its load circuit.

8. Poor interlacing denotes failure of odd-field lines to fall precisely half-way between even-field lines.

9. Low frequencies are associated with the ramp portion of a peaked-sawtooth waveform; high frequencies are associated with the peaking-pulse portion.

10. If a peaked-sawtooth waveform passes through a differentiating circuit that has a suitable time-constant, a pulse output waveform is generated.

11. Tube-type vertical-sweep systems operate at considerably higher AC and DC voltages than solid-state systems, and have higher internal resistance.

12. Pincushion-correction circuitry operates to shape deflection waveforms as required to eliminate curvature in raster edges.

13. Top-bottom pincushioning and left-right pincushioning are encountered.

14. An exponentially decaying waveform falls to 37% of its original amplitude at the end of one time-constant, a rising waveform acquires 67% of its final amplitude; a decaying waveform falls to 14% amplitude, and a rising waveform attains 86% amplitude at the end of two time-constants.

15. A sine wave that is processed by an integrating circuit has its waveform unchanged; its amplitude becomes decreased, and its output voltage lags the input voltage.

True-False

1.	True	9.	True
2.	False	10.	True
3.	False	11.	True
4.	False	12.	False
5.	True	13.	False
6.	True	14.	False
7.	True	15.	False
8.	True		

Multiple Choice

1. sawtooth
2. peaked-sawtooth
3. 0.5
4. good interlacing
5. linearize the scanning action
6. a linear ramp
7. poor interlace

8. 1
9. 5
10. peaking pulse
11. ramp
12. pulse
13. higher voltage
14. curved raster edges
15. reluctance variation

Problems

1. 0.017 sec and 0.032 sec
2. 726 watts, approximately
3. 320 ohms, approximately
4. 62 ohms, approximately
5. 364 ohms, approximately
6. 18 ohms, approximately
7. 90 ohms, approximately
8. 0.0094 sec
9. 0.42, approximately
10. 1 watt

CHAPTER 13

Questions

1. Picture-signal and sound-signal channels are common only through the IF section.
2. A double-conversion superheterodyne system employs two mixers and two IF frequencies.
3. A sound converter develops the intercarrier sound signal, whereas the video detector develops the picture signal in a color receiver.
4. The sound-IF signal is heterodyned with the picture-IF signal in the sound converter.
5. A sound converter is usually driven by the last IF stage in a color receiver.

6. Because the picture- and sound-IF carriers are separated by 4.5 MHz, the intercarrier sound-IF frequency becomes 4.5 MHz.

7. An intercarrier sound-IF signal has a bandwidth of approximately 50 kHz.

8. Neutralization is often required in sound-IF amplifiers to avoid regeneration or oscillation.

9. Tube-type intercarrier-sound systems operate at higher DC potentials than solid-state systems.

10. Ratio-detector action is similar to discriminator action, except that a ratio detector has substantial inherent limiting action also.

11. Preemphasis consists of a boost in the higher audio frequencies, whereas deemphasis consists of an attenuation in the higher audio frequencies.

12. Unless an audio-output stage is properly loaded, the output transistor is likely to overheat and fail.

13. A typical integrated sound-IF amplifier, detector, and audio preamplifier contains 12 transistors.

14. Sync buzz is produced when an IF stage, for example, does not operate in class A; overload causes nonlinear operation and modulation of the vertical sync pulse into the sound signal.

15. Noisy sound output in a color receiver is often caused by faulty limiting action.

True-False

1. False	9. True
2. True	10. False
3. False	11. False
4. True	12. True
5. True	13. True
6. False	14. False
7. True	15. False
8. True	

Multiple Choice

1. 4.5 MHz	4. 60 Hz
2. 20 Hz to 10 kHz	5. 4.5 MHz
3. notch or pulse on the 4.5-MHz FM signal	6. 50 kHz
	7. 80

8. completely limit high percentages of AM signal

9. into saturation and into collector-current cutoff

10. attenuation of the higher audio frequencies

11. low collector potential

12. weak or no sound output

13. S curve

14. diode limiters

15. output transistor damage

Problems

1. 90, approximately

2. 56 kHz, approximately

3. 41.225 MHz and 41.275 MHz

4. 4.475 MHz and 4.525 MHz

5. 300 k

6. 65 k, approximately

7. 650 k, approximately

8. 87 MHz

9. 5 μH

10. 150 ohms, approximately

CHAPTER 14

Questions

1. A television remote-control unit provides adjustment of the channel-selector switch, volume control, intensity control, and hue control (plus throwing the on-off switch) at distances up to 50 feet.

2. Automatic tint control consists of judicious hue distortion so that flesh tones appear to remain more nearly constant in a color image.

3. CATV tuners provide 10 more channels for cable reception in addition to the 12 VHF channels and 70 UHF channels provided by a conventional tuner.

4. Ultrasonic waves are radiated by a remote-control transmitter to produce signals at the receiver for actuating control relays.

5. A transducer converts one form of energy into another form of energy; for example, a microphone converts sound energy into electrical energy.

6. A typical channel-change relay utilizes a signal frequency of 40 kHz.

7. Memory capacitors are designed to hold a charge indefinitely, for biasing the gate of an IGFET in a remote-control receiver.

8. Seven control coils are provided in a typical remote-control unit.

9. One method of tint control employs a chroma demodulation angle of $124°$ instead of $90°$. Another method utilizes gating circuitry so that only the chroma axes in the vicinity of the orange hues are changed in phase.

10. An ATC phase-shift capacitor increases the phase angle between the R—Y and B—Y chroma demodulation axes by $34°$.

11. A preference control permits the viewer to vary the amount of change in chroma-demodulation phase.

12. The high-voltage regulator tube is a possible source of X-radiation.

13. 0.5 milliroentgens per hour at a distance of 2 inches from a receiver is the maximum permissible X-radiation level.

14. Normal background radiation produces an average count from 40 to 150 per minute.

15. An X-ray field of 0.5 mR/hr produces a count of 4000 per minute, approximately.

True-False

1. False	9. True
2. True	10. True
3. True	11. True
4. True	12. True
5. False	13. False
6. False	14. True
7. False	15. True
8. True	

Multiple Choice

1. stabilize reproduction of flesh tones

2. make a record of TV programs

3. provide judicious distortion of the color spectrum

4. 13

5. 70

6. 31

7. 4

8. ultrasonic

9. 35.5 to 44.5 kHz

10. microphone

11. 50 feet

12. positive and negative clamp diodes

13. 124° chroma demodulation axes

14. high-voltage regulator tube

15. Geiger-Mueller counters

Problems

1. 0.0089 meter, approximately

2. 47.2 kHz, approximately

3. 0.0032 μF, approximately

4. 1300 ohms, approximately

5. 1300 ohms, approximately

6. 32 ohms, approximately

7. 32 ohms, approximately

8. 15 seconds

9. 30 seconds

10. 0.2 watt

CHAPTER 15

Questions

1. Basic color TV troubleshooting instruments include a VOM or TVM with a high-voltage probe, oscilloscope, signal generator, transistor tester, tube tester, sweep-and-marker generator, color-bar generator, and white-dot/crosshatch generator.

2. A color picture-tube tester or test jig and a TV analyzer are often utilized in addition to the basic test instruments.

3. Unless an oscilloscope has vertical-amplifier response to at least 4 MHz, it cannot be used effectively in troubleshooting chroma circuits.

4. Keyed-rainbow generators provide a spectrum of color-difference signals with a progressive phase advance of 30° per bar.

5. An NTSC color-bar generator provides color bars comprising the primary and complementary colors at full brightness and saturation.

6. White-dot and crosshatch generators are utilized in convergence procedures.

7. Sweep and marker generators are employed in VHF, UHF, IF, and chroma alignment procedures.

8. The two basic types of transistor testers are the out-of-circuit type and the in-circuit type.

9. Typical color picture-tube testers check for interelectrode leakage, for short-circuits, and measure the emission of each electron gun.

10. Absorption markers produce dip indications along frequency-response curves, whereas beat markers produce "pips" along the curves.

11. High-voltage probes, low-capacitance, and demodulator probes are employed in general troubleshooting procedures.

12. Nearly all conventional oscilloscopes can be utilized as vector-scopes.

13. Ten chroma bars are displayed on the picture-tube screen by a keyed-rainbow generator.

14. Post injection markers mix the marker signal with the sweep signal after the latter has passed through the receiver circuits.

15. Double markers may be displayed on a response to curve to indicate the picture- and sound-carrier points, or the picture-carrier and color-subcarrier points. Triple markers may be displayed to indicate the picture-carrier, sound-carrier, and color-subcarrier points.

True-False

1. False	9. False
2. True	10. True
3. True	11. True
4. True	12. False
5. False	13. True
6. True	14. True
7. True	15. True
8. True	

Multiple Choice

1. 25 kV	3. signal injection
2. convergence procedures	4. 4.5 MHz

5. calibrated time bases

6. 10

7. horizontal sync pulse

8. 30°

9. align tuned circuits

10. 70.7%

11. turn on normal semiconductor junctions

12. 1.5 to 1

13. collector-junction leakage

14. FM

15. CW

Problems

1. 3.595295 MHz

2. 3.3 μs, approximately

3. 12

4. 3,150

5. 10

6. 41 and 42

7. 76, 77, 78, and 79

8. 18 MHz

9. 18 MHz

10. 24 MHz

CHAPTER 16

Questions

1. If an antenna is sharply directional, and broadcast stations are located in various directions from the antenna, a rotor is desirable or necessary.

2. A preamplifier should be mounted at the antenna, so that the noise voltages picked up by the lead-in do not impair the signal-to-noise ratio unnecessarily.

3. During installation, the technician generally adjusts the AGC, horizontal-hold, vertical-hold, centering, height, noise-gate, and width controls.

4. A balun is an impedance transformer utilized in an antenna-input system.

5. A color picture tube is set up by adjusting the high-voltage value, checking the bias and screen controls for the picture tube, making purity adjustments, and converging the picture tube if necessary.

6. Typical log periodic antennas have fifteen elements.

7. If an antenna has an unsatisfactory signal-to-noise ratio, it should be replaced by a more highly directional antenna, with equal or greater gain, and usually supplemented by a rotor.

8. An attenuation pad would be installed in an antenna lead-in if the prevailing signal level overloaded the receiver.

9. Not all color receivers provide the same picture-tube adjustments.

10. A degaussing coil consists of a large number of turns of wire in a doughnut shape about one foot in diameter; the coil is energized from a 117-volt 60-Hz outlet.

True-False

1. False	6. True
2. True	7. True
3. False	8. False
4. True	9. False
5. True	10. True

Multiple Choice

1. 15
2. 60
3. separate VHF and UHF signals
4. avoid receiver overloading
5. match a 75-ohm single-ended source to a 300-ohm double-ended load

6. incorrect adjustment of the convergence controls
7. demagnetize the picture-tube assembly
8. AC voltage/current

Problems

1. 13%, approximately	5. 3 watts
2. 420 ohms, approximately	6. 60 ohms
3. 212 ohms, approximately	7. 75 ohms
4. zero	8. 75 ohms

CHAPTER 17

Questions

1. Color-TV troubleshooting procedures start with analysis of picture and sound trouble symptoms.

2. In the case of a no-picture symptom, the screen glows but no image is reproduced; a no-raster symptom is the same as a dark-screen symptom.

3. Basic signal-tracing procedures involve waveform checks through the signal channel with an oscilloscope supplemented by demodulator and low-capacitance probes.

4. Basic signal-injection procedures entail application of RF, IF, or VF voltages at suitable points along the signal channel.

5. A sine wave has an rms value that is equal to 0.707 of its peak value, and a peak-to-peak value that is equal to twice its peak value.

6. A sine wave has a single frequency, whereas a square wave has an array of frequencies in odd-harmonic relationship.

7. Vectorgrams are developed by application of an R—Y signal to the vertical-input channel, and a B—Y signal to the horizontal-input channel of a scope, usually with a keyed-rainbow signal applied to the color receiver.

8. A distorted IF frequency-response curve can cause separated picture and sound.

9. Typical picture trouble symptoms include weak image, overload, negative picture, smear, sound interference, and poor sync.

10. Defective components are generally pinpointed by DC voltage measurements, often supplemented by resistance measurements.

11. A turn-off test is made by short-circuiting the base and emitter terminals of a transistor to determine whether its collector voltage then rises to the supply-voltage value.

12. A turn-on test is made by bleeding collector voltage into the base of a transistor to determine whether its collector potential decreases.

13. A lo-pwr ohmmeter has an advantage in transistor circuit testing in that the applied test voltage is insufficient to turn on the transistor junctions.

14. The resistance of a coil is unrelated to its inductance value because a given inductance value can be obtained by winding a reference coil with larger or smaller wire.

15. A color-TV receiver can sometimes be serviced without the aid of servicing data if the technician is very familiar with the particular model.

True-False

1. False	9. True
2. True	10. True
3. True	11. True
4. True	12. False
5. False	13. False
6. True	14. False
7. False	15. False
8. False	

Multiple Choice

1. color-killer setting	6. independent parameters
2. ratio detector	7. lags
3. AGC section	8. 100 kHz
4. normal waveforms with p-p voltages	9. Lissajous figure
5. rms, peak, and peak-to-peak values	10. R—Y and B—Y demodulators

Problems

1. The second harmonic is passed at full amplitude; the third harmonic is passed at reduced amplitude.

2. 2.8 MHz, approximately

3. The 39th harmonic is passed at full amplitude; the 42nd harmonic is passed at reduced amplitude.

4. 0.5 MHz, approximately

5. 0.66 μs

6. 0.87 μs, approximately

7. 100 pF, approximately

8. 3200 ohms, approximately

9. 3200 ohms, approximately

10. 320 ohms, approximately

CHAPTER 18

Questions

1. CATV systems are used in fringe and far-fringe areas to provide normal signals and in areas where multipath propagation is a serious problem.

2. A preamplifier steps up the signal level of one channel; a combiner mixes signals from a number of channels.

3. A translator heterodynes UHF signals into the VHF band.

4. Trunk cables run from a combiner to a bridger.

5. Typical coaxial cable has a loss of 1 dB per 100 ft at 150 MHz.

6. A head end comprises preamplifiers, translators, and combiners.

7. Bridgers are basically signal splitters; they generally include amplifiers also.

8. Typical feeder amplifiers provide a gain of 25 dB.

9. A tilt control imposes a low-end VHF loss, to equalize the signal amplitude over the VHF band.

10. The final amplifier at the end of a trunk line is called a distribution amplifier.

11. A subscriber tap is a tap point along a feeder line.

12. A CATV system maintains a signal under 2000 μv to avoid radiation problems and amplifier overload difficulties.

13. Feeder lines can be run up to 1000 feet without a line extender.

14. Approximately 1500 microvolts are supplied to a TV receiver in a CATV system.

15. A line from a tap point along a feeder cable is called a subscriber drop.

True-False

1. False		9. False	
2. True		10. False	
3. True		11. True	
4. False		12. True	
5. True		13. False	
6. True		14. False	
7. True		15. True	
8. True			

Multiple Choice

1. multipath propagation
2. coaxial cable
3. convert high frequencies to low frequencies
4. greater at high frequencies
5. trunk cables
6. 50 dB
7. boosters, translators, and combiners
8. pedestals
9. 75
10. bridger
11. wide-band VHF amplifiers
12. low-frequency loss
13. distribution cables
14. 1500 microvolts
15. signal splitters

Problems

1. 5.36 microseconds, approximately
2. 2 inches, approximately
3. 700 MHz
4. 3.2×10^{-3}, approximately
5. 10^{-5}
6. 3.162×10^{2}
7. 10^{5}
8. 10
9. 100
10. 0.5, approximately

Index